8th International Conference on Aeronautical, Aerospace and Mechanical Engineering (AAME 2025)

AA007375

Journal of Physics: Conference Series Volume 3041

Suzhou, China
28-30 March 2025

ISBN: 979-8-3313-2376-9
ISSN: 1742-6588

Printed from e-media with permission by:

Curran Associates, Inc.
57 Morehouse Lane
Red Hook, NY 12571

Some format issues inherent in the e-media version may also appear in this print version.

Institute of Physics
Dirac House, Temple Back
Bristol BS1 6BE UK

Phone: 44 1 17 929 7481
Fax: 44 1 17 920 0979

techtracking@iop.org

Additional copies of this publication are available from:

Curran Associates, Inc.
57 Morehouse Lane
Red Hook, NY 12571 USA
Phone: 845-758-0400
Fax: 845-758-2633
Email: curran@proceedings.com
Web: www.proceedings.com

TABLE OF CONTENTS

Journal of Physics: Conference Series **3041** (2025) 011001 doi:10.1088/1742-6596/3041/1/011001

Preface

It is with great pleasure that we present the proceedings of the 8th International Conference on Aeronautical, Aerospace and Mechanical Engineering (AAME 2025). Building on the success of previous editions, AAME 2025 was held in Suzhou, China, from March 28 to 30, 2025, adopting a hybrid format to accommodate global participation while fostering in-person collaboration. The conference brought together leading researchers, engineers, and industry experts to share cutting-edge advancements and explore future directions in aeronautical, aerospace, and mechanical engineering.

This year's conference featured two distinguished keynote speakers who set the tone for the event. Prof. Zheng Hong (George) Zhu from York University, Canada, a renowned expert in space robotics and dynamics, shared groundbreaking insights into the latest developments in aerospace technologies. Prof. Lixi Huang from The University of Hong Kong, China, delivered an inspiring talk on innovations in mechanical engineering and their applications in sustainable systems. Additionally, six invited speakers from prestigious institutions contributed their expertise, further enriching the conference with diverse perspectives on emerging trends and challenges in the field.

AAME 2025 attracted participants from several countries, including China, Canada, Russia, Japan and other countries. The conference provided a vibrant platform for delegates to present their research, engage in stimulating discussions, and forge collaborations. All submitted papers underwent a rigorous peer-review process by the International Technical Committee, ensuring the highest standards of quality and originality. The selected papers reflect the conference's commitment to advancing knowledge and addressing real-world engineering problems.

The proceedings are organized into nine thematic chapters, covering a wide range of topics such as:

Aircraft Structure Design and Optimization

Aviation Electronics System Design and UAV Cooperative Control

Supersonic Aircraft Design and Fluid Dynamics Simulation

Dynamic Modelling and Fluid Dynamics Analysis in Aerospace Engineering

Operation Status Monitoring and Heat Analysis of Aviation Equipment and Components

Aerospace System and Engine Performance Simulation and Evaluation

Mechanical Equipment Design, System Model and Reliability Analysis in Aerospace Engineering

Complex System Control and Function Simulation Analysis

Aerospace Material Mechanical Performance Analysis and Modelling

Each chapter showcases the ingenuity and dedication of researchers pushing the boundaries of their disciplines. The diversity of topics and the depth of analysis underscore the dynamic nature of aeronautical, aerospace, and mechanical engineering today.

We extend our heartfelt gratitude to the conference committees, technical reviewers, keynote and invited speakers, and all participants for their invaluable contributions to AAME 2025. Special thanks to the local organizing team in Suzhou for their exceptional efforts in hosting this event. Their hard work ensured a seamless and enriching experience for all attendees.

We hope this volume serves as a valuable resource for researchers, practitioners, and students alike, inspiring further innovation and collaboration in the field. The success of AAME 2025 reaffirms the importance of international dialogue and knowledge exchange in driving engineering progress. We look forward to welcoming you to the next edition of AAME, where we will continue to explore the frontiers of engineering excellence.

John P.T. Mo

Conference Chair

Professor of Manufacturing Engineering

BSc(Engg), MSc(Engg), PhD(Loughborough), FIEAust CPEng NER APEC Engineer IntPE(Aus), CEng, FIMechE, MHKIE, FEF

School of Engineering, RMIT University

Australia

March 2025

Journal of Physics: Conference Series **3041** (2025) 011001 doi:10.1088/1742-6596/3041/1/011001

Conference Committees

Conference Chairs

Prof. Yuehong Qian, Soochow University, China

Prof. John Mo, Royal Melbourne Institute of Technology, Australia

Conference Co-chairs

Prof. Lixi Huang, The University of Hong Kong, Hong Kong, China

Prof. Yan Wang, Nanjing University of Aeronautics and Astronautics, China

Technical Program Chairs

Prof. Xin Zhang, Hong Kong University of Science and Technology, Hong Kong, China

Prof. Cees Bil, RMIT University, Australia

Prof. Zheng Hong Zhu, York University, Canada

Technical Program Co-chairs

Prof. Liming Yang, Nanjing University of Aeronautics and Astronautics, China

Prof. Chengwei Fei, Fudan University, China

Dr. Zhaolin Chen, Nanjing University of Aeronautics and Astronautics, China

Local Committees

Assoc. Prof. Jialin Lou, Soochow University, China

Dr. Yuxuan Chen, Soochow University, China

Technical Committee Members

Prof. Gang Sun, Fudan University, China

Prof. Wenlin Wang, Dongguan University of Technology, China

Prof. Anika Zafiah Mohd Rus, Universiti Tun Hussein Onn Malaysia, Malaysia

Prof. Tai Yan Kam, National Yang Ming Chiao Tung University, Taiwan, China

Prof. Yuan Kang, Chung Yuan Christian University, Taiwan, China

Prof. Shyh-Chour Huang, National Kaohsiung University of Science and Technology, Taiwan, China

Prof. Madhan Arumugam Kumar, KFUPM campus Dhahran, Saudi Arabia

Assoc. Prof. Zhouteng Ye, Beihang University, China

Assoc. Prof. Dr. Kwek-Tze Tan, The University of Akron, USA

Assoc. Prof. Fufu Wang, Chinese Academy of Sciences, China

Assoc. Prof. Chrystal Zhang, Royal Melbourne Institute of Technology, Australia

Assoc. Prof. Jiang Ding, Guangxi University, China

Assoc. Prof. Cao Zhongchen, Tianjin University, China

Assoc. Prof. Baharudin B.T.H.T, Universiti Putra Malaysia, Malaysia

Assoc. Prof. Oscar Philander, Cape Peninsula University of Technology - Adaptronics Advanced Manufacturing Technology Laboratory, South Africa

Asst. Prof. Kuo Liu, Dalian University of Technology, China

Journal of Physics: Conference Series **3041** (2025) 011001 doi:10.1088/1742-6596/3041/1/011001

Journal of Physics: Conference Series **3041** (2025) 011001 doi:10.1088/1742-6596/3041/1/011001

STATEMENT OF PEER REVIEW

All papers published in this volume of Journal of Physics: Conference Series have been peer reviewed through processes administered by the proceedings Editors. Reviews were conducted by expert referees to the professional and scientific standards expected of a proceedings journal published by IOP Publishing.

Journal of Physics: Conference Series **3041** (2025) 011002 doi:10.1088/1742-6596/3041/1/011002

Peer Review Statement

All papers published in this volume have been reviewed through processes administered by the Editors. Reviews were conducted by expert referees to the professional and scientific standards expected of a proceedings journal published by IOP Publishing.

- **Type of peer review:** Double Anonymous
- **Conference submission management system:** Morressier
- **Number of submissions received:** 99
- **Number of submissions sent for review:** 99
- **Number of submissions accepted:** 65
- **Acceptance Rate (Submissions Accepted / Submissions Received × 100):** 65.7
- **Average number of reviews per paper:** 1.02
- **Total number of reviewers involved:** 11
- **Contact person for queries:**
  Name: Amanda WU
  Email: amanda@saise.org
  Affiliation: SAISE

Towards Autonomous Aerial Docking of Multirotor UAVs: Integrating Vision and RTK in Outdoor Scenarios

Xufei Dong[1], Fujian Cai[1], Haojie Liu[1,*], and Rui Huang[1]

[1]State Key Laboratory of Mechanics and Control of Aerospace Structures, Nanjing University of Aeronautics and Astronautics, Nanjing 210016, China.

Email: liuhj@nuaa.edu.cn

Abstract. This research investigates the autonomous aerial docking of multirotor UAVs in an outdoor environment. Initially, a novel docking mechanism utilizing NbFeB magnets is proposed for a Multirotor UAV, allowing for misalignment errors of up to 5 cm. Subsequently, the relative position estimation between the Multirotor UAV and the docking mechanism is obtained by using onboard visual and Real-Time-kinematics (RTK) sensing. Eventually, a cascade PID-based autonomous docking controller is developed to ensure precise alignment. Experiment results demonstrate that the quadrotor platform can successfully dock with the mechanism via the proposed method.

1. Introduction

Multirotor UAVs have been widely used in remote sensing[1]-[2], rescue operations[3]-[4], aerial photography[5]-[6], and other fields. However, their limitations in payload capacity and endurance restrict their applications[7]. The modular UAVs can address these shortcomings of multirotor UAVs in terms of payload capacity and endurance, thereby expanding their application scope. Modular UAVs encompass two technological pathways: ground assembling and aerial assembling configurations. The representative example of ground modular configuration is the lift project of boeing[8]. In contrast, the representative example of aerial modular configuration is Modular Multirotor by David et al[9]-[12]. Compared to ground modular UAVs, aerial modular UAVs can be divided into multiple small modules, enabling them to pass through narrow gaps[13]-[14]. Additionally, aerial modular UAVs can replace damaged modules in mid-air. Therefore, aerial modular UAVs offer greater flexibility compared to ground modular UAVs.

Aerial autonomous docking stands as a pivotal technology within multirotor, posing challenges in the relative pose estimation between the multirotor and its target, as well as in the design of the docking mechanism and the development of control algorithms. In recent years, scholars world-wide have delved extensively into this technology. For instance, Choi et al pioneered a method employing the YOLO algorithm and depth cameras for relative pose estimation, alongside the creation of a two stable equilibrium docking mechanism enabling successful docking within a certain margin of error[15]-[17]. David et al utilized the Whycon method for pose estimation and engineered a framework-based docking mechanism leveraging permanent magnets[18]-[21]. Jain et al harnessed ArUco for pose estimation and introduced a parent-child UAV system[22]. These pioneers predominantly rely on visual methods for pose estimation, leading to docking failures if markers are outside the camera's field of view or are unidentifiable. Furthermore, the docking mechanisms in these studies exhibit insufficient redundancy or lack shear optimization. Additionally, these studies were conducted indoors, neglecting the impacts of outdoor factors like weather and lighting. In response to the aforementioned issues, a combination of visual and RTK methods is employed for position estimation. The docking mechanism is optimized to enhance its redundancy, and improved its shear force resistance.

The remainder of the paper is organized as follows: Section 2 presents the design of docking system, including the docking mechanism and the quadrotor platform. Section 3 establishes the dynamics of quadrotor platform and coordinate systems, and presents the flying controller system. Section 4 demonstrates the quadrotor platform is able to complete docking by itself under the guidance of RTK and OpenMV. Conclusions are drawn in Section 5.

2. The modular aerial system

2.1. Quadrotor Platform

As shown in Figure 1, the main structure of quadrotor platform is made of carbon fiber, and the wheelbase of the multirotor is 220mm. The Velox V2207 kv1750 is used as the motors of the quadrotor platform, when it compares with the propeller of Ts146 its maximum force is 1636g, while the quadrotor platform weights 1458.6g. The battery capacity is 1550mAh, which provide a flight time of five minutes. To balance the weight of docking mechanism, the battery and OpenMV are moved to the tail of the quadrotor platform. The mini-pix is chosen as the flight controller for the quadrotor platform, which features a built-in magnetometer and barometer. It has two telemetry ports for receiving RTK differential data and OpenMV information. Its size is compact, measuring only 40*40*12mm.

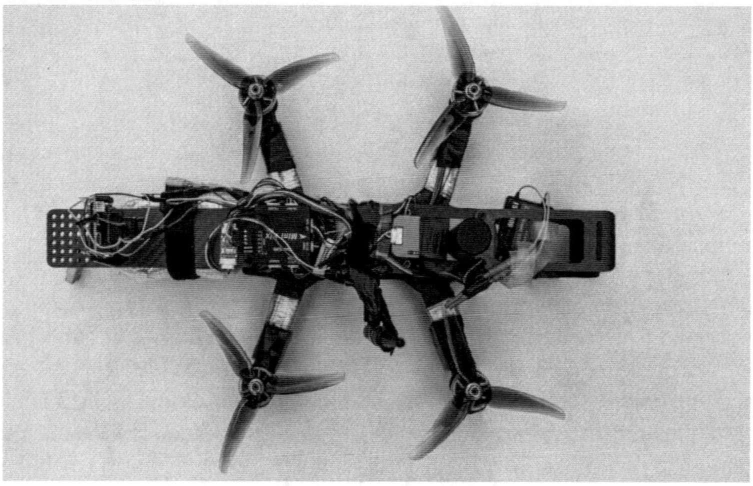

Figure 1. The quadrotor platform

Figure 2. The OpenMV

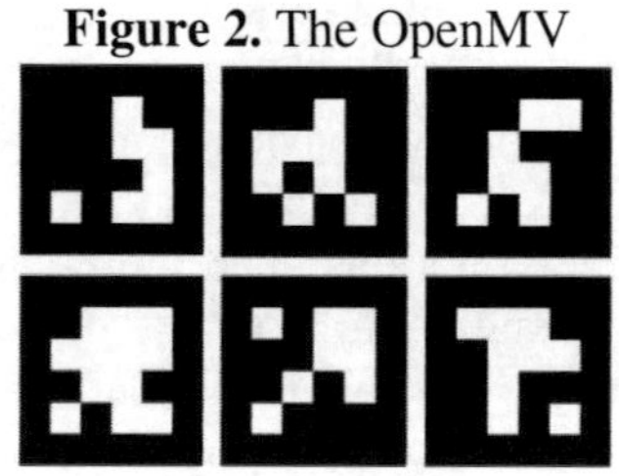

Figure 3. The Apriltag

2.2. Onboard Sensors

The onboard sensors include OpenMV and RTK. As shown in Figure 2 and Figure 3, the Apriltag+OpenMV method is employed as vision method for relative pose estimation between the quadrotor platform and the Apriltag[23]. When the OpenMV camera aligns with the Apriltag, it can fit the edges of objects in the image using least squares method, then identify the corresponding Apriltag through a depth-first algorithm. Finally, with the corresponding rotation matrix, the three-dimensional coordinates of the Apriltag can be determined. Its accuracy is estimated at millimeter level. To avoid obstructing the field of view of the OpenMV, the quadrotor platform adopts the X configuration.

The RTK is used as supplement of Apriltag+OpenMV. When Apriltag is not be seen by OpenMV, The RTK is used as pose estimation. As shown in Figure 4 and Figure 5, RTK positioning technology uses both a reference station and a mobile station to simultaneously receive positioning data from satellites. The reference station sends the differentially corrected information it has computed to the mobile station. With the differentially corrected information and positioning data received by itself, the mobile could determine its own position, and the accuracy is at centimeter level. By fusing vision and RTK, the robustness and accuracy of relative position estimation of docking process are improved.

Figure 4. The reference station

Figure 5. The mobile station

2.3. Formatting author affiliations Docking Mechanism

The docking mechanism is made by 3-D print, the material is ABS. The docking mechanism consist of two parts, one is installed on the quadrotor platform and the other is installed on the ground support, which is shown in Figure 6 and Figure 7. The one on the ground support weights 92g, the other one on the quadrotor platform weights 81g. Because of high strength-to-weight ratio of Neodymium Iron Boron (NbFeB) magnets[24], the magnet is chosen for initial alignment, capture and retention. The magnets are cylinder, the height of them are 10mm, the radius of them are 10mm too. Sixteen magnets are installed into the groove of docking mechanism.

Figure 6. The docking mechanism on the quadrotor platform

Figure 7. The docking mechanism on the ground support

A docking cone is installed on the docking mechanism at the quadrotor platform and a docking slot on the docking mechanism at the ground support for final alignment. The docking cone is chamfered, so that only one docking cone is needed to securely fix both the quadrotor platform and the ground support. This design improves the success rate of the docking process. The addition of the docking cone and docking slot enhances the shear resistance of the docking mechanism, allowing for a more stable connection. Figure 8 and Figure 9 illustrates the polarity of the magnets installed in the docking mechanism. This installation method ensures that if the quadrotor platform docks in the wrong orientation, the magnets will repel it, thereby ensuring the correct orientation after docking.

Figure 8. The magnet polarity of docking mechanism on the quadrotor platform

Figure 9. The magnet polarity of docking mechanism on the ground support

Journal of Physics: Conference Series **3041** (2025) 012001 doi:10.1088/1742-6596/3041/1/012001

2.4. Magnetic Force Simulation

COMSOL 6.1 is utilized for the magnetic field simulation of the docking mechanism's magnets. Given the rapid lateral decay of magnetic force in permanent magnets, two magnets are modeled in opposition. The impact of the remaining magnets on magnetic force is relatively negligible. Hence, they were excluded to streamline computational efficiency. Illustrated in Figure 10 are the two cylindrical magnets employed in the docking mechanism. The 3D model's simulation domain is a sphere with a radius of 500mm. The maximum mesh element size is 100mm, with a minimum element size of 0.1mm, as shown in Figure 11. The simulation progresses in steps of 1mm, commencing at 2mm and concluding at 52mm. As depicted in Figure 12, the electromagnetic force diminishes with displacement. The peak electromagnetic force occurs at 2mm, measuring 85.71N, while the minimum electromagnetic force is registered at 52mm, amounting to 1.32N.

Figure 10. The polarity of the magnet

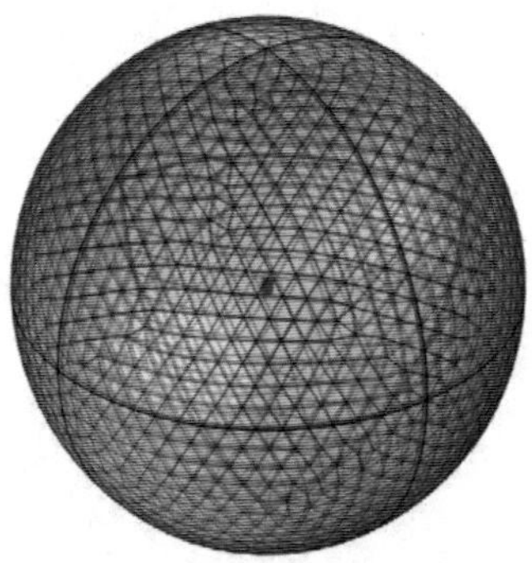

Figure 11. The mesh generation of magnetic simulation

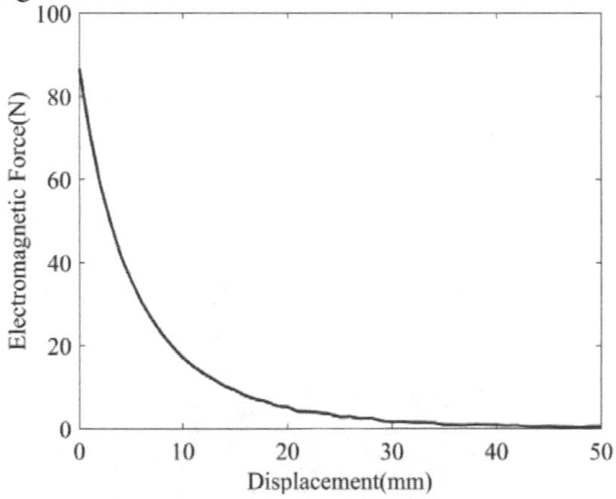

Figure 12. The electromagnetic force displacement curve

3. Dynamic modelling and control law synthesis

3.1. Dynamic Modeling

To establish the control law of docking, three coordinate systems are established as follows:

The world coordinate system W : it is fixed to the Earth, with the origin at the takeoff point, the X-axis pointing towards true north, the Y-axis pointing towards true east, and the Z-axis pointing towards the ground.

The body coordinate system B : it is fixed on the quadrotor platform. The origin is at the mass center of quadrotor platform the X-axis points towards the nose direction within the quadrotor platform symmetric plane, the Z-axis is within the quadrotor platform symmetric plane, perpendicular to the X-axis pointing downwards, and the Y-axis can be determined by the right-hand rule.

The target coordinate system T : it is fixed on the ground support. The origin is at the center of docking mechanism, the X-axis is in the ground support symmetric plane, deviating from the docking mechanism towards the direction of the support leg, the Z-axis is in the ground support symmetric plane, perpendicular to the X-axis pointing downwards, and the Y-axis can be determined by the right-hand rule.

Journal of Physics: Conference Series **3041** (2025) 012001 doi:10.1088/1742-6596/3041/1/012001

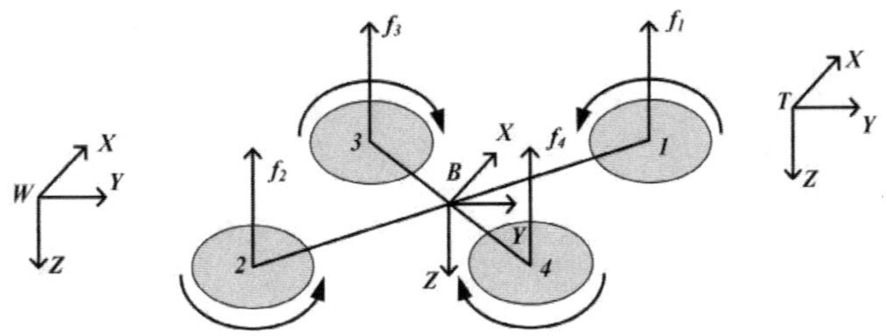

Figure 13. The frames and the motors include numbers and directions

Figure 13 shows the world frame W , body frame B , and target frame T . The position of the Body frame is $\boldsymbol{R}_b$, with velocities and accelerations of $\boldsymbol{V}_b$, $\boldsymbol{a}_b$. The position of the Target frame is $\boldsymbol{R}_t$. The attitude of the Body frame is $\boldsymbol{\theta}_b = (\theta_b, \phi_b, \psi_b)$, with angular velocities $\boldsymbol{\omega}_b = (\omega_\theta, \omega_\phi, \omega_\psi)$ of and angular accelerations of $\boldsymbol{\alpha}_b = (\alpha_\theta, \alpha_\phi, \alpha_\psi)$. The tension, roll, pitch, and yaw moments of Body frame are represented as F_T , M_{xb} , M_{yb} , M_{zb} respectively. Each motor i at the quadrotor platform provides an angular velocity of ω_i , resulting in forces and moments as follows:

$$f_i = k_f \omega_i^2 \tag{1}$$

$$M_i = k_m \omega_i^2 \tag{2}$$

Where k_f , k_m are determined experimentally. The external forces and moments at the quadrotor platform are determined by the forces and moments generated by each motor, as well as gravity.

The motor numbers and directions are as shown in Figure 12. With the motor numbers and directions, the F_T , M_{xb} , M_{yb} , M_{zb} can be written as follows:

$$\begin{pmatrix} F_T \\ M_{xb} \\ M_{yb} \\ M_{zb} \end{pmatrix} = \begin{pmatrix} 1 & 1 & 1 & 1 \\ -l_x & l_x & l_x & -l_x \\ l_y & -l_y & l_y & -l_y \\ -\frac{k_m}{k_f} & -\frac{k_m}{k_f} & \frac{k_m}{k_f} & \frac{k_m}{k_f} \end{pmatrix} \begin{pmatrix} f_1 \\ f_2 \\ f_3 \\ f_4 \end{pmatrix} \tag{3}$$

The following equations can be derived from the Newton-Euler equations:

$$m\boldsymbol{a}_b = \begin{pmatrix} 0 \\ 0 \\ mg \end{pmatrix} + \boldsymbol{R}_b^w \begin{pmatrix} 0 \\ 0 \\ -F_T \end{pmatrix} \tag{4}$$

$$\boldsymbol{I}_b \boldsymbol{\alpha}_b = \boldsymbol{M}_b - \boldsymbol{\omega}_b \times \boldsymbol{I}_b \boldsymbol{\omega}_b \tag{5}$$

Where $\boldsymbol{I}_b$ is the mass moment of inertia of quadrotor platform, $\boldsymbol{M}_b = (M_{xb}, M_{yb}, M_{zb})$ is the resultant external moment of quadrotor platform, $\boldsymbol{R}_b^w$ is the rotation matrix that transforms from the body coordinate system B to the world coordinate system W .

3.2. Control Law Synthesis

Ardupilot is opted for developing a flight control system. Its flight mode and position control modules are chosen for further development. The flying control process is shown in Figure 14.

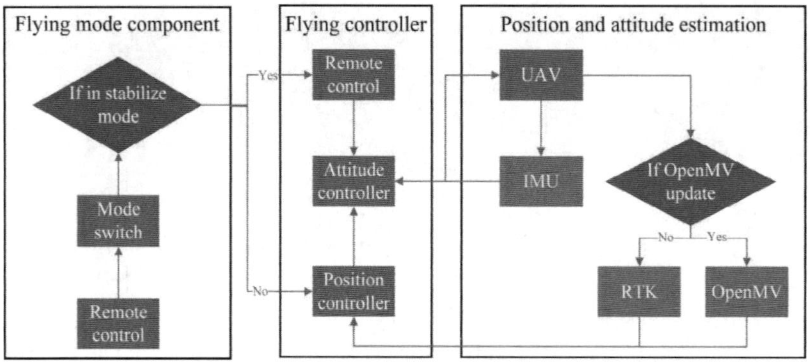

Figure 14. The flying control process

In order to improve the docking accuracy and robustness, two new flying modes have been incorporated: hover mode and docking mode. Initially, the quadrotor platform takes off in stabilize mode. Subsequently, the hover mode is engaged, causing the quadrotor platform to hover approximately one meter in front of the ground support. When the OpenMV identifies the Apriltag, the docking mode is activated. The quadrotor platform automatically aligns itself with the docking system and completes the docking process autonomously.

The aerial position estimation relies on RTK and OpenMV. When utilizing OpenMV data as the primary source for position estimation, certain positional transformations are necessary. The $\boldsymbol{R}_o$ represents the position vector of the OpenMV in relation to the quadrotor platform. The position between the AprilTag and OpenMV is $\boldsymbol{R}_v$. The $\boldsymbol{R}_a$ denotes the position vector of the AprilTag with respect to the docking system. The position of the quadrotor platform $\boldsymbol{R}_b$ can be calculated as follows:

$$\boldsymbol{P}_b = \boldsymbol{R}_a + \boldsymbol{R}_t - \boldsymbol{R}_b^w(\boldsymbol{R}_v + \boldsymbol{R}_o) \tag{6}$$

The position control in ArduPilot consists of the XY axis and the Z axis. For the XY axis position control, if OpenMV has identified the AprilTag, OpenMV data will be used for calculation $\boldsymbol{P}_b$.

Otherwise, it will be calculated using RTK data. The position error of the XY axis $\boldsymbol{P}_e^{xy}$ can be derived as:

$$\boldsymbol{P}_e^{xy} = \boldsymbol{P}_T^{xy} - \boldsymbol{P}_b^{xy} = (P_e^x, P_e^y) \tag{7}$$

Where $\boldsymbol{P}_T^{xy}$ is the target position of the XY axis.

Inputting $|\boldsymbol{P}_e^{xy}|$ into the square root controller yields the corrected velocity magnitude of the XY axis V_c^{xy}. The desired velocity $\boldsymbol{V}_T^{xy}$ of XY axis can be derived as:

$$\boldsymbol{V}_T^{xy} = \frac{V_c^{xy}}{|\boldsymbol{P}_e^{xy}|}\boldsymbol{P}_e^{xy} + \boldsymbol{V}_F^{xy} \tag{8}$$

Where $\boldsymbol{V}_F^{xy}$ is the feedforward velocity of XY axis.

The time interval $\mathrm{d}t$ is inputted into a low-pass filter to yield $\mathrm{d}t_1$, $\mathrm{d}t_2$, $\mathrm{d}t_3$, $\mathrm{d}t_4$ and $\mathrm{d}t_5$, each with a distinct cutoff frequency. The velocity error of XY axis $\boldsymbol{V}_e^{xy}$ can be derived as:

$$\boldsymbol{V}_e^{xy} = (\boldsymbol{V}_T^{xy} - \boldsymbol{V}_b^{xy} - \boldsymbol{V}_l^{xy})\mathrm{d}t_1 + \boldsymbol{V}_l^{xy} \tag{9}$$

Where $\boldsymbol{V}_l^{xy}$ is the previous the velocity error of XY axis and $\boldsymbol{V}_b^{xy}$ is the current velocity of the XY axis.

The differential term of acceleration along XY axis $\boldsymbol{a}_d^{xy}$ can be derived as:

$$\boldsymbol{a}_d^{xy} = k_{d1}[(\frac{\boldsymbol{V}_e^{xy} - \boldsymbol{V}_l^{xy}}{\mathrm{d}t} - \boldsymbol{a}_{dl}^{xy})\mathrm{d}t_2 + \boldsymbol{a}_{dl}^{xy}] \tag{10}$$

Where k_{d1} is coefficient of differential term of XY axis and $\boldsymbol{a}_{dl}^{xy}$ is the previous differential term of XY axis.

The integral term of acceleration along XY axis $\boldsymbol{a}_i^{xy}$ and the proportional term of acceleration along XY axis $\boldsymbol{a}_p^{xy}$ can be derived as:

$$\boldsymbol{a}_i^{xy} = k_{i1}\boldsymbol{V}_e^{xy}\mathrm{d}t + \boldsymbol{a}_{il}^{xy} \tag{11}$$

$$\boldsymbol{a}_p^{xy} = k_{p1}\boldsymbol{V}_e^{xy} \tag{12}$$

Where k_{i1} is the coefficient of integral term of XY axis, k_{p1} the coefficient of proportional term of XY axis and $\boldsymbol{a}_{il}^{xy}$ is the previous integral term of the XY axis.

The target acceleration of XY axis $\boldsymbol{a}_T^{xy}$ can be derived as:

$$\boldsymbol{a}_T^{xy} = \boldsymbol{a}_d^{xy} + \boldsymbol{a}_i^{xy} + \boldsymbol{a}_p^{xy} + \boldsymbol{a}_F^{xy} \tag{13}$$

Where $\boldsymbol{a}_F^{xy}$ is the feedforward acceleration of XY axis.

The current position of the Z-axis is obtained through an extended Kalman filter using RTK and a barometer. Due to the high accuracy of this method, the OpenMV data of the Z-axis is not integrated. The position error of Z axis P_e^z could be derived as:

$$P_e^z = P_T^z - P_b^z \tag{14}$$

Where P_T^z is the target position of Z axis and P_b^z is the current position of Z axis.

Inputting P_e^z into square root controller to get target velocity of Z axis V_T^z. The correct target velocity of Z axis V_c^z can be calculated as:

$$V_c^z = V_T^z + V_F^z \tag{15}$$

Where V_F^z is the feedforward velocity of Z axis.

The velocity error of Z axis V_e^z can be derived as:

$$V_e^z = (V_T^z - V_b^z - V_l^z)\mathrm{d}t_3 + V_l^z \tag{16}$$

Where V_l^z is the previous the velocity error of Z axis and V_b^z is the current velocity of the Z axis.

Then uses the same methods of XY axis, the target acceleration of Z axis a_T^z can be calculated. Different from XY axis, the Z axis need to calculate target throttle T_t^z .the acceleration error of Z axis a_e^z can be calculated as:

$$a_e^z = (a_T^z - a_b^z - a_l^z)\mathrm{d}t_4 + a_l^z \tag{17}$$

Where a_l^z is the previous the acceleration error of Z axis and a_b^z is the current acceleration of the Z axis

The differential term of throttle along Z axis T_d^z can be computed as:

$$T_d^z = k_{d2}[(\frac{a_e^z - a_l^z}{dt} - T_{dl}^z)\mathrm{d}t_5 + T_{dl}^z] \tag{18}$$

Where k_{d2} is the coefficient for the differential term of the Z axis and T_{dl}^z is the previous differential term of the Z axis.

The integral term of throttle along Z axis T_i^z and the proportional term of throttle along Z axis T_p^z can be calculated as:

$$T_i^z = k_{i2}a_e^z\mathrm{d}t + T_{il}^z \tag{19}$$

$$T_p^z = k_{p2}a_e^z \tag{20}$$

Where k_{i2} is the coefficient of integral term of Z axis and k_{p2} is the coefficient of proportional term of Z axis and T_{il}^z is the previous throttle integral term of the Z axis.

The target throttle of Z axis T_{Tz} could be calculated as:

$$T_t^z = T_d^z + T_i^z + T_p^z + T_F^z \tag{21}$$

Where T_F^z is the feedforward throttle of Z axis.

4. Docking experiment

During the aerial docking experiment, the OpenMV is positioned in front of the quadrotor platform, while the Apriltag is positioned above the ground support, the whole docking process is shown in Figure 15. The quadrotor platform initiated take off in stabilize mode, then hovered in front of the Apriltag. When the OpenMV identified the Apriltag at 23.33s, the docking mode was engaged, prompting the quadrotor platform to align with the docking mechanism and successfully completed the docking procedure. The position error data is shown in Figure 16, which illustrates OpenMV did not detect the Apritag after 26.61 seconds, the quadrotor platform still successfully docked under the guidance of RTK, which highlighting the quadrotor platform's robustness. The controller parameters are as follows: $k_{p1} = 1.5$, $k_{i1} = 0.1$, $k_{d1} = 0.005$, $k_{p2} = 0.7$, $k_{i2} = 1$, $k_{d2} = 0.01$. It is worth noting that in Figure 17, the acceleration of the XYZ axis exhibits multiple peaks after 27.12s, indicating the completion of aerial docking at that time.

Figure 15. The process of docking

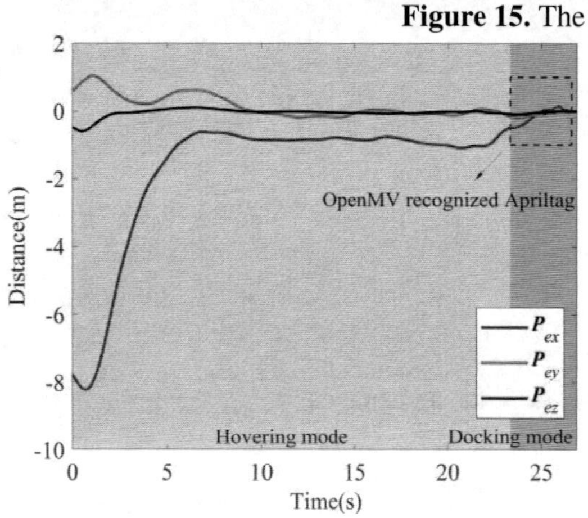

Figure 16. Time history of position error

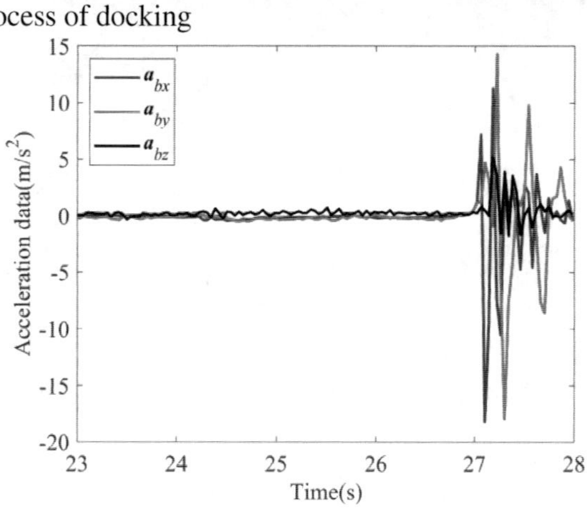

Figure 17. The acceleration of docking time

5. Conclusion

In conclusion, a multirotor UAV developed the capable of autonomously docking under the guidance of RTK and vision systems. This project emphasizes the design of the docking system, flight dynamics, flying control systems. The experiment shows that the multirotor can successfully dock using the docking mechanism with the assistance of RTK and OpenMV. The OpenMV identifies the Apriltag at 23.33s, and then switched to the docking mode. The acceleration data indicates the quadrotor platform completed docking at 27.12s. Moving forward, the docking which under complex environments and the cooperative control strategies for multiple multirotor will be explored.

6. References

[1] Burri M, Nikolic J, Hurzeler C, Caprari G and Siegwart R 2012 *2nd Int. Conf. on Applied Robotics for the Power Industry* (Switzerland: ETH Zurich) pp 70-5

[2] Khattak S, Papachristos C and Alexis K 2019 *Int. Conf. on Robotics and Automation* (Montreal, QC: Palais des congres de Montreal) pp 3563-9
[3] Tranzatto M, et al 2022 CERBERUS: Autonomous Legged and Aerial Robotic Exploration in the Tunnel and Urban Circuits of the DARPA Subterranean Challenge *Preprint* gr-qc/07067
[4] Tomic T, Schmid K, Lutz P, Domel A, Kassecker M, Mair E, Grixa I, Ruess F, Suppa M and Burschka D 2012 *J. Automat. Mag* **19** 46-56
[5] Gurtner A, Greer D G, Glassock R, Mejias L, Walker R A and Boles W W 2009 *J. Geosci. Remote Sensing* **47** 709-21
[6] Klosterman S, et al. 2018 *J. Agricultural and Forest Meteorology* **248** 397–407
[7] Wu Y and She H 2023 *35th Chinese Control and Decision Conf.* (Yichang: Xindao International Hotel) pp 3362-9
[8] Duffy M J and Samaritano A 2015 *33rd AIAA Applied Aerodynamics* Conference (Dallas: American Institute of Aeronautics and Astronautics) pp 3013-33
[9] Naldi R, Forte F and Marconi L 2011 *50th IEEE Conf. on Decision and Control and European Control Conf.* (Orlando: Hilton Orlando Bonnet Creek) pp 3584-9
[10] Fitch R, Stoy K, Kernbach S, Nagpal R and Shen W-M 2014 *J. Robotics and Autonomous Systems* **62** 943-4
[11] Naldi R, Forte F, Serrani A and Marconi L 2015 *J. IEEE Trans. Contr. Syst. Technol* **23** 1869-85
[12] Oung R and D'Andrea R 2014 *J. The International Journal of Robotics Research* **33** 375–400
[13] Zhao M, Kawasaki K, Anzai T, Chen X, Noda S, Shi F, Okada K and Inaba M 2018 *J. The International Journal of Robotics Research* **37** 1085–112
[14] Yang H, Park S, Lee J, Ahn J, Son D and Lee D 2018 *IEEE Int. Conf. on Robotics and Automation* (Brisbane: IEEE Int. Conf. on Robotics and Automation) pp 7017–23
[15] Choi A J, Park J and Han J-H *AIAA SCITECH 2022 Forum* (San Diego: American Institute of Aeronautics and Astronautics) 2022-0883
[16] Choi A J, Park J and Han J-H 2021 *Active and Passive Smart Structures and Integrated Systems XV* (United States: Online Only) p 38
[17] Choi A J, Yang H-H and Han J-H 2021 *J. Systems and Signal Processing* **154** 107579
[18] Nitsche M, Krajnık T, Mejail M and Duckett T *IROS Workshop on Open Source Aerial Robotics* (Hamburg: Congress Center Hamburg) pp 1-6
[19] Li G, Gabrich B, Saldana D, Das J, Kumar V and Yim M 2019 *Int. Conf. on Robotics and Automation* (Montreal, QC, Canada: IEEE) pp 346–52
[20] Saldana D, Gabrich B, Li G, Yim M and Kumar V 2018 2018 IEEE Int. Conf. on Robotics and Automation (Brisbane: Brisbane Convention & Exhibition Centre) pp 691–8
[21] Xu J, D'Antonio D S and Saldaña D 2024 Modular Multi-Rotors: From Quadrotors to Fully-Actuated Aerial Vehicles *Preprint* gr-qc/00788
[22] Jain K P, Park M and Mueller M W 2020 Docking two multirotors in midair using relative vision measurements *Preprint* gr-qc/05565
[23] Olson E *2011 IEEE Int. Conf. on Robotics and Automation* (Shanghai: Shanghai International Conference Center) pp 3400–7
[24] Duan H, Li H, Luo Q, Zhang C, Li C, Li P and Deng Y 2016 *J. Sci. China Inf.* **59** 053201

Acknowledgments
This work was supported by the National Natural Science Foundation of China (Grant No. 12472013), the foundation of National Key Laboratory of Aircraft Configuration Design and Research(Grant No. JBGS-2024-07) and the Aeronautical Science Foundation of China (Grant No. 2024M005052002).

High-fidelity and aero-structural design optimization of a martian propeller

N A Benai-dara, Z L Chen*, L K Kanama, Y Liu and A Rahman

College of Aerospace Engineering, Nanjing University of Aeronautics and Astronautics, Nanjing, China

*zhaolin_chen@nuaa.edu.cn

Abstract. The propeller is an important component of Mars Helicopter as it generates thrust and lifting force and acts as a control system during flight. Due to the low density and low temperature on Mars, the propeller performance is limited. The present research aims at improving the Figure of Merit FM and evaluating the factor of safety of a Martian blade by combining Response Surface Methodology RSM and High-Fidelities CFD and FEA. The surrogate model RSM is reported to be more suitable for identifying the effect of individual variables and locating the optimum combination. As appropriate to Mars operating conditions, NACA 5605 is used as the onboard airfoil and the design variables are the twist angle and chord length distributions at stations $\mathrm{r/R} = 0.19$, $\mathrm{r/R} = 0.86$ and $\mathrm{r/R} = 1$ resulting in unconventional blade planform. The optimized propeller generates thrust of $\mathrm{T} = 5.178\mathrm{N}$ and reaches a FM of 0.69 at 2737RPM with minimum factors of safety for compressive and tensile of 5.34 and 11.08. The flow physics around the optimized blade reveals that the root section does not experience flow separation when the twist angle is 9~12deg.

1. Introduction

Mars Helicopter was proposed to increase the ground exploration due to various challenges encountered by previous means such as Rovers and Landers and Satellites [1, 2]. Among the components of MH, the rotor system is vital as it generates propulsion and lifting forces as well as controlling the vehicle body during flight. However, the rotor performance is limited due to a low density ($\rho \approx 0.0179 kg/m^3$) and low temperature ($T \approx -60°\mathrm{C}$) resulting in low Reynolds number compressible flows. These flows are characterized by the viscous effect—which allows the formation of large-scale vortices due to boundary layer separation—and the compressibility effect that suppresses the onset shear layer instabilities and result in wake vortices [3].

The chord-based Reynolds number of the current blade is around $Re_c \approx 10^4$ (see figure 1). The airfoil boundary layer at this condition doesn't carry much near-surface momentum and thus it is susceptible to flow separation when it encounters a strong adverse pressure gradient. As a result, effective camber is reduced, and the pressure drag component increases. If the momentum is too feeble, the boundary layer stays unattached while the transition delays forming a wake with periodic shedding. The separated boundary layer may act as a trip and ingest Kelvin-Helmholtz vortices in the layer which divide into turbulent spots ending up in boundary layer reattachment to process to turbulent flow and therefore, creating Laminar Separation Bubble (LSB) [4, 5]. LSBs have detrimental effects on the airfoil efficiency. Besides the low Reynolds number effects is the compressibility effect where the shock waves cause further flow complexities. Johnson *et al.* [6] and Munday *et al.* [7] investigated the effect of highly subsonic Mach number on CLF 5605 airfoil and triangular airfoil respectively. Their findings reveal that the airfoil performance tends to diminish as Mach number augments.

Generally, performance of a rotor blade can be improved by selecting an appropriate cross-sectional airfoil. Koning [8] recapitulated the information needed to back up the selection of airfoil for a Martian propeller. Cambered and thick airfoils can be used to delay the flow separation. Thus, NACA 5605 was selected as the cross-sectional airfoil of the blade.

Numerical optimization using 3D RANS have been applied to blade shape optimization recently. The majority of these efforts to improve efficiency rely on the gradient-based Optimization GBO method [9, 10]. The method works as a population-based evolutionary and needs a significant reprogramming in the routine analysis. On the other hand, Response Surface Methodology introduced more than two decades ago has many advantages over GBO methods. It can work with data collected from various tools and does not require the calculation of the design parameters. Moreover, RSM can quiet down the high frequency disturbance of the output function providing an optimum solution and can consider more than one objective function [11]. RSM has been extended to various fields beyond its original use and is reported to be more suitable for identifying the effect of individual variables and locating the optimum combinations [12, 13].

The current study takes as baseline the rotor system of Ingenuity with the aim of increasing further the thrust and Figure of Merit. The payload of the vehicle was 1.8kg and the rotor of 1.21m diameter were designed to function up to 2800RPM reaching a tip Mach number and speed of $Ma_{tip} = 0.76$ and $V_{tip} = 177m/s$ [14]. In the current study, the Faced Central Composite Design FCCD within RSM associated with 3D Navier-Stokes and Finite Element Analysis is employed to optimize a single Martian propeller where the design variables are the airfoil chord lengths and its twist angles. The blade radius is the same as that of MH Ingenuity and the tip Mach number is constrained to $Ma_{tip} = 0.75$ $(a = 233.13m/s)$. The work constitutes the first use of the combination RSM-CFD&FEA for the Martian blade shape optimization.

2. Surrogate Model Optimization

2.1. Optimization Strategies

The factors affecting blade efficiency are the airfoil chord distribution, twist angle of individual airfoil and blade radius [15, 16]. The current blade radius was constant. The chord length and twist angle distribution were optimized separately by first considering the untwisted blade planform as shown in figure 2. Through the blade span, NACA 5605 was used. The blade was divided into three sections: root, mean and tip. The root, mean and tip airfoils were placed at 19%, 86% and 1% of the blade radius respectively.

There were in total three design variables for each optimization phase as summarized in table 1 with their range. For each optimization phase, twenty combinations were obtained and the total area A_{total} of each combination was approximately the same as or less than that of Ingenuity. Based on the fact that the velocity at the root is lower than the velocity at tip [17], the root airfoil needs to be inclined at high angle of attack meanwhile avoiding major flow separation. Therefore, the ranges of [9deg, 12deg] and [1deg, 3deg] were used for the root and tip airfoils respectively. The pitch axis was placed at 40% of chord length [18]. The aim of the different combinations was to obtain the thrust force that offsets the required force ($T_{req} = 1.674N$ per blade) as well as maximizing the hovering performance and examining the factor of safety FoS defined respectively as follows:

$$FM = \sqrt{\frac{2}{\pi}\frac{C_T^{1.5}}{C_P}} \quad and \quad FoS = \frac{Yield\ stress}{Working\ stress} \tag{1}$$

where C_T *and* C_P are thrust and power coefficients respectively.

Table 1. Lower and upper bounds of the design variable values.

Chord length			Twist angle		
Variable	Lower	Upper	Variable, deg	Lower	Upper
L_2	0.13	0.17	θ_2	9	12
L_3	0.07	0.11	θ_3	5	7
L_4	0.012	0.018	θ_4	1	3

Mars has an ultra-low average temperature which can cause corrosion of materials and reduces their fatigue life. Due of its capacity to withstand corrosion and its lightweight, Carbon Fiber was selected and applied to the blades.

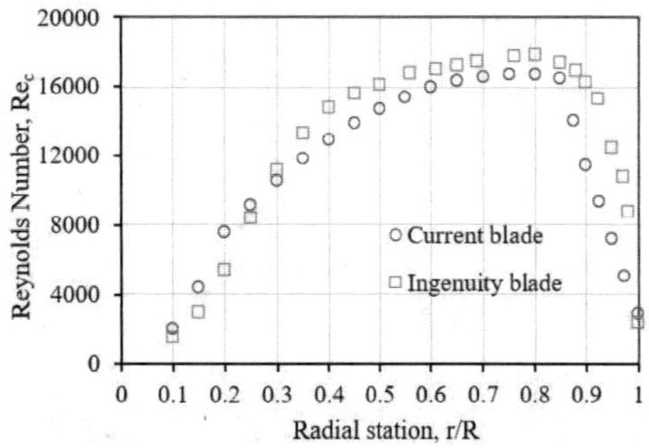

Figure 1. Approximate chord-based Reynolds number of the blade along the spanwise at maximum design parameters (chord lengths).

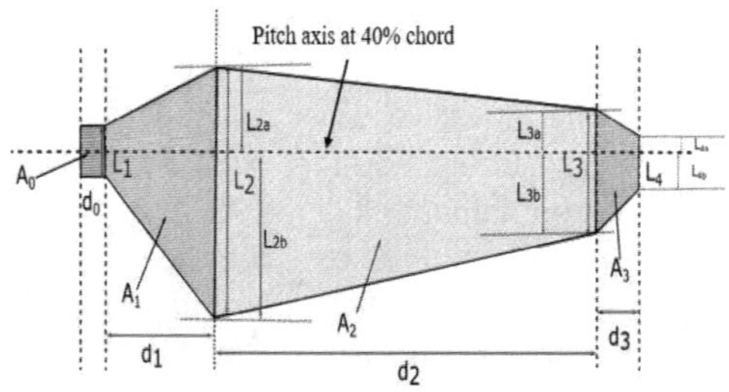

Figure 2. Parameters characterizing the blade.

2.2. Physical Modelling

Fluent solver was employed to evaluate the aerodynamic characteristics of the blades. It is built on the integral form of the governing equations which guarantees the enforcement of the conservation laws. Koning *et al.* [19] tested the performance of a fully turbulent model with that of a transition model in predicting FM of the rotors of Ingenuity and it was found that the transition model predicts a well-agreed results with the experimental curve of FM. Therefore, RANS transition model $SST\,\gamma - Re_\theta$ was utilized.

The aerodynamic characteristics of blades were to be evaluated, and it is important to find out a number of grid elements which does not affect the results. To verify the grid sensitivity, a total of four grids were considered to evaluate for the thrust and torque at 2737RPM and since the flight regime is hover, the incoming velocity was zero. Figure 3 showcases the static thrust coefficient C_T and power coefficient C_P of the four grids. As the number of elements increased, the result arrived to be relatively unchanging. Grids 1 and 2 were inappropriate as the results of C_T and C_p were unstable. Grids 3 and 4 generated a nearly unchanged C_T and C_p. Therefore, considering the computational cost, the number of elements for Grid 3 was applied to all the blades.

Table 2. Verification of Grid Sensitivity.

Grids	No. of spanwise nodes	No. of chordwise nodes	No. of leading edge nodes	No. of elements
1	111	45	12	1.05×10^5
2	138	54	18	2.07×10^6
3	201	67	22	3.97×10^6
4	238	80	34	5.32×10^6

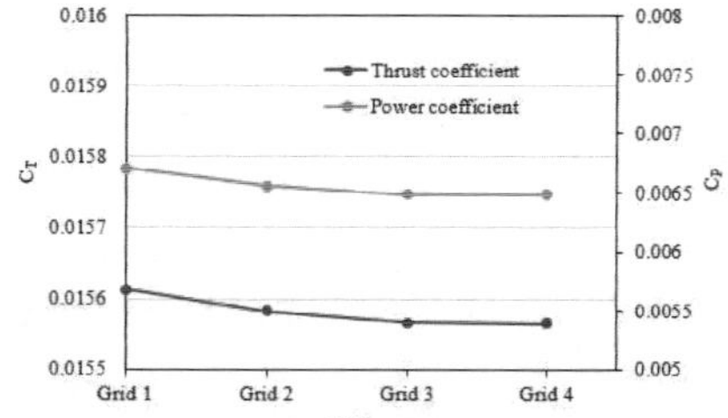

Figure 3. Comparison of Thrust and Power coefficient for the four grids.

3. Results

RSM provides a complet approach to design and evaluation which includes analysis of variables, building of the model and adequacy and uncertainty estimates. Table 3 presents the ANOVA of FM for the significance test. The global p-value of the model was significant as p-value was less than 0.0001. P-value is a measure of how significance a model or an input parameter is. When the p-value is small than 0.05, the parameter or the model is said to be significant otherwise it is insignificant. Factor C was not significant but its interaction with other factors were significant. The significant factors were A and B. The parameters A, B and C were related to the individual effect of L_2 (chord length of root airfoil), L_3 (chord length of mean airfoil), L_4 (chord length of tip airfoil) respectively on FM response. AB, AC and BC represented the second order interaction terms of L_2 and L_3; L_2 and L_4, L_3 and L_4 respectively. The mathematical model of FM in terms of coded factors was given by:

$$FM = 0.2764 + 0.9321A + 0.5445B + 0.3757C - 3.2546AB - 14.2335AC + 20.2875BC. \quad (2)$$

From the sign of the coefficients of the above equation, the factors A, B, C and the interaction BC had positive effects whereas AB and AC had negative effects on FM response. The magnitude of these coefficients showed the quantity of their influence.

Table 4 summarizes the fit statistics of the current model. The correlation factors R^2, adjusted and predicted R^2 were close to 1 and the difference between adjusted R^2 and predicted R^2 was less than 0.2. Guinta *et al.* [20] showed that if the value of adjusted R^2 is between 0.9 and 1, the current model can predict with some accuracy the variation seen in the objective function. Moreover, Adeq Precision measuring the signal to disturbance ratio was above 4 which is desirable. Therefore, this model was reliable and could be used to navigate the design space.

Table 3. ANOVA for the study of significance of design variables on Figure of Merit.

Source	Sum of Squares	df	Mean Square	F-value	p-value	
Model	0.0475	6	0.0079	19.90	< 0.0001	significant
A-L2	0.0056	1	0.0056	6.26	<0.0001	
B-L3	0.0391	1	0.0391	102.02	< 0.0001	
C-L4	0.0003	1	0.0003	0.6478	0.2293	
AB	0.0001	1	0.0001	0.1222	0.5715	
AC	0.0013	1	0.0013	5.26	0.0135	
BC	0.0012	1	0.0012	5.06	0.0166	
Residual	0.0001	13	5.728E-06			
Lack of Fit	0.0000	8	5.009E-06	0.7282	0.6720	not significant
Pure Error	0.0000	5	6.879E-06			
Cor Total	0.0008	19				

Table 4. Fit statistics for Figure of Merit.

Fit statics	Value
R^2	0.9583
Adjusted R^2	0.9390
Predicted R^2	0.9255
Adeq Precision	24.43

Relying on the model result, a numerical optimization was carried out to identify the optimum parameters while maximizing thrust and minimizing torque. The Desirability approach within the FCCD-RSM in Design Expert software was used. Table 5 summarizes the optimum arrangements along with predicted and computed Figures of Merit. The optimum tip chord length corresponded to the minimum value within the range which allows for a sweep resulting in reduction of compressibility effect and performance enhancement. The Fluent solver result for FM agreed quite well with the predicted value by Design-Expert with an discrepancy error of 2.52%.

Table 5. Blade performance at optimum chord distribution.

Variable	Value	$FM_{predicted}$	$FM_{computed}$
L_2	0.16		
L_3	0.08	0.457	0.4685
L_4	0.012		

The twist angle optimization was performed to balance the lift distribution over the spanwise in order to enhance the performance metrics. In addition, the evaluation of the thickness distribution for each blade sample was examined by considering the factor of safety. Mars Planet has an ultra-low average temperature ($T \approx -60°C$) and Carbon Fiber whose properties are compatible to such temperature (see table 6) was applied to the blades while conducting the structural analysis. The hub was constrained as fixed support and the aerodynamics and inertia loads were applied on the blades. The highest performance metrics for the blade were obtained when $\theta_2 = 9deg$, $\theta_3 = 7deg$ and $\theta_4 = 3deg$ with the thrust of 2.589 and FM of 0.69. The compressive and tensile FoS of this blade were largely above the requirement—5.34 and 11.08.

Table 6. Properties of Carbon Fiber.

Property	Value
Density, kg/m^3	1800
Young modulus E_x and E_y, Pa	2.3×10^{11}
Tensile Strength, MPa	2231
Compressive Strength, MPa	1082

In order to understand the effect of flow feature on FM, the flow separation along the spanwise of optimized blade was evaluated using X-Shear Stress plot. The separation point is located when the upper surface curve of the cross section crosses the X-axis towards the negative Y-axis. Around the root section, the flow remained attached over the cross-section until $r/R = 0.3$. From $r/R = 0.3$ to $r/R = 0.6$, the separation point moved towards the leading edge. Towards the blade tip where the compressibility effect began to be pronounced ($r/R = 0.7{\sim}0.9$), the separation point remained somewhat constant. And whereas at the blade tip where the compressibility was pronounced, it shrank back. The tip Mach number of the Martian blade yielded a value around 0.766 as can be seen in figure 5, less than the sonic Mach number.

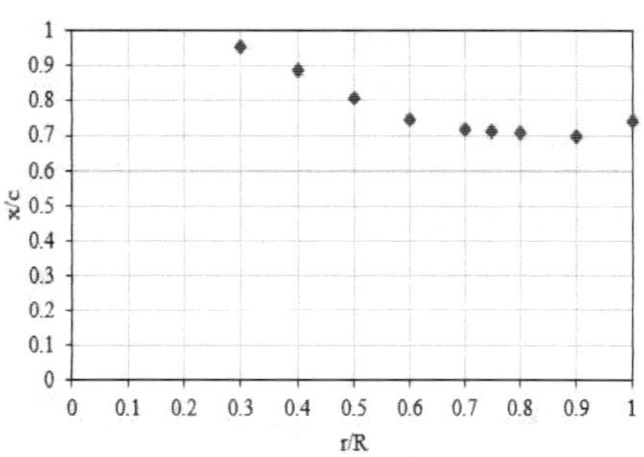

Figure 4. Flow separation points along the span.

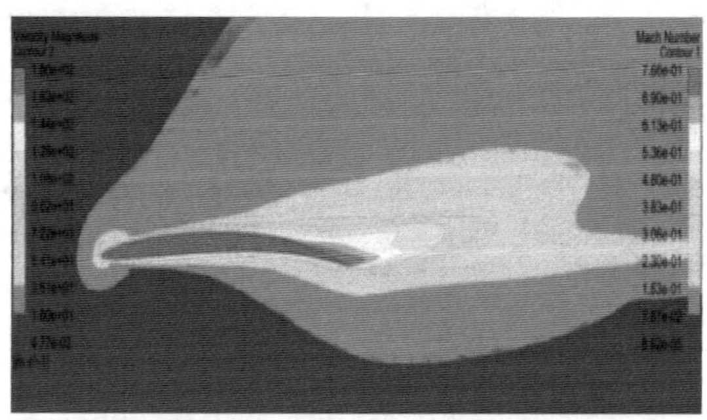

Figure 5. Velocity contour and Mach number at blade tip.

4. Conclusion

The combination of the surrogate model RSM and High Fidelities CFD&FEA was carried out to obtain optimized blade which generates thrust above 1.674N per blade as well as improving FM. It was found that the optimized propeller attained a thrust of $T = 5.178N$ and Figure of Merit of FM = 0.69 with minimum compressive and tensile FoS of 5.34 and 11.08 respectively. The flow physics revealed that the root section did not experience flow separation when the root airfoil was inclined at $AoA = 9{\sim}12deg$ and the separation moved towards the leading edge along the radial with no reattachment. The current study constitutes the first use of the combination RSM-CFD&FEA for the Martian blade shape optimization.

5. References

[1] Balaram J, Daubar I J, Bapst J and Tzanetos T 2019 *9th Int. Conf. On Mars.*
[2] O'Brien P C 2003 *AIAA* 2993-6275 23 - 25 September, Long Beach, California.
[3] Suwa T, Nose K, Numata D, Nagai H and Asai K 2012 *30th AIAA Appl. Aerod. Conf.* 25 - 28 June New Orleans, Louisiana
[4] Alam M M, Zhou Y, Yang H X, Guo H and Mi J 2010 *Exp. Flu.* **48** 81-103.
[5] Windte J, Scholz U and Radespiel R 2006 *Aero. Sci. Technol.* **10** 484-94.
[6] Johnson W *et al.* 2020 Mars science helicopter conceptual design. *NASA/TM*–2020-220485.
[7] Munday P, Taira K, Suwa T, Numata D and Asai K 2015 *J. Airc* **52** 924-31.
[8] Koning W J F 2019 *NASA/CR*–2019-220236.
[9] Ning A 2021 *Structural and Multidisciplinary Optimization* **64** 991-1014.
[10] Yang X, Ma D, Zhang L, Yu Y, Yao Y and Yang M 2023. *Aeros. Sci.Technol.* **133** 108142.
[11] Myers R H, Montgomery D C and Anderson-Cook C M 2016 *Response Surface Methodology* 4th edn (John Wiley & Sons).
[12] Landman D, Burnette D C, Konuk E and Baris E 2022 *AIAA Aviation Forum.*
[13] Kim K 2004. *J. Flu. Eng.* **126** 735-42.
[14] Grip H F, Johnson W, Malpica C, Scharf D P, Mandić M, Young L, Allan B, Mettler B and Martin M S 2017 *43rd Euro. Rot. For. Coun. of Euro. Aeros. Soc.*
[15] Mian H H, Wang G, Zhou H and Wu X 2021 *Aeros. Sci.Technol.* **111** 106563
[16] Kwon H Il, Yi S, Choi S and Kim K 2015 *J. Propulsion and Power* **31** 1057-72.
[17] Conlisk A T 2001 *J. Aeros. Sci.* **37** 419–76.
[18] Lumba R, Chi C, Datta A, Koning W, Perez Perez N and Cummings H 2023 *J. Amer. Helic. Soc.* **68** 42003-42018.
[19] Koning W J F, Johnson W and Grip H F 2019. *AIAA* **57** 3969-79.
[20] Guinta A A 1997 PhD thesis, Department of Aerospace Engineering, Virginia Polytechnic Institute and State University, Blacksburg, VA.

Funding

The authors disclosed receiving the following financial aid for the research, authorship, and/or publication of this article. The present work is supported by the National Natural Science Foundation of China under Grant No. 12002161, 12032011, and the National Key Laboratory of Helicopter Aeromechanics, Funding No. 2023-HA-LB-067-06.

A three-dimensional path planning method of UAV formation based on deep Q-network

Chao Zhang[1,2], Boyuan Chen[1], Chunshi Fan[3], Linghui Yu[3] and Zhiwen Wang[3]

[1] Department of Integrated Technology and Control Engineering, School of Aeronautics, Northwestern Polytechnical University, Xi'an 710072, China
[2] National Key Laboratory of Aircraft Design, Xi'an 710072, China
[3] DFH Satellite Co. Ltd, Beijing 100094, China

caec_zc@nwpu.edu.cn; boyuan.chen@mail.nwpu.edu.cn

Abstract. Nowadays, unmanned aerial vehicles (UAV) are playing an increasingly important role in both military and civilian fields. However, previous studies have mostly focused on a single drone or 2D route planning, and they have not combined Deep-Learning with route planning. We have made some efforts to make up this vacancy. In this paper, a Deep Q-Network (DQN) based UAV formation path planning method is proposed to overcome the limitations of previous research. Firstly, a new modeling scheme for the state space and action space of drone formations in discrete space is proposed by this article. Secondly, this article introduces deep learning algorithms into path planning and compares them with traditional algorithms, achieving good results. Thirdly, based on the previous work, the complex functions of deep learning algorithms were utilized to classify risk areas and enable intelligent agents to make decisions, achieving good results.

1. Introduction

Along with the continuous development of unmanned aerial vehicle design, aircraft control, information engineering, artificial intelligence and other fields, the application scenarios of unmanned aerial vehicles in military and daily life have also become increasingly diverse. In recent years, UAV technology has become more mature, and UAV formations have become a hot research topic due to their high task efficiency and more functions. As a critical technology reflecting the autonomous control capability of drones, the essence of path planning is to design a collision-free optimal path in the three-dimensional space environment from the starting position to the destination [1-3].

The selection of path planning algorithms is also an important part. We roughly divide path planning algorithms into three categories: traditional optimization algorithms, intelligent algorithms, and algorithms based on deep reinforcement learning. Traditional optimization algorithms include A* algorithm, Dijkstra algorithm, etc. [4-5]. Hernändez-Hernández et al. proposed a multi-objective A* algorithm for finding unmanned aerial vehicle paths from the starting point to the destination, taking into account travel time, path angle, fuel consumption, and area deviation [6]. Intelligent algorithms include genetic algorithm, particle swarm optimization algorithm, etc. [7-8]. Wang et al. proposed an improved genetic algorithm for solving the path planning problem of unmanned aerial vehicle formations facing heterogeneous targets. It combines the knapsack problem, shortest path problem, and scheduling problem, and has good practicality in large-scale instances [9]. Li et al. establish an interaction model between UAVs, terrain, and enemy threats. Based on this, we introduce an ant colony algorithm enhanced with a fuzzy logic memory mechanism (FLM-IACO) to obtain lower-cost flight routes [10]. Classic reinforcement learning algorithms include DQN algorithm and Deep Deterministic Policy Gradient (DDPG) algorithm [11-12]. Hu et al. proposed a route planning algorithm for unmanned aerial vehicles based on REL-DDPG, which effectively solves the route planning problem in complex three-dimensional terrain environments [13]. Chao et al. proposed a new bio-inspired path planning method using E-DQN (event-based deep Q-network), referring to introducing event stream to reinforcement learning network in order to help the drone to find the goal in the environments [14].

However, most drone path planning only focuses on a single drone, and the maneuverability and physical performance of drone formations differ significantly from those of a single drone. Following the original path planning methods can lead to problems such as collision with obstacles and collisions between drones. Modeling the state space and action space of drone formation can effectively solve the above problems.

The advantages of various deep reinforcement learning algorithms are combined to re model unmanned aerial vehicle formations, endowing them with new action and state spaces. A Double-DQN network model is constructed and applied to the path planning problem of unmanned aerial vehicle formations in complex environments, achieving good results [15]. The average calculation speed in a 100 * 100 * 100 data grid is 0.4 seconds each time.

The structure of this paper is as follows. In Section 2, modeling of the battlefield environment and UAV formation was completed. Section 3 builds the DQN network and defines the reward function. Section 4 describes the simulation results. The conclusion is presented in Section 5.

2. Modelling of environment and UAV formation

2.1. Modelling of environment

The environment in which drones perform tasks includes but is not limited to complex terrains such as cities, suburbs, and mountains. For environment modeling, we divide it into basic terrain modeling and risk zone modeling. They are both divided into two parts for model implementation and graphic visualization.

For undulating terrain, it is impossible to accurately obtain terrain data for each point. In this article, we adopt the method of two-dimensional cubic spline interpolation to construct a denser terrain grid by interpolating the sampling points. This method can simulate terrain finely while completing the task of terrain discretization and facilitating subsequent visualization operations. The principle of two-dimensional interpolation is shown in Figure 1.

In the battlefield environment, battlefield threat is one of the important constraints in drone route planning, mainly including radar detection, anti-aircraft missiles, anti-aircraft guns, etc. When executing route planning, drones need to avoid the threat area on the battlefield and complete designated tasks without causing damage to themselves. This article takes the location of air defense weapons as the center of the circle and sets a circular threat zone based on the combat radius of air defense weapons. The visualization result of the final task environment modeling is shown in Figure 2.

2.2. Modelling of UAV formation

The flight performance of a UAV formation shares similarities and differences with individual UAVs. The main difference lies in the action that UAV takes. We take a step size unit of 100m for example, in the action space of "left ahead", the drone formation needs to first turn left and then return the nose to reach the next point. This is difficult for unmanned aerial vehicle formations of several tens of meters in size. Moreover, this complex action directly increases the possibility of collisions within the drones in a formation, bringing certain risks.

Journal of Physics: Conference Series **3041** (2025) 012003 doi:10.1088/1742-6596/3041/1/012003

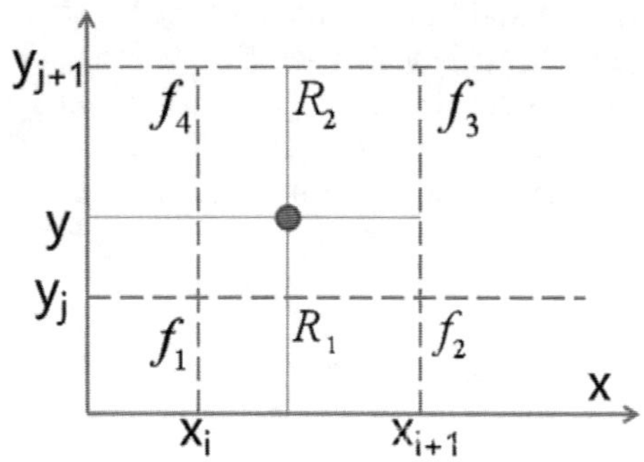

Figure 1. The principle of two-dimensional interpolation.

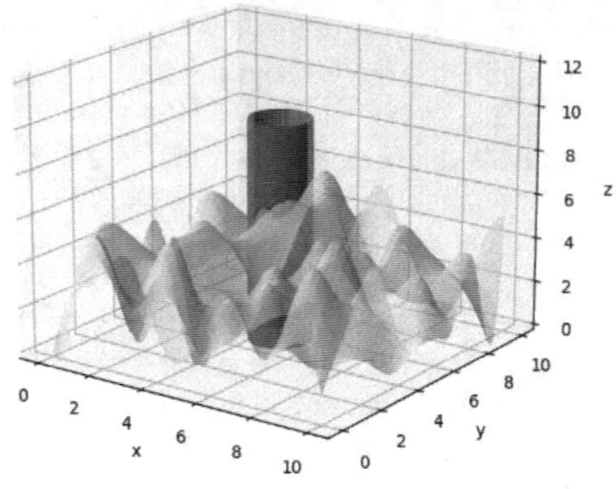

Figure 2. Modeling diagram of terrain and risk areas.

In order to improve the safety redundancy of drone formation flight, this paper re models the action space of drone formation, and designs a total of nine action spaces as shown in Figure 3.

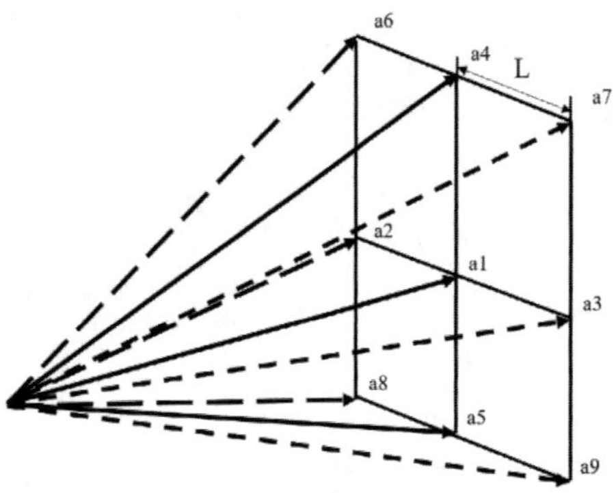

Figure 3. The action space of UAV Formation

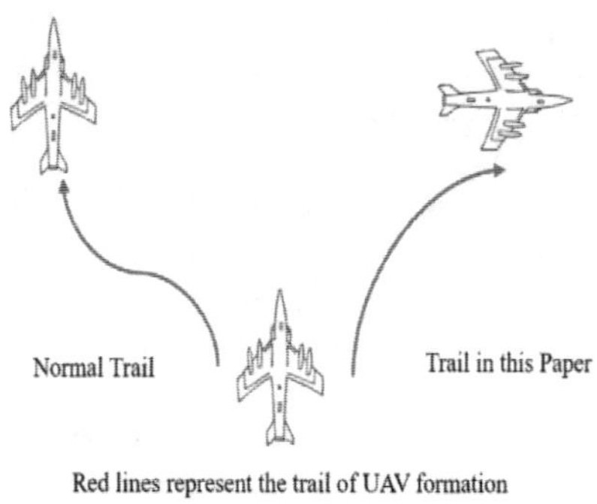

Figure 4. Difference between normal trail and that in this paper.

Unlike the conventional action space, the intelligent perception also changes its orientation after moving left/right, as shown in Figure 4. The advantage of this action state space design is that it is more in line with the actual flight situation of the drone formation, ensuring safe flight of the formation while minimizing step size and improving the accuracy of path planning.

The modeling of action space brings about the problem of state space modeling. The action space before and after changing the orientation of UAV formation is reflected differently in computer programs, which means that the same action needs to perform different operations on the x, y, z coordinates in different orientations. Therefore, based on other similar studies, this article introduces a fourth state space besides the x, y, and z three-dimensional state spaces: orientation. It has four values, representing the positive and negative directions of the x-axis and y-axis, respectively.

3. Route planning algorithm based on DQN

3.1. Deep Q-network

Reinforcement learning is a machine learning method based on Markov decision processes. Define the interaction methods and benefits between intelligent agents and the environment, and continuously try and learn through the interaction between intelligent agents and the environment. Ultimately, a sound decision-making strategy is formed.

DQN is based on Q-Learning and uses Deep Neural Networks (DNN) to replace the Q-table, which is the action value function. We represent it as: $Q(s_t, a_t|\theta)$. s_t and a_t represent the state of the intelligent agent and the selected action at time t. θ represents network parameters. DQN updates the network by minimizing the loss function, thereby making the output decision of the network the optimal solution. The loss function is shown in Equation 1.

$$L(\theta)=\frac{1}{2}[y-Q(s_t,a_t\,|\,\theta)]^2 \tag{1}$$

The neural network of the agent in the DQN algorithm suffers from overfitting during the update process due to strong correlation between training samples. Double DQN introduces Target-Net on the basis of DQN to solve the problem of overfitting. After selecting actions through Evaluate-Net, update the Evaluate-Net based on the Q-value in target net. Finally, after a period of calculation, use Evaluate-Net to replace the old Target-Net. The entire process is shown in Figure 5.

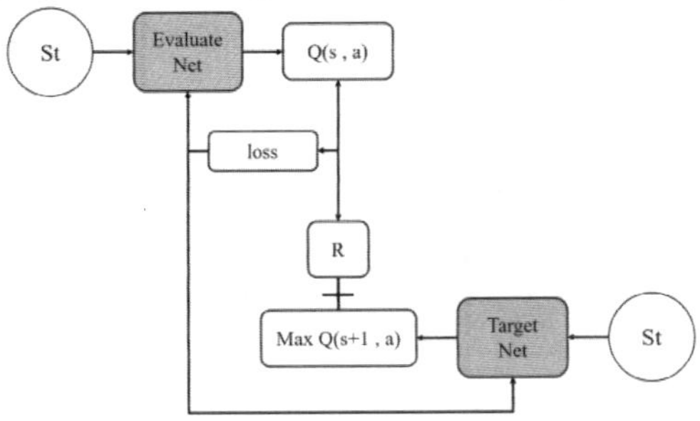

Figure 5. Structural framework of Double-DQN

This paper uses Double-DQN to solve the route planning problem of UAV formations and constructs an algorithm model. By selecting the best action based on the current state s of the drone, the maximum benefit can be obtained, ultimately achieving global route planning for the UAVs in the environment.

3.2. Implementation of algorithms

3.2.1. State space and action space. In the battlefield environment, the state space of the UAV formation, as described in Chapter 2, includes x, y, z three-dimensional coordinates and the orientation of the current drone formation's nose.

We assume that a terrain sampling point near the starting point is the origin of the coordinate axis. And define the center point of the drone formation as the coordinates of the intelligent agent. The drone formation is difficult to maintain a fixed shape during flight, so we model it as a sphere based on its maximum flight range, which facilitates subsequent calculations and programming.

The x, y, and z coordinates of the drone formation can be obtained by the relative distance between the coordinates of the drone formation and the origin of the coordinate axis, and the corresponding orientation can be obtained after each action. In summary, the expression for the state vector of the drone is:

Journal of Physics: Conference Series **3041** (2025) 012003 doi:10.1088/1742-6596/3041/1/012003

$$s_t = \left[s_t^x, s_t^y, s_t^z, s_t^o \right] \tag{2}$$

The action space is a collection of actions that drones may take. The action space set in this article includes the following nine actions: a1, flying forward; a2, turn left; a3, turn right; a4, fly forward and upwards; a5, fly forward and down; a6, fly left and up; a7, fly right and up; a8, fly downwards to the left, a9, Fly downwards to the right. State space A can be shown as:

$$a_t = \left[a_1, a_2, a_3, a_4, a_5, a_6, a_7, a_8, a_9 \right] \tag{3}$$

3.2.2. Reward function. The profit function directly affects the Q value, thereby affecting the outcome of decision-making. This article defines the benefit function from five aspects: terrain obstacle avoidance, risk avoidance, flight guidance, task completion, and UAV height limitation.

- Task completion reward function

When the drone reaches the endpoint, a positive reward is given. Due to the large size of the formation, the reward determination is set within a certain range of the endpoint, taking a sphere. When the sphere representing the drone formation intersects with it, it is judged that the task is completed, and the profit expression is:

$$r_{dest} = \begin{cases} 0 & dist\left(s_{now}, s_{tar}\right) > R \\ k_{dest} & dist\left(s_{now}, s_{tar}\right) < R \end{cases} \tag{4}$$

In the equation, k_{dest} is a positive constant. Function dist () refers to the destination between agent and the target. R is a constant which is decided by actual situation.

- Flight guidance reward function

In order to help drones reach the destination faster, a drone flight guidance benefit function is set up. The difference in absolute distance from the distance before and after the drone formation movement is used to measure the effectiveness of guidance. It is intuitive and can achieve the best effect, and its expression is:

$$r_{lead} = k_{lead} (dist\left(s_{now}, s_{tar}\right) - dist\left(s_{now*}, s_{tar}\right) \tag{5}$$

- Terrain obstacle avoidance reward function

Drones should avoid collisions with the ground. When the distance between the formation and the ground is less than the safe distance, negative benefits are given. The expression is:

$$r_{land} = \begin{cases} 0 & h_{now} - h_{land} > h_{safe} \\ k_{land} & h_{now} - h_{land} < h_{safe} \end{cases} \tag{6}$$

In the equation, k_{land} is a negative constant. h is the height of drones and terrine.

- UAV height limitation reward function

Drones have their static lift limit, making it difficult to ensure safety when flying to a certain altitude. Provide negative benefits when the drone swarm exceeds a certain height. The expression is:

$$r_{height} = \begin{cases} 0 & h_{now} < h_{max} \\ k_{height} & h_{now} > h_{max} \end{cases} \tag{7}$$

Journal of Physics: Conference Series **3041** (2025) 012003 doi:10.1088/1742-6596/3041/1/012003

- Risk aversion return function

The secondary purpose of path planning is to avoid various risk zones, and this article mainly designs reward functions for circular and square risk zones. When the distance between the intelligent agent and the center of the risk zone is less than a certain value, negative returns are given. The expression for a circular risk zone is:

$$r_{dan} = \begin{cases} 0 & dist\left(xy_{now}, xy_{dan}\right) < r + r_d \\ k_{dan} & dist\left(xy_{now}, xy_{dan}\right) < r + r_d \end{cases} \tag{8}$$

xy_{now} represents the x and y coordinates of the current intelligent agent, xy_{dan}represents the x and y coordinates of the center of the risk zone, and $r + r_d$ represents the distance margin of the risk zone, which is determined by the actual situation.

The expression for the square risk zone is:

$$r_{dan} = \begin{cases} 0 & other \\ k_{dan} & abs\left(xy - xys\right) < a \end{cases} \tag{9}$$

In the equation, xys represents the center coordinates of a rectangle. Abs () is an absolute value function.

Based on the definitions of the five benefit functions mentioned above, we can ultimately determine that the final expression for the total benefit for each specific state and action is:

$$R_{total} = r_{dest} + r_{lead} + r_{land} + r_{height} + r_{dan} \tag{10}$$

After confirming the above information, the algorithm network model is constructed as shown in the Figure 6.

4. Experiment and result analysis

This article uses the PyTorch framework to build a Double DQN network in a Python programming environment. The computer used for training algorithms and testing is configured with an Intel i5-13500H CPU, NVIDIA RTX 4050 Laptop GPU, 16GB of running memory, and GPU acceleration is not used.

Table 1. Algorithm parameters

Parameter	Value
Learning Rate	0.02
Batch Size	512
Epsilon	0.9
Memory Capacity	10000

4.1. Basic experimental analysis

In order to accelerate learning speed and avoid falling into local optima or not finding the endpoint, this article has made certain improvements to the main program. In the first two hundred cycles, the greedy strategy EPSILON value is set to 0, which means that in the first 200 planning cycles, the agent will explore in a completely random manner, accumulating as much data as possible until the task is completed or danger is encountered. And in the last 200 planning sessions, set EPSILON to 1 and stop updating the network. Plan the route completely according to the size of the Q value, and use the trained network to plan the path, observe its success rate, average path length and calculation time.

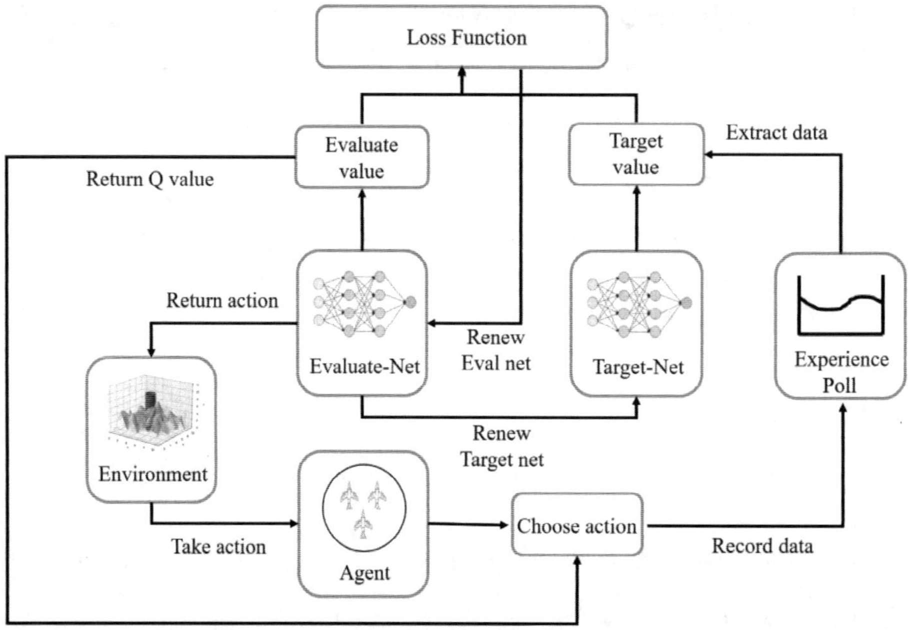

Figure 6. Route planning algorithm framework.

The success rate of route planning is the ratio of the number of times the EPSILON is adjusted to 1 after 1000 rounds of learning, executed completely according to the network's output actions, and successfully completed route planning to the total number of tests. The calculation formula is:

$$G_p = \frac{C_S}{C_A} \tag{11}$$

In the equation, C_S represents the number of successful planning attempts. C_a represents the number of total planning attempts.

The average planned path length is the average length of the intelligent agent's path in each successful plan, and its formula is:

$$S_a = \frac{1}{C_A}\left(\sum_1^n \sum_1^m dist\left(s_t, s_{t+1}\right)\right) \tag{12}$$

The average planning time is the ratio of the total time required to run the entire program to the number of planning iterations, and its calculation formula is:

$$T_a = \frac{1}{C_A}\left(\sum_1^n t_i\right) \tag{13}$$

This experiment randomly generated 10 task environment maps, including terrain and risk zone environments. Simulation calculations were conducted for fixed starting and ending points, and a random learning curve was selected as shown in Figure 7. The horizontal axis represents the number of simulations, and the vertical axis represents the average profit per step of the simulation. The learning curve only records the middle 800 EPSILON=0.9 learning processes for the purpose of observing the learning effect. Through the learning curve, we can see that the neural network reaches initial convergence around 50 iterations and complete convergence around 300 iterations.

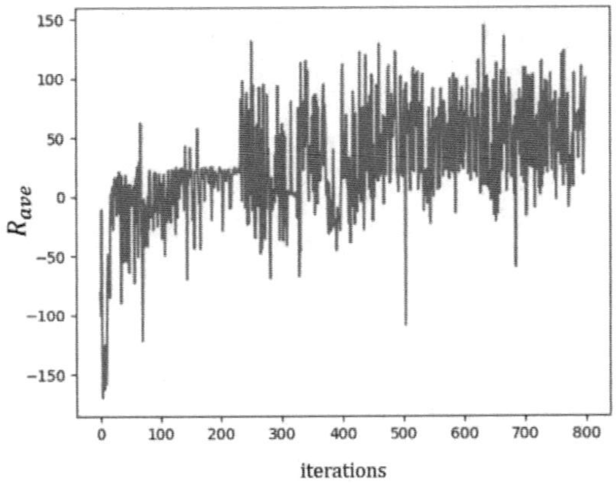

Figure 7. The average reward curve of the algorithm

Through a total of ten maps and a total of two thousand calculations, the average success rate of the planning is 62.8%. For the task in Figure 9, the average planned path length is 12.03km (the straight-line distance between two points is 8.54km), and the average planning time is 0.4s.

At the same time, we compared the calculation results of the same task with A* algorithm and Ant Colony Optimization, and the results are shown in the following table.

Table 2. Formatting sections, subsections and subsubsections.

Algorithm	Average time	Success rate	Shortest route
DQN	0.4s	62.8%	11.61km
A*	5.3s	100%	10.82km
ACO	3.8s	24.6%	11.03km

The partial route planning results obtained are shown in Figures 8-9.

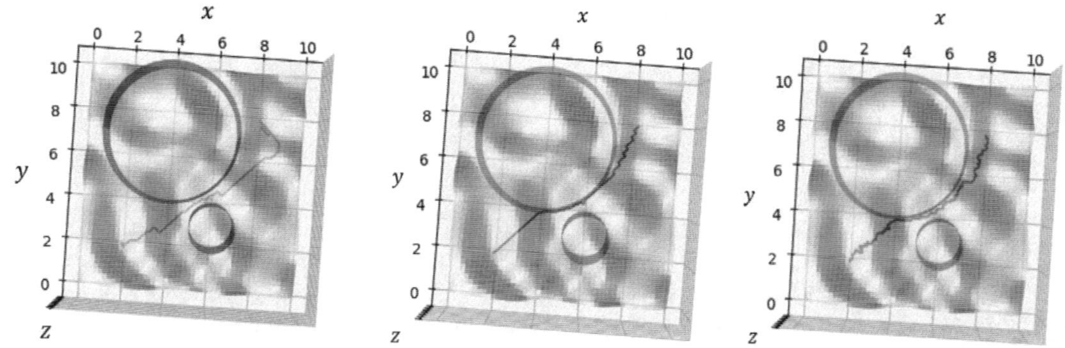

Figure 8. Route planning results carried out by DQN, A*, ACO.

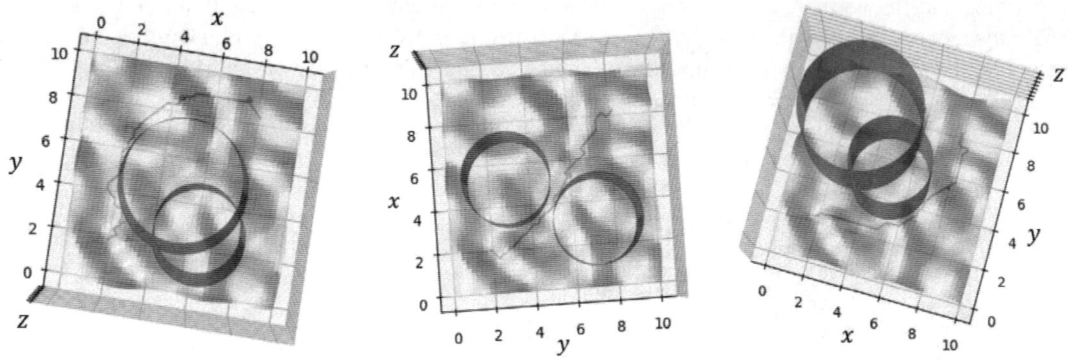

Figure 9. Some of route planning results carried out by DQN.

Obviously, although the A * algorithm has the highest success rate and a shortest planning path, the computation time is too long. ACO has a shorter computation time but often requires a high number of iterations, and the success rate in the previous iterations is extremely low, resulting in poor results. Both algorithms above cannot handle the more complex problems presented later through simple adjustments. In contrast, the DQN algorithm used in the article is faster and has a higher success rate, making it a more suitable algorithm for the current model. In addition, DQN can solve more complex problems in order to adjust the real environment.

4.2. More complex risk zone experimental analysis

4.2.1. Experimental analysis of graded risk zones. In the actual task process, not all tasks have completely risk-free areas within the task scope, and not all risks need to be avoided. Therefore, this article conducted multiple experiments to cover almost all flight routes in the risk zone, but the degree of danger varies among different risk zones. The lower risk zone is larger, and a lower absolute value of k is set. Ten maps were randomly generated for planning and calculation, and the final partial route planning results are shown in the figure 10.

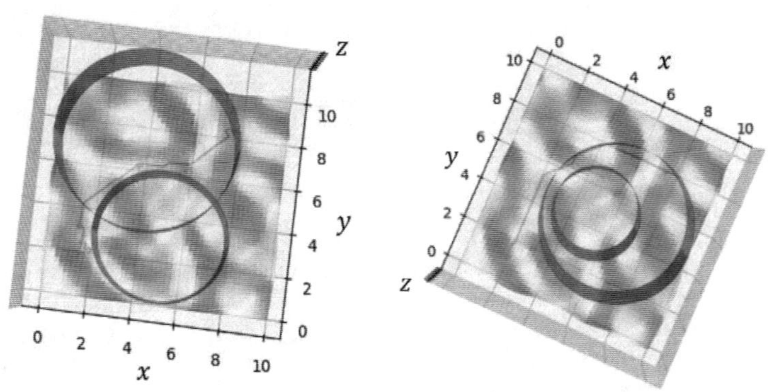

Figure 10. Result of multiple risk areas route planning.

4.2.2. Experimental analysis of square risk zones. In reality, it is often necessary to consider the situation of rectangular risk areas, and the same method of 3D drawing is used to represent the rectangular risk areas. The final experimental results obtained by randomly generating 5 maps for calculation are shown in the figure 11.

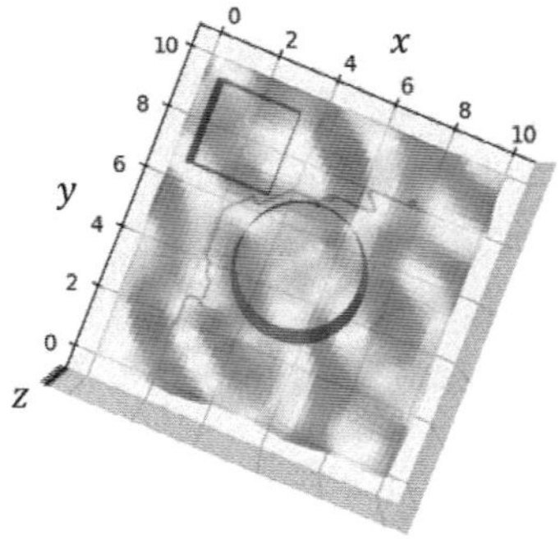

Figure 11. Result of square risk areas route planning.

4.2.3. Experimental analysis of larger map. The previous experiments were conducted on a 10km * 10km map. In order to demonstrate the feasibility of the method on a larger map, a new 20km * 20km map was designed. In theory, the use of the flight guidance benefit function in Chapter 4 is independent of map size, step size, etc., as it only depends on how much one step can approach the target. Due to the change in map size and the number of data points, it is necessary to adjust the parameters in Chapter 5 accordingly. The sampling data size (BATCH-SIZE) needs to be changed from 512 to 1024, and the memory capacity needs to be changed from 10000 to 40000. During the experiment, the square risk area and graded risk area in front were combined with a large map to randomly generate 5 maps. Some route planning results are shown in the figure 12.

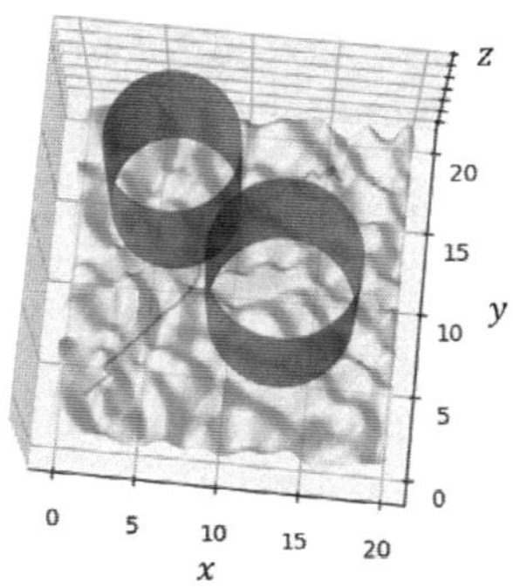

Figure 12. Result of larger map route planning.

However, due to the increase in data volume, the average path planning success rate decreased to 53.5% and the average path planning duration increased to 2.1 seconds.

5. Conclusion

This paper introduces the concepts and research background of unmanned aerial vehicle (UAV) path planning and deep learning, especially DQN. It summarizes and analyzes the current research status of UAV path planning problems, especially in three-dimensional environments. Based on the existing research, a model for planning UAV flight paths in complex mountainous environments with multiple shapes and hazards is constructed by combining deep Q-Network technology. Intended to quickly assist drones in planning a shorter path to their mission destination in complex environments. Unlike previous studies, this article focuses on intelligent agents for large drone clusters. A new model is proposed for the flight dynamics performance of drone clusters, optimizing the algorithm function, and adding a new value of danger to the risk zone, which increases the selection method of intelligent agents.

In the future, we will continue to optimize the algorithm and adopt more advanced improvement algorithms compared to Double DQN to improve computing speed and planning success rate. Secondly, real-time improvements should be made to the entire program to enable it to continuously plan paths during the intelligent agent's flight process. Finally, optimize the current battlefield model to better fit the actual formation flight and battlefield scenarios.

Journal of Physics: Conference Series **3041** (2025) 012003 doi:10.1088/1742-6596/3041/1/012003

References

[1] Zhou C, Gu S, Wen Y, The review unmanned surface vehicle path planning: Based on multi-modality constraint, vol.200. Ocean Engineering. 2020. pp 107043.

[2] Mu D, Li T, Han X, et al. Global path planning for Unmanned Surface Vehicles in complex maritime environments considering environmental interference, vol 310. Ocean Engineering. 2024. pp 118580

[3] Clerk Maxwell, A Treatise on Electricity and Magnetism, 3rd ed., vol. 2. Oxford: Clarendon, 1892. pp 68–73.

[4] Zhang A, Li C, Bi W, Rectangle expansion A* pathfinding for grid maps. vol.29. Chinese Journal of Aeronautics, 2016. pp 1385-1396.

[5] Bulut Okan, Shin Jinnie, Cormier Damien, C. Learning Analytics and Computerized Formative Assessments: An Application of Dijkstra ’ s Shortest Path Algorithm for Personalized Test Scheduling, vol.10. Mathematics, 2023. pp 2230.

[6] L. Hernändez-Hernändez, A. Tsourdos, H.-S. Shin, A. Waldock, “Multi objective UAV routing,” in Proc. Int. Conf. Unmanned Aircraft Syst., Orlando, FL, USA, 2014. pp 534–542.

[7] Albalawneh Da’ad Ahmad, Mohamed M.A, A new federated genetic algorithm-based optimization technique for multi-criteria vehicle route planning using ArcGIS network analyst, vol.20. International Journal of Pervasive Computing and Communications. 2024. pp 206-227.

[8] Lu C, Yang J, Bernt Johan Leira, et al. Three-Dimensional Path Planning of Deep-Sea Mining Vehicle Based on Improved Particle Swarm Optimization. vol.11. Journal of Marine Science and Engineering. 2023.

[9] Wang X, Liu Z, Li X, Optimal delivery route planning for a fleet of heterogeneous drones: A rescheduling-based genetic algorithm approach. vol.179. Computers & Industrial Engineering. 2023. pp 1-13.

[10] Li Y, Zhang Z, Sun Q , Huang Y, An improved ant colony algorithm for multiple unmanned aerial vehicles route planning. vol.361. Journal of the Franklin Institute. 2024, pp. 107060-107060.

[11] Volodymyr Mnih, Koray Kavukcuoglu, David Silver, et al. Playing Atari with Deep Reinforcement Learning. Computer Science, 2013.

[12] Timothy P. Lillicrap, Jonathan J. Hunt, Alexander Pritzel, et al. Continuous control with deep reinforcement learning. Proceedings of the 4th International Conference on Learning Representations, San Juan, Puerto Rico, 2016.

[13] Hu Z, Gao X, Wan K, et al. Relevant experience learning: A deep reinforcement learning method for UAV autonomous motion planning in complex unknown environments. vol.34. Chinese Journal of Aeronautics, 2021. pp 187-204.

[14] Chao Y, Rüdiger Dillmann, Arne Roennau, Xiong Z, E-DQN-Based Path Planning Method for Drones in Airsim Simulator under Unknown Environment. vol.9. Biomimetics (Basel, Switzerland), 2024. pp 238.

[15] H. Jang, S. Hao, P.M. Chu, P. K. Sharma, Y. Sung, K. Cho. Deep Q-network-based multi-criteria decision-making framework for virtual simulation environment, vol.33. Neural Comput. & Applic. 2021. pp 10.

Acknowledgements

This work is partially supported by the National Key Scientific Research Projects of China (JSZL2022607B002, JCKY2022203C027), the Fundamental Research Funds for the Central Universities (HYGJXM202310 and HYGJXM202311). The authors also gratefully acknowledge the helpful comments and suggestions of the reviewers which have improved the presentation.

Research on Two-dimensional Torsion Resistance Characteristics of GEO Telecommunication Satellite Extravehicular Motion Cable

Siqiao Ge[1*], Beifei Sheng[1], Yu Fu[1], Fengping He[1], Dongyang Luo[1] and Xuyang Du[1]

[1]Institute of Telecommunication and Navigation Satellites, China Academy of Space Technology, 100094, Beijing, China

[*]Email: gesiqiao2008@163.com

Abstract: The connecting cable between the inter-satellite link (ISL) antenna and the deployment arm of geosynchronous Earth orbit (GEO) telecommunication satellites undergoes continuous two-dimensional twisting motion, with up to 30,000 rotations during the entire on-orbit lifetime. By establishing a two-dimensional twisting mechanical performance verification platform, we obtained the resistance torque and lifetime characteristics of the extravehicular cable installed using traditional methods. Experiment results showed that the satellite extravehicular cable installed using traditional methods cannot effectively resist mechanical damage caused by antenna twisting throughout the entire life cycle. Therefore, this paper proposes a strength-enhanced cable based on a combination of metal hoses and polyimide films to resist two-dimensional twisting motion. Accelerated lifetime tests have verified the mechanical strength and twisting characteristics of this cable, which can be applied to the external cable of the ISL antenna of GEO telecommunication satellites, effectively improving satellite reliability.

1. Introduction

The number and variety of spacecraft launched into Earth's orbit are increasing yearly. At the same time, the demand for satellite communication from ground telecommunication network equipment is rapidly increasing. More and more satellites or ground equipment rely on telecommunication satellite networks to provide high-dynamic, wide-range, high-speed data relay services, enabling ground stations to continuously track and transmit information data to various spacecraft operating in medium and low orbits, such as satellites, spacecraft, and space stations. As the volume of information transmission from various spacecraft increases, geosynchronous Earth orbit (GEO) telecommunication satellites, as the primary transmission carrier for space-based data relay services, not only face increasing demands for their on-orbit operational services but also face new requirements for the reliability of their on-orbit services. This poses new challenges for the on-orbit application of high-orbit communication satellites [1-3].

Telecommunication satellites involved in building satellite networks are generally equipped with adjustable ISL antennas, which require a two-dimensional antenna drive mechanism to rotate the antenna around the X and Y axes at a certain angle [4]. On one hand, according to the current task demand, the number of rotations of the ISL antenna during the entire life cycle of on orbit telecommunication satellites can reach more than 30000 times. On the other hand, satellites usually require control cable connections between the ISL antenna and the satellite, therefore the cables outside the satellite cabin undergo a large amount of two-dimensional torsional motion throughout the satellite's entire life cycle.

Due to the exposure of satellite cables outside the cabin to the space environment, the total radiation dose borne by the cables far exceeds the level of cables inside the cabin. At the same time, as this section of cable moves with the ISL antenna, a large number of movements throughout its entire life cycle will further exacerbate the aging of the cables. Therefore, existing on orbit satellites have carried out a series of protective designs to address the radiation resistance issue of this cable. As for LEO communication satellites, by studying the radiation resistance characteristics of cables and researching new cable protection measures, metal braided mesh sheaths or tantalum film coatings are usually used for radiation reinforcement [5]. There are also studies on new materials through ground experiments, which aim to improve the radiation resistance of satellite cabin cables by modifying the materials themselves [6]. However, most of the above cable protection measures are aimed at protecting static cables [7,8]. Ground tests have shown that their effectiveness is very limited for cables that undergo a large amount of torsional motion throughout their entire life cycle [9]. Therefore, it is necessary to carry out work to improve the reliability of extravehicular motion cables.

Regarding the enhancement of the reliability of the extravehicular motion cable, this paper conducts research in the following aspects: Firstly, through experimental methods, the reliability and working life of traditional coated cables within the current activity cycle of the ISL antenna are obtained. Furthermore, aiming at the defects of traditional coated cables, a protective measure for external cables based on stainless steel soft sleeves is proposed. Through conducting full-life cycle reliability verification tests, its ability to resist space radiation is proven, and it is further demonstrated that this cable protection measure can adapt to the working characteristics of the ISL antenna, effectively improving the reliability of GEO telecommunication satellites in orbit service.

2. Composition of the drive mechanism for the ISL antenna

The ISL antenna of a GEO telecommunication satellite performs two-dimensional rotation during on-orbit operation to achieve tracking and scanning of user targets. The connection relationship between the antenna and the satellite is shown in figure 1.

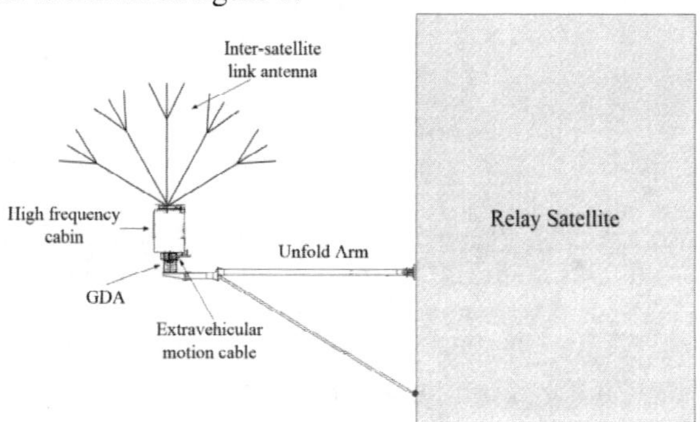

Figure 1. Schematic diagram of the ISL antenna of a GEO telecommunication satellite.

The ISL antenna system comprises an ISL antenna, an extravehicular equipment, an antenna drive assembly (ADA), and a deployment arm. The ADA is composed of an antenna drive assembly mechanism (ADAM) and antenna drive assembly electronics (ADAE). The relative position of the ISL antenna and its extravehicular equipment remains fixed. The deployment arm is deployed in place once and then fixed perpendicular to the satellite surface. The ADA facilitates two-dimensional rotation between the satellite's extravehicular equipment and the deployment arm in orbit.

The extravehicular equipment of the ISL antenna is connected to the ADA via cables. The ADA needs to follow the ISL antenna for frequent two-dimensional rotation. The connecting cables are divided into three bundles: 1) The cable bundle from the deployment arm to the ADAM pitch axis. Since the relative position of the deployment arm and the ADAM pitch axis remains unchanged, this bundle of cables remains stationary; 2) The cable bundle from the deployment arm to the ADAM azimuth axis. Both terminals of this cable bundle are fixed on the satellite connection frame and the azimuth axis, respectively. When the antenna moves in pitch, the cables twist, and when the antenna moves in azimuth, the cables swing with the ADAM; 3) The cable bundle from the antenna deployment arm into the extravehicular equipment. This part of the cables twists with the antenna for a long time in orbit.

During the entire life cycle of GEO telecommunication satellites, the drag torque generated by the movement of the extravehicular cables exposed to the space environment, along with the effects of space radiation, can have a serious impact on the lifespan of the cables. Therefore, it is necessary to explore the impact of frequent twisting movements on the performance of the extravehicular moving cables throughout their entire life cycle. On the one hand, the movement of the cables will generate a certain drag torque on the rotating mechanism, which is also one of the main loads of the two-dimensional drive mechanism ADA. It is necessary to collect baseline data on the drag torque. On the other hand, the reciprocating bending and twisting of the cables require an investigation into whether their performance meets the operational requirements throughout the entire life cycle of the satellite.

3. Resistance torque and lifespan test of external cables

3.1. Extravehicular cable resistance torque test and accelerated lifespan test system

A two-dimensional rotational resistance torque experiment mechanical and servo control system are established, as shown in figure 2 and figure 3, respectively. In order to ensure that the extravehicular motion cable can fulfil the full-life cycle mission requirements of high-orbit communication satellites in the cosmic environment with alternating high and low temperatures and high radiation intensity, ground tests are conducted on cables for long-term motion (high-frequency cables, low-frequency cables).

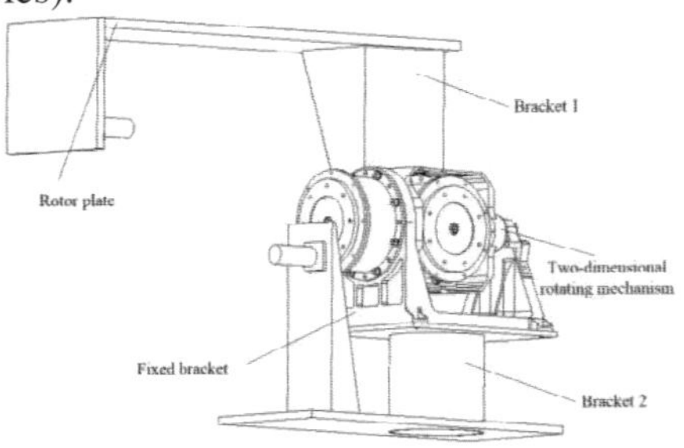

Figure 2. Mechanical system for two-dimensional rotational resistance torque test of extravehicular motion cable.

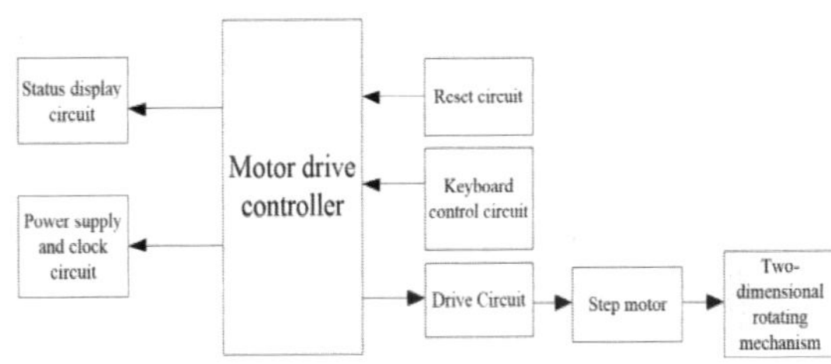

Figure 3. Servo control system for two-dimensional rotation mechanism.

The experimental setup comprises a two-dimensional rotating mechanism, two supports, a rotating plate, and a fixed support. During the experiment, the cable is installed at the fixed points on the rotating plate and the fixed support. The two-dimensional rotating experimental setup is controlled by a newly developed servo control system, which allows manual control of the mechanism to rotate to a specified position. The cable resistance torque test is measured using a torque sensor, which is installed on the two rotating shafts of the rotating mechanism. The torque on the rotating shafts can be collected in real time through a strain testing system.

3.2. Conditions for accelerated lifespan test of cables

Conduct an accelerated motion life test on the test cable under the following conditions: Under normal pressure, the test temperature points are set at -60 °C, -40 °C, -20 °C, -10 °C, 0 °C, 20 °C, 50 °C, 70 °C and 90 °C, and maintain for about 20 minutes after reaching the temperature measurement point, and then conduct a torque test. There are three working conditions for the torque test of the cable: the azimuth axis rotates ± 21 ° separately; The pitch axis rotates -21 °~+47 ° separately; Both axes rotate simultaneously ± 10 °.

The testing process is divided into two scenarios: major motion and minor motion scenarios, based on different motion conditions of the cable outside the cabin. The dynamic combination ranges of azimuth and pitch angle changes under the two scenarios are shown in table 1 and table 2, respectively.

Table 1. Combination of dynamic angles of azimuth and pitch axes under minor motion conditions.

Serial number	Azimuth axis angle	Pitch angle
1	0° →12°	12° →0°
2	12° →0°	0° →-12°
3	0° →-12°	-12° →0°
4	-12° →0°	0° →12°

Table 2. Combination of dynamic angles of azimuth and pitch axes under major motion conditions.

Serial number	Azimuth axis angle	Pitch angle
1	0° →21°	47° →0°
2	21° →0°	0° →-21°
3	0° →-21°	-21° →0°
4	-21° →0°	0° →47°

During the experiment, the cable performance is tested in stages, including: (1) using a digital multimeter to test the conductivity and insulation performance of low-frequency cables; (2) using a network analyser to test the insertion loss and standing wave performance of high-frequency cables.

3.3. Resistance torque and lifespan characteristics of extravehicular motion cables

Based on different fixing schemes or cable states of the extravehicular cables, the test is divided into two stages: (1) The first stage is the resistance torque and lifespan baseline test stage. The cable conductors are bundled and installed into a radiation-resistant braided reinforcement sheath, which is then wrapped with thermal control layers. The cable is clamped and compressed, and it bends twice within the extravehicular equipment space to reach ADAM, with a moving part length of approximately 180mm. (2) The second stage is the cable lifespan characteristic test. The moving cable is wrapped with a metal flexible steel tube instead of the radiation-resistant braided sheath used in the first stage, and the cable directly bends once outside the extravehicular equipment to reach ADAM.

3.3.1. Results of resistance torque and lifespan baseline test. Based on actual testing and calibration results, excluding the interference torque from the test mechanism itself, the trend of the absolute values of the maximum resistance torque on the azimuth and pitch axes as a function of temperature is shown in figure 4.

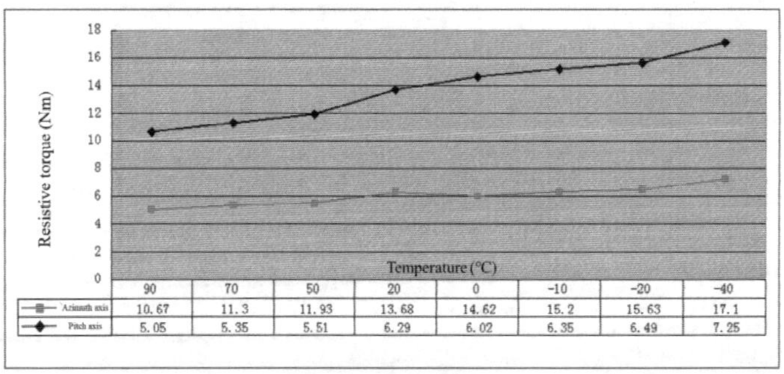

Figure 4. Trend chart of absolute value of maximum resistance torque in the first stage.

As can be seen from figure 4, the resistance torque experienced by both axes increases as the temperature decreases. Specifically, when both axes are linked and the azimuth axis is at -21° and the pitch axis is at -21°, the maximum resistance torque occurs on the azimuth axis. When both axes are linked and the azimuth axis is at +21° and the pitch axis is at +47°, the maximum resistance torque occurs on the pitch axis.

Under the control of the servo control system, the test mechanism alternates between major and minor operating conditions at a rotation ratio of 1:9, completing a total of 4,672 rotations under major operating conditions and 42,048 rotations under minor operating conditions. After the experiment, the cable performance was inspected, and the following test conclusions were drawn:

(1) In terms of appearance, the cable fixing point is slightly loose, and there is a phenomenon of delamination and edge warping at the multi-layer bonding point of thermal control;

(2) In terms of electrical performance, the technical indicators of high-power high-frequency flexible cable have significantly deteriorated. When measured with a multimeter, both the inner and outer conductors of the cable are non-conductive, indicating that the high-power flexible cable has broken.

The aforementioned experimental phenomena indicate that cables with thermal control multilayer coating cannot meet the high-intensity motion requirements of extravehicular cables, necessitating the adoption of adaptive processes for reinforcement.

3.3.2. Experiment results of lifespan characteristics. The second stage of life characteristic test involves three performance tests during the testing phase, with the following test phenomena:

(1) After conducting 941 tests under major operating conditions and 8490 tests under minor operating conditions, the high-power high-frequency cable indicators were tested. It was found that the port standing wave indicator was abnormal. Upon dissection, it was discovered that the cable had broken;

(2) When conducting a separate test on the high-power high-frequency cable, the cable sample became non-conductive after operating under both major and minor conditions for a total of 10,600 cycles, and the cable broke again;

(3) Tested other cables in the cable bundle under high operating conditions of 10512 rpm and low operating conditions of 94608 rpm. After the test, all indicators of the tested cables were found to be normal.

The aforementioned experimental phenomena indicate that the conductor experiences significant twitching within the radiation-resistant steel tube during movement, suggesting the need for improvement and optimization of the cable bundling and fixing process. Measures should be taken to prevent the polyimide film covering the conductor from cracking during movement.

4. Protective measures for extravehicular motion cables based on the collaboration of metal tubes and insulated multilayers

4.1. Reinforcing cable structure with anti-motion characteristics

Considering that the performance of high-power high-frequency cables is severely affected during the lifespan test of extravehicular cables, and that the actual extravehicular cables also contain heating copper wires, in order to enhance the cable strength and improve the life characteristics during movement, a protective measure for extravehicular moving cables based on the collaboration of metal tubes and insulation layers is proposed, as shown in figure 5.

The motion cable is divided into low-frequency and high-frequency cables according to the frequency level of the transmitted signal, bundled and split into 'L' and 'H' wire bundles as shown in figure 5. The heater copper wire is distributed around the bundled cable, as shown in the 'C' wire bundle in figure 5. The cable bundle is wrapped with multiple layers of insulation film and then installed inside a metal tube.

The cable directly passes through a bend from outside the cabin equipment to reach ADAM. The radiation-resistant steel hose is directly fixed with nuts to the outside of the cabin equipment and the connecting flange of ADAM. At the connecting flange of ADAM, the low-frequency cable is divided into three bundles and fixed in the shape of an "Ω" ring. The ADA cable is bridged between the mechanism base and the tail of the azimuth axis motor and is tightly fixed with a clamp.

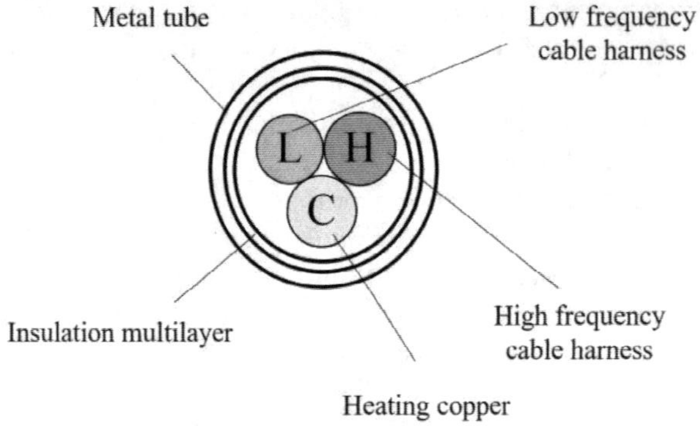

Figure 5. Schematic diagram structure of an extravehicular motion cable based on the collaboration of metal tubes and insulated multilayers.

4.2. Characteristics of lifespan test for reinforced cables

Conduct a lifespan experiment on the reinforced structure of the extravehicular cable with anti-motion characteristics mentioned above, involving a total of 100,000 cycles of rotational testing. During the test, periodically inspect the port standing wave of high-frequency cables, measure the conductivity of constantan heating wires, and observe the damage to the multi-layer appearance of the thermal control system.

After the experiment, the low-frequency cable exhibited normal conductivity, insulation performance, and appearance. Compared with the high-frequency cable samples before and after the test, all indicators were within the normal fluctuation range, and the cable appearance was intact. The experiment indicators are shown in table 3 and table 4.

Table 3. Parameters comparison of testing small-signal flexible cables.

Test items		Before the experiment		After the experiment	
Frequency range		1500MHz~ 2500MHz		1500MHz~ 2500MHz	
		2#	3#	2#	3#
Standing wave 1	S11	1.62	1.59	1.67	1.63
Standing wave 2	S22	1.64	1.66	1.61	1.69
Central insertion loss	S21	3.07	3.3	3.06	3.30
In band fluctuation	ΔS21	1.01	1.03	1.02	1.0
Center frequency phase		75.93°	102.34°	71.67°	98.74°

Table 4. Parameters comparison of testing high-frequency, high-power flexible cables.

Test items		Before the experiment	After the experiment
Frequency range		1500MHz~ 2500MHz	1500MHz~ 2500MHz
Standing wave 1	S11	1.51	1.47
Standing wave 2	S22	1.38	1.29
Central insertion loss	S21	0.47	0.56
In band fluctuation	ΔS21	0.15	0.19
Center frequency phase		-25.29°	-27.11°

Journal of Physics: Conference Series **3041** (2025) 012004 doi:10.1088/1742-6596/3041/1/012004

From the lifespan test conducted on the extravehicular cable with the aforementioned counter-motion characteristics, it can be seen that the extravehicular motion cable structure based on the collaboration of metal tubes and insulation multilayers can effectively enhance the mechanical strength of the cable, and can adapt to the two-dimensional torsional motion of the ISL antenna of GEO telecommunication satellites throughout their entire life cycle in orbit.

5. Conclusions

This paper conducts resistance torque and lifespan experiments on the external motion cables of GEO telecommunication satellites, obtaining the mechanical characteristics of traditional external-coated cables and verifying their inadequacy to meet the operational requirements of the entire lifespan of GEO telecommunication satellites. A torsion-resistant cable structure based on metal tubes and multi-layer insulation is proposed. The effectiveness of this structure in resisting mechanical torsion is verified through experiments, and its lifespan characteristics are obtained. This cable structure can be used for external applications of GEO satellites in orbit, effectively enhancing the reliability of satellite in-orbit service.

6. References

[1] Liu H, Wang Y, Qiu X 2022 Satellite relay task scheduling based on dynamic antenna setup time and splittable task *Proc. of 2022 IEEE Global Communications Conf.* (Rio de Janeiro: Brazil/ IEEE) pp 3917-3922

[2] Wei Z, Yong W and Hua DA 2022 Coverage analysis of conical beam antenna for LEO satellite relay terminal *Proc. of 2022 IEEE International Conf. on Electrical Engineering, Big Data and Algorithms* (Changchun: China/ IEEE)

[3] Winterstein A and Greda LA 2018 An adaptive calibration and beamforming technique for a GEO satellite data relay *International Journal of Satellite Communications and Networking* **36**(2) pp 207–219

[4] Pesyna KM, Humphreys TE, Heath RW, Novlan TD and Zhang JC 2017 Exploiting antenna motion for faster initialization of centimeter-accurate GNSS positioning with low-cost antennas *IEEE Trans. on Aerospace and Electronic Systems* **53**(4) pp 1597–1613

[5] Min J, Tao ZZ, Song XP, Zhang J, Wang ZH, Ji YL, Zhai GF and Ye XR 2018 Study on electron beam cross-linked ethylene-tetrafluoroethylene copolymer-insulated cables for aerospace applications: Mechanical performance, crystallization kinetics, and fluoride precipitation *Polymers for Advanced Technologies* **29**(5) pp 1497-1506

[6] Xue ZY, Ye YL, Yan RJ, Wang W, Li WP, Yang W, Shi XN 2022 The Effect of X-ray Radiation Dose on the Performance of Power Cable Insulation Materials *Proc. of 2022 4th International Conf. on Smart Power & Internet Energy Systems* (Beijing: China/ IEEE)

[7] Fothergill JC, Montanari GC, Stevens GC, Laurent C, Teyssedre G, Dissado LA, Nilsson UH and Platbrood G 2003 Electrical, microstructural, physical and chemical characterization of HV XLPE cable peelings for an electrical aging diagnostic data base *IEEE Trans. on Dielectrics and Electrical Insulation* **10**(3) pp 514-527

[8] Garros B 1999 Ageing and reliability testing and monitoring of power cables: diagnosis for insulation systems: the ARTEMIS program *IEEE Electrical Insulation Magazine* **15**(4) pp 10-12

[9] Song X, Xie J, Wang W, Guo L, Zhang Q, Cheng Y and Liu X 2021 Study of development & application about assembly technology of optical fiber in high-capacity satellite Journal of physics. Conference series. 1846(1) pp 12031

Journal of Physics: Conference Series **3041** (2025) 012005 doi:10.1088/1742-6596/3041/1/012005

Sine wave motion control of the helicopter rescue simulator based on CSP mode

Jitao Si[1], Wei Xiong[1]*, Hongwang Du[1] and Zheng Li[1]

[1] Naval Architecture and Ocean Engineering College, Dalian Maritime University, No.1 Linghai Road, Dalian 116026, China

Corresponding author: Wei Xiong Address: xiongwei@dlmu.edu.cn

Abstract. In order to improve the position reproduction accuracy and motion trajectory smoothness of the Helicopter Rescue Simulator during sine wave motion, a real-time control system based on Cyclic Synchronous Position Mode (CSP Mode) is developed. A Stewart platform control system based on Speedgoat real-time target machine is constructed, and a signal generation module meeting the requirements of CSP Mode is designed. A time-step-based position signal transmission mode is established to achieve sine wave control of the Helicopter Rescue Simulator. The results show that the sine wave motion control system can successfully control the Helicopter Rescue Simulator and has complete control functionality. The sine motion control system based on CSP Mode offers advantages such as strong signal tracking ability, low position error, and smooth motion trajectory, addressing the drawbacks of traditional Profile Position Mode with poor dynamic characteristics and insufficient precision in Profile Velocity Mode during sine wave motion.

1. Introduction

The helicopter rescue operation, known for its high mobility and rapid response, is a key component of rescue forces. The Helicopter Rescue Simulator is a rescue simulation device in which an inverted Stewart platform is mounted at a high position, with the static platform fixed at the top and the dynamic platform below simulating the helicopter's posture [1-3]. Currently, the mainstream control systems in related fields are mostly based on Profile Position Mode and Profile Velocity Mode, with the relevant technologies being highly mature. However, these control systems often struggle to balance the smoothness of velocity changes and the accuracy of position reproduction during multi-point continuous motion. Although methods such as linear interpolation, spherical quadrilateral interpolation, and algorithms for position error compensation have alleviated some of these shortcomings, they have also introduced new issues such as increased system complexity and difficulty in parameter adjustment [4-5].

Cyclic Synchronous Position Mode (CSP Mode) is a high-precision control mode in servo control systems that controls the position of servo motors by receiving external synchronization signals. It is primarily applied in CNC machines and automated production lines. This control mode ensures high-precision signal reproduction and smooth motion trajectories without the need for complex algorithms, though it requires high demands on communication and hardware. With the development and application of bus technologies such as EtherCAT, which improve data communication capabilities, the use of CSP technology to control the Helicopter Rescue Simulator has become a new possibility and direction [6-8].

In view of this, based on CSP Mode, this paper establishes a new sine wave motion control system for the Helicopter Rescue Simulator through EtherCAT real-time Ethernet protocol for communication, utilizing Speedgoat real-time target machine as the hardware foundation and Simulink Real-Time for system development. This system improves the platform's motion accuracy, trajectory smoothness, and signal tracking ability, and corresponding experimental validation and data comparison are conducted.

2. Coordinate system and kinematic inverse solution

Establishing an appropriate coordinate system and performing kinematic inverse solution are fundamental for controlling the Helicopter Rescue Simulator.

2.1. Establishment of the coordinate system

The Helicopter Rescue Simulator is based on an inverted electric Stewart platform, and the schematic diagram of the coordinate system of the Simulator is shown in figure 1. In figure 1, the upper part represents the static platform, where the static coordinate system $O_bX_bY_bZ_b$ is established, and the lower part represents the dynamic platform, where the dynamic coordinate system $O_aX_aY_aZ_a$ is established. Both coordinate systems are right-handed orthogonal coordinate systems. The dynamic coordinate system is fixed on the dynamic platform, with its origin coinciding with the center of the dynamic platform, moving with the motion of the platform. The static coordinate system is fixed on the static platform, with its origin coinciding with the center of the static platform, remaining stationary in space. When the Simulator is in the initial working position, the directions of the axes X_a and X_b, Y_a and Y_b, and Z_a and Z_b are the same, but the origins of the two coordinate systems, O_a and O_b, are offset by an initial distance and do not coincide. In figure 1, A_1 to A_6 represent the six hinge points on the dynamic platform, and B_1 to B_6 represent the six hinge points on the static platform. The corresponding hinge points are connected by electric actuators [9].

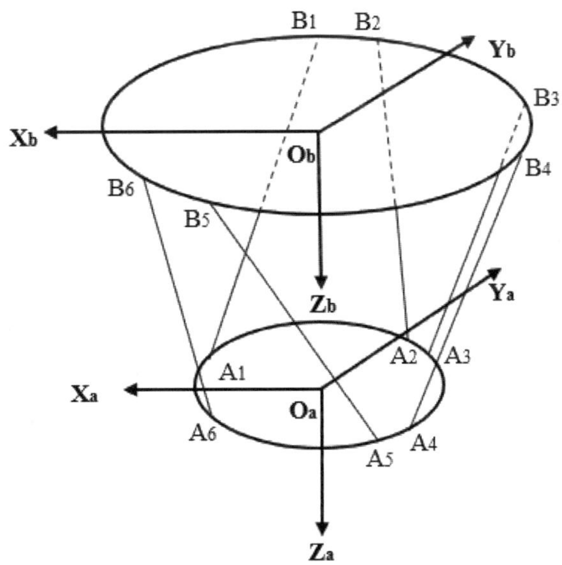

Figure 1. Coordinate System of the Helicopter Rescue Simulator

2.2. Kinematic inverse solution

Kinematic inverse solution refers to the process of determining the lengths of the six electric actuators when the translational displacements of the dynamic platform of the Helicopter Rescue Simulator along the x, y, and z axes, as well as the rotational angles about the three coordinate axes, are given.

Mathematically, this problem can be transformed into finding the distances between two points in space, with the key being to obtain the coordinates of the 12 hinge points in the same coordinate system [10]. The coordinates of the hinge points on the dynamic platform in the dynamic coordinate system can be transformed into the coordinates in the static coordinate system using equation (1).

$$P_i^* = TP_i + P_0 \tag{1}$$

Where P_i^* is a 3×1 matrix representing the coordinates of a hinge point on the dynamic platform in the static coordinate system; P_i is a 3×1 matrix representing the coordinates of the corresponding hinge point on the dynamic platform in the dynamic coordinate system; P_0 is a 3×1 matrix representing the translational displacement of the center of the dynamic platform along the X_b, Y_b, and Z_b axes in the static coordinate system; and T is a 3×3 rotation matrix, as shown in equation (2).

$$T = \begin{bmatrix} \cos(\beta)\cos(\gamma) & -\cos(\beta)\sin(\gamma) & \sin(\beta) \\ \sin(\alpha)\sin(\beta)\cos(\gamma)+\cos(\alpha)\sin(\gamma) & \cos(\alpha)\cos(\gamma)-\sin(\alpha)\sin(\beta)\sin(\gamma) & -\sin(\alpha)\cos(\beta) \\ \sin(\alpha)\sin(\gamma)-\cos(\alpha)\sin(\beta)\cos(\gamma) & \cos(\alpha)\sin(\beta)\sin(\gamma)+\sin(\alpha)\cos(\gamma) & \cos(\alpha)\cos(\beta) \end{bmatrix} \tag{2}$$

Where α is the rotation angle of the moving platform around axis X_a, β is the rotation angle of the moving platform around axis Y_a, and γ is the rotation angle of the moving platform around axis Z_a.

After obtaining the coordinates of the 12 hinge points of the Simulator in the static coordinate system from equation (1), the lengths of the electric actuators can be determined by calculating the distances between two points in space.

3. Hardware and algorithmic structure of the simulator

The entire control system communicates based on EtherCAT technology and consists of four hardware components: a PC responsible for control functions, a Speedgoat real-time target machine for algorithm execution, six DELTA ASD-B3-0421-E servo drives, and a Stewart platform consisting of six ECM-B3M-C20604SS1 servo motors. Their relationship is shown in figure 2.

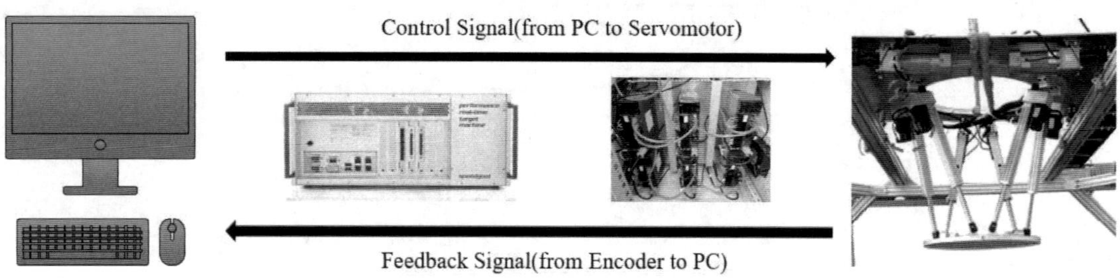

Figure 2. Hardware Composition of the Simulator

Unlike traditional master-slave control modes, the PC in this system only performs simple control and interaction functions. After the PC issues a command, the core program written in Simulink Real-Time runs on the Speedgoat real-time target machine, which then sends motion data to the servo drives, controlling the motion of the servo motors. The encoders on the servo motors provide feedback on the motion, which is processed through the servo drives and within Speedgoat, and finally displayed and stored on the PC. This approach reduces the negative impact of the distributed clock of the Windows system on the control system.

The software part of the control system is built using Simulink Real-Time for the underlying program, and the UI control interface is developed using MATLAB/AppDesigner. The flow diagram of the program execution is shown in figure 3. The user issues control commands through the User

Interface, which are passed to the State Machine. The Timer and Trajectory Generator inside the Position Signal Generator produce position signals under the command of the State Machine. These signals are processed by the Inverse-Solution Calculator and then sent to the Servo-Driver Group via the RxPDO Group to control the motion of the servo motors. The encoders on the servo motors provide feedback on the motor's movement to the Servo-Driver Group, and this information is transmitted to the Data Processor via the TxPDO Group for further calculation. The results are then fed back to the State Machine and displayed and stored on the User Interface.

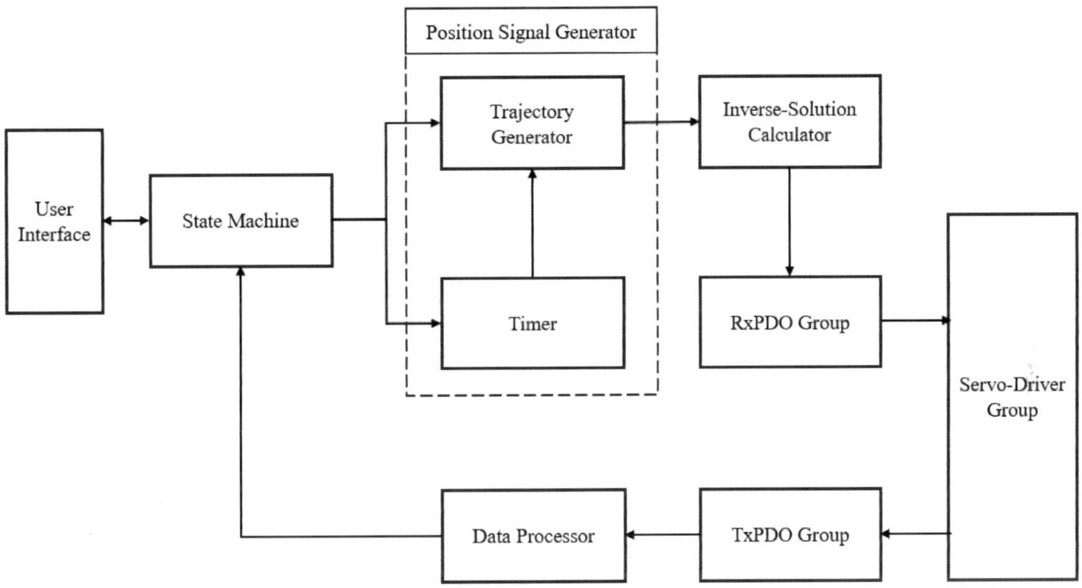

Figure 3. Software Framework of the Control System

The operation of the Position Signal Generator is the key part of the program. Its function is to output controllable sinusoidal signals, and the internal principle is shown in figure 4.

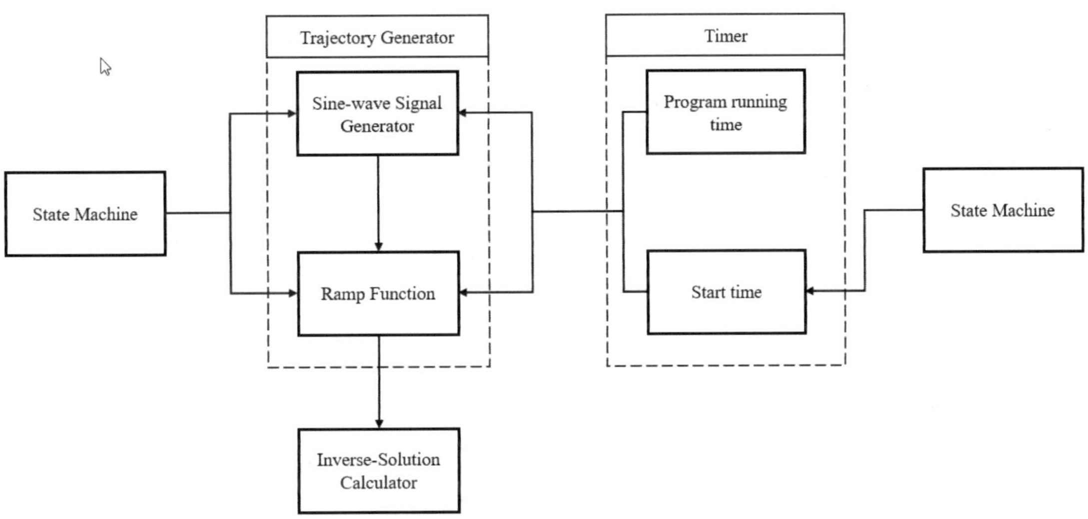

Figure 4. Schematic of the Position Signal Generator

The control information from the State Machine simultaneously affects both the Trajectory Generator and the Timer. The output of the Timer is the Program running time minus the Start time, which records the time elapsed since the State Machine command was issued. This result is simultaneously output to both the Sine-wave Signal Generator and the Ramp Function. The sine wave signal generated by the Sine-wave Signal Generator under the control of the State Machine is processed by the Ramp Function to reach the set amplitude after some cycles. According to the computation step size set in Simulink Real-Time, this signal is periodically sent to the Inverse-Solution Calculator.

4. Experiment and result analysis

Due to the absence of displacement sensors directly measuring the pose of the moving platform in the simulator, and the potential errors in pose data from commonly used kinematic forward solutions for the Stewart platform, such as the Newton-Raphson method, which may introduce inaccuracies, encoder feedback from the six servo motors was directly collected in the experiment. A comparison and analysis of the motor data were then conducted to avoid unnecessary errors affecting the control result analysis.

In order to observe the signal reproduction errors of the control system under the same amplitude but different frequencies, and under the same frequency but different amplitudes, a total of eight control experiments were conducted, with amplitudes of 10 mm and 20 mm, and frequencies of 0.1 Hz, 0.5 Hz, 1.0 Hz and 2.0 Hz. These experiments were designed to evaluate the signal reproduction accuracy and signal tracking ability of the sine wave motion control system based on CSP Mode. As shown in table 1, the average experimental results of the six servo motors are presented, where the Helicopter Rescue Simulator vibrates along the z-axis with an amplitude of 10 mm and frequencies of 0.1 Hz, 0.5 Hz, 1.0 Hz and 2.0 Hz.

Table 1. Error results at different frequencies for an amplitude of 10 mm.

Frequency (Hz)	Peak error (mm)	Trough error (mm)	Average error (mm)
0.1	0.0013	0.0005	0.00090
0.5	0.0358	0.0345	0.03515
1.0	0.1378	0.1365	0.13715
2.0	0.5148	0.5140	0.51440

Table 2 presents the average experimental results of the six servo motors, where the Helicopter Rescue Simulator vibrates along the z-axis with an amplitude of 20 mm and frequencies of 0.1 Hz, 0.5 Hz, 1.0 Hz and 2.0 Hz.

Table 2. Error results at different frequencies for an amplitude of 20 mm.

Frequency (Hz)	Peak error (mm)	Trough error (mm)	Average error (mm)
0.1	0.0029	0.0027	0.00280
0.5	0.0699	0.0687	0.06930
1.0	0.2744	0.2727	0.27355
2.0	1.0309	1.0267	1.02880

The experimental data in table 1 and table 2 show that, at low frequencies, the signal reproduction accuracy of the motors is very high, with the error being less than 1 micron at its smallest. For the same frequency, as the amplitude increases, the error also increases. Similarly, for the same amplitude, as the frequency increases, the error increases. These results validate the signal reproduction accuracy of the sine wave motion control system for the Helicopter Rescue Simulator built using the CSP Mode.

In addition, the input and output signals of each experiment were monitored to verify the sine trajectory tracking performance of the control system. It was found that the time difference between the input and output signals was maintained at approximately 43 milliseconds.

To evaluate whether there has been an improvement in signal reproduction accuracy compared to the original control system in the laboratory, the Helicopter Rescue Simulator was controlled to perform a sine motion along the z-axis with an amplitude of 50 mm and a frequency of 0.1 Hz. Figure 5 shows the motor position trajectory curves for both the current system (CSP_System_Output) and the original system (Original_System_Output) in the laboratory. In the enlarged view on the right side of figure 5, the data '188.2982' represents the peak value of the input sine signal.

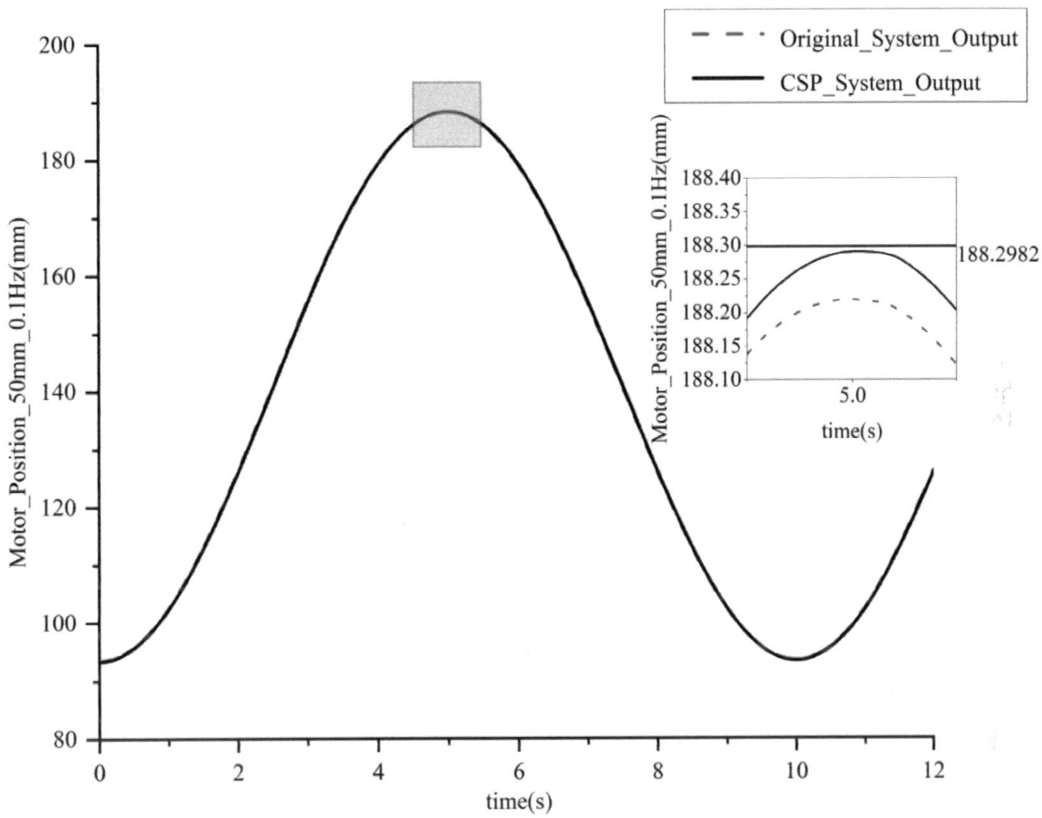

Figure 5. Motor trajectory curve for an amplitude of 50 mm and frequency of 0.1 Hz.

Based on the experimental results and analysis, it can be concluded that the control system has high signal reproduction accuracy, with position errors reduced compared to the original system in the laboratory. The system demonstrates strong signal tracking ability and smooth motion trajectories, meeting the predicted and design requirements for applying the CSP Mode to the Helicopter Rescue Simulator sine control system.

5. Conclusion

The author of this paper has developed a sine motion control system based on CSP Mode for the Helicopter Rescue Simulator. The high-rate data transmission via EtherCAT meets the real-time interaction requirements of CSP Mode. The sine motion control system, built using Simulink Real-Time, runs on the Speedgoat real-time target machine. Under the interaction and control of the PC, it performs kinematic calculations, generates sine signals, and gradually sends position signals to the servo drives in time steps. The servo drives then control the servo motors under CSP Mode and read and feedback encoder data, which is sent back to the target machine for calculation and control, thus achieving real-time control of the Helicopter Rescue Simulator. The results show that the real-time control system based on CSP Mode has strong signal-following capability, high position reproduction accuracy, and smooth sine motion trajectories. These characteristics align with the initial research

Journal of Physics: Conference Series **3041** (2025) 012005 doi:10.1088/1742-6596/3041/1/012005

predictions and design requirements, significantly enhancing the functionality of the Helicopter Rescue Simulator's sine motion control system.

6. References

[1] Stewart D A 1965 *Proc. Inst. Mech. Eng.* A platform with six degress of freedom **180** 371-386
[2] Dong Y, Baohong L and Boding Z 2005 *Flight Dyn.* Analysis of the helicopter simulator qualification **23** 9-12
[3] Shengye W, Haitao W, Wei X and Guangfeng G 2022 *Mach.Tool Hydraul.* Workspace Analysis of Helicopter Hovering Rescue Simulator **50** 130-134
[4] Kuoyang T, Tungchung W and Tsutian L 2004 *IEEE Int.Conf.on Syst., Man and Cybern.* vol. 4 (The Hague, Netherlands) (IEEE) pp. 3950-55
[5] Xiaoqing L 2023 *Electron.Sci.Technol.* Research Progress on Control Technology of Multi-Degree of Freedom Parallel Robot **36** 89-94
[6] Stefan N and Aurelian S 2024 *Int. Conf. on Appl. and Theor. Electr.* (Craiova, Romania) (IEEE) pp.1-5
[7] Hasan M.K. 2021 *Int. Conf. on Electron., Commun. and Inform. Technol.* (Khulna, Bangladesh) (IEEE) pp. 1-4
[8] Wenzheng Z 2022 *Mach.Tool Hydraul.* Development of dVRK Manipulator Control System Based on Speedgoat Machine **50** 91-95
[9] K. L, Fitzgerald J.M. and Lewis F.L. 1993 *IEEE Trans. Ind. Electron.* vol40 (IEEE) pp.282-293
[10] Hamidreza H, Hasan M and Mehdi M 2010 *2nd Int. Conf. on Comput. and Autom. Eng.* (Singapore) (IEEE) pp. 472-477

Journal of Physics: Conference Series **3041** (2025) 012006 doi:10.1088/1742-6596/3041/1/012006

Analysis of 3:1 internal resonance of a rigid-flexible coupling antenna system with three degrees of freedom

H Jiang, X Gao, Y Sheng, W Tang, P Shi

School of Physical and Mathematical Sciences, Nanjing Tech University, Nanjing 211816, China

gaoxm@njtech.edu.cn

Abstract. This paper investigates the nonlinear dynamic behavior of a spatially deployable antenna's rigid-flexible coupling structure. Firstly, a three-degree-of-freedom T-beam structure model consisting of a rigid manipulator and a symmetrical antenna is established. By applying the Lagrange equation and the assumed modes method, the governing nonlinear dynamics equations for the T-beam's in-plane motion are formulated. Subsequently, the system's nonlinear dynamic behavior under 3:1 internal resonance conditions is investigated employing the matrix-form multi-scale methodology. The general expression of the analytical solution of the system under 3:1 the internal resonance is given and the stability of the approximate solution is analyzed. The criteria governing both periodic and aperiodic motions are established. The amplitude-frequency formulation enables derivation of bifurcation characteristics for the system's nonlinear mode interactions. Finally, the validity of the approximate solution is confirmed through numerical verification. The results show that there are multiple modal solutions, the numerical solution and the analytical solution are in good agreement with each other. In addition, the effectiveness of the matrix-form multi-scale method for nonlinear dynamics analysis of rigid-flexible coupling systems with multiple degrees of freedom is proved.

1. Introduction

Following recent progress in aerospace engineering, large deployable truss antenna structure has been widely used in satellite antenna because of its foldable, deployable and lightweight structure characteristics, which has improved the effective utilization rate of space. However, owing to the structural complexity inherent in large deployable truss antennas, large flexibility and many gaps, it is easy to produce nonlinear vibration in the microgravity environment of outer space. In the face of huge demands and challenges, many experts and scholars have conducted research on space truss antenna structure [1-5].

Figure 1. Schematic diagram of the ring truss structure.

Journal of Physics: Conference Series **3041** (2025) 012006 doi:10.1088/1742-6596/3041/1/012006

As illustrated in Figure 1, the ring truss antenna system primarily consists of three key components: the extension arm, the ring truss and the paraboloidal cable net structure. These structures are coupled with each other and have many degrees of freedom, so it is difficult to model. It is very important to simplify the ring truss reasonably and establish an equivalent model which can be analyzed analytically. Numerous scholars have conducted extensive studies on the development of equivalent models for truss-based antenna systems. The ring truss structure is often simplified as a cylindrical shell model or a ring beam model. Armaghan Salehian et al. [6] derived the dynamic and strain energy expressions based on the spatial and temporal derivatives of the central displacement component of the truss element, and used the one-dimensional homogenization model to find the inherent vibrational frequency of the framework. In [7], Fushou Liu's team transformed a jointed truss with damping properties into an equivalent anisotropic viscoelastic beam through energy-based homogenization techniques, and derived the motion equation of the equivalent beam model. In reference [8], Chen's research team examined the nonlinear breathing oscillations of a simplified circular truss antenna's cylindrical shell under thermal loading. T. Liu et al. [9] studied the hopf bifurcation and intermittent chaos characteristics of space antennas under thermal excitation. Previous research by Zhang et al. [10,11] established an equivalent cylindrical shell model for ring truss structures using a simplified periodic element methodology to facilitate nonlinear dynamics examination.

Concerning the beam-model simplification of antenna systems, scholars have carried out the research on the in-plane vibration and out-of-plane vibration systems of the beam. For the in-plane vibration of the beam model, Mingchang Niu et al. [12] used the Hamiltonian principle to investigate the vibrational behavior of a circular structure incorporating multiple discrete mass elements. In their foundational work [13], Lee et al. implemented the displacement function methodology to model and analyze the static response of Timoshenko beams undergoing in-plane bending with highly nonlinear boundary constraints. Wenhua Hu et al. [14] obtained the Z-beam frequency and analytical mode shape through the Hamiltonian principle. Xiumin Gao et al. [15-17] used the multi-scale method to derive closed-form solutions for the nonlinear oscillatory behavior of the simplified multi-DOF antenna, with numerical verification confirming the result, which proved the good applicability of the matrix form multi-scale method in the study of the nonlinear vibration of the antenna system. Regarding three-dimensional out-of-plane beam vibrations, Lee [18] employed the displacement function approach to determine static deformations in Timoshenko beams subjected to nonlinear out-of-plane bending conditions. Yong Zhou et al. [19] established an analytical model of a three-dimensional variable curvature piezoelectric laminated bending beam. Rajasekaran [20] employed the differential transformation technique to derive shape functions for nodal variables in non-uniform bending beam elements incorporating axial functionally graded material properties. Applying the Hamiltonian principle, Zhang et al. [21-23] formulated the nonlinear partial differential governing equations and corresponding boundary conditions for beam-ring structures, subsequently investigating their nonlinear dynamic behavior under periodic thermal loading in three-dimensional ring truss antenna configurations. Wu et al. [24] studied the influence of thermoelastic effects on spatial structure expansion, and proposed a new method based on flexible multi-body systems. By adopting absolute nodal coordinate formulation, the governing equations for an Euler-Bernoulli beam under transient thermal excitation were formulated. Later, Liu's team [25] quantitatively analyzed how thermal gradients alter the deformation patterns of thin flexible beams subjected to rigid-body motions. Cao et al. [26] conducted thermal structure coupling analysis of radiation-disturbed orientation control of agile space vehicles.

The present work investigates the nonlinear dynamic behavior of a hybrid rigid-flexible antenna structure. The paper is structured as follows: In Section 2, the space model of the planar rigid-flexible coupling antenna is established by using the floating coordinate method. In Section 3, the 3:1 internal resonance phenomenon is systematically investigated through a matrix-formulation of the method of multiple scales, enabling rigorous characterization of the nonlinear modal interactions. Then, the analytical solution of the system under 3:1 internal resonance is derived. Section 4 presents numerical validation of the derived analytical solutions through systematic computational experiments, while Section 5 synthesizes the principal findings and their engineering implications.

2. Dynamic modeling

The antenna model is approximated by a 3-degree-of-freedom T-shaped configuration, featuring a central rigid arm with two identical flexible beams symmetrically attached, and the in-plane vibration nonlinear response of the structure is studied. Assuming that the flexible beam is an Euler-Bernoulli beam structure, it is modeled as shown in figure 2.

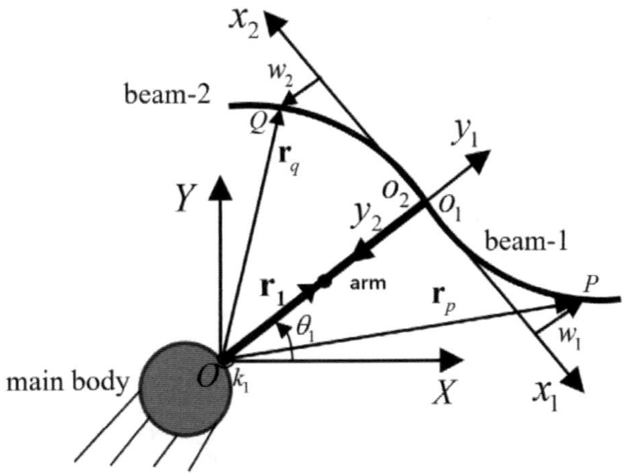

Figure 2. The schematic diagram of the three-DOF T-shaped antenna structure.

The global reference frame OXY and the local body-fixed coordinate system $o_1x_1y_1$ and $o_2x_2y_2$ are established based on the satellite ontology. The position vectors $\mathbf{r}_1$, $\mathbf{r}_p$, and $\mathbf{r}_q$ represent the center of mass of the rigid arm, point *P* on flexible beam 1, and point *Q* on flexible beam 2, respectively. The expressions for $\mathbf{r}_1$, $\mathbf{r}_p$, and $\mathbf{r}_q$ are given as

$$\mathbf{r}_1 = \begin{bmatrix} \cos\theta_1 & -\sin\theta_1 \\ \sin\theta_1 & \cos\theta_1 \end{bmatrix} \begin{bmatrix} \frac{l_1}{2} \\ 0 \end{bmatrix} \tag{1}$$

$$\mathbf{r}_p = \begin{bmatrix} \cos\theta_1 & -\sin\theta_1 \\ \sin\theta_1 & \cos\theta_1 \end{bmatrix} \begin{bmatrix} l_1 \\ 0 \end{bmatrix} + \mathbf{A}_0 \begin{bmatrix} x_1 \\ w_1(x_1,t) \end{bmatrix} \tag{2}$$

$$\mathbf{r}_q = \begin{bmatrix} \cos\theta_1 & -\sin\theta_1 \\ \sin\theta_1 & \cos\theta_1 \end{bmatrix} \begin{bmatrix} l_1 \\ 0 \end{bmatrix} + \mathbf{B}_0 \begin{bmatrix} x_2 \\ w_2(x_2,t) \end{bmatrix} \tag{3}$$

$\mathbf{A}_0$ and $\mathbf{B}_0$ are the coordinate transformation matrix from the body coordinate system $o_1x_1y_1$ and $o_2x_2y_2$ to the inertial coordinate system OXY, respectively, expressed as

$$\mathbf{A}_0 = \begin{bmatrix} \cos(\theta_1 - \pi/2) & -\sin(\theta_1 - \pi/2) \\ \sin(\theta_1 - \pi/2) & \cos(\theta_1 - \pi/2) \end{bmatrix} \tag{4}$$

$$\mathbf{B}_0 = \begin{bmatrix} \cos(\theta_1 + \pi/2) & -\sin(\theta_1 + \pi/2) \\ \sin(\theta_1 + \pi/2) & \cos(\theta_1 + \pi/2) \end{bmatrix} \tag{5}$$

Base on the assumed modes method, taking the first-order mode into consideration, there is

$$w_1(x_1,t) = \varphi_1(x_1)p_1(t), w_2(x_2,t) = \varphi_1(x_2)q_1(t) \tag{6}$$

where $\varphi(x_i)$ is the cantilever beam mode function, $p_1(t)$ and $q_1(t)$ are the corresponding modal coordinates.

Ignoring gravitational effects, the total mechanical energy of the 3-DOF T-shaped hybrid rigid-flex beam system consists solely of kinetic energy and elastic potential energy, expressed as:

$$T = \frac{1}{2}m_1\dot{\mathbf{r}}_1^T\dot{\mathbf{r}}_1 + \frac{1}{2}J_1\dot{\theta}_1^2 + \frac{1}{2}\int_0^L \rho A\dot{\mathbf{r}}_p^T\dot{\mathbf{r}}_p dx + \frac{1}{2}\int_0^L \rho A\dot{\mathbf{r}}_q^T\dot{\mathbf{r}}_q dx \tag{7}$$

$$V=\frac{1}{2}k_1\theta_1^2+\frac{1}{2}\int_0^L EI\left(\frac{\partial^2 w_1}{\partial x_1^2}\right)^2 dx_1+\frac{1}{2}\int_0^L EI\left(\frac{\partial^2 w_2}{\partial x_2^2}\right)^2 dx_2 \tag{8}$$

where m_1 and l_1 are the mass and length of the rigid arm, J_1 is the moment of inertia of the rigid arm, ρ, A, L and EI represent the density, cross-sectional area, length and flexural stiffness of the flexible beam, respectively, and k_1 is the spring stiffness of torsional spring.

3. 3:1 internal resonance analysis

The dimensionless parameters are introduced as follows

$$\begin{cases} d_1=\dfrac{m_1}{\rho AL}, b_1=\dfrac{l_1}{L}, \kappa_1=\dfrac{k_1}{\rho AL^3}, \kappa_2=\dfrac{EI\int_0^L\left(\dfrac{d^2\varphi_1(x)}{dx^2}\right)^2 dx}{\rho AL} \\ \kappa_3=\kappa_2, \bar{q}_1=\dfrac{p_1}{L}, \bar{q}_2=\dfrac{q_1}{L} \end{cases} \tag{9}$$

By choosing the generalized coordinates as $\boldsymbol{\eta}=[\theta_1,\bar{q}_1,\bar{q}_2]^T$, through systematic application of Lagrange's equations, the normalized dynamic equations characterizing the T-beam's three-degree-of-freedom system are derived as:

$$\mathbf{M}(\boldsymbol{\eta})\ddot{\boldsymbol{\eta}}+\mathbf{K}\boldsymbol{\eta}-\mathbf{P}(\boldsymbol{\eta},\dot{\boldsymbol{\eta}})=\mathbf{0} \tag{10}$$

The mass matrix $\mathbf{M}$, stiffness matrix $\mathbf{K}$, and the vector $\mathbf{P}$ of the sum of centrifugal and Coriolis forces in the equation are defined in the appendix.

By defining the time scale $T_r=\varepsilon^r t, r=0,1,2...$, where ε is any small parameter, the solution of equation (10) is approximated via third-order multiple-scale expansion as:

$$\boldsymbol{\eta}(t,\varepsilon)=\boldsymbol{\eta}_0+\varepsilon\boldsymbol{\eta}_1(T_0,T_1,T_2)+\varepsilon^2\boldsymbol{\eta}_2(T_0,T_1,T_2)+\varepsilon^3\boldsymbol{\eta}_3(T_0,T_1,T_2)+O(\varepsilon^4) \tag{11}$$

By substituting equation (11) into equation (10), expanding the matrix and vector, and comparing the coefficients of the same power of ε on both sides of the equation, the formula can be obtained as follows

$$\mathbf{K}_0\boldsymbol{\eta}_0-\mathbf{P}_0=\mathbf{0} \tag{12}$$

$$\mathbf{M}_0 D_0^2\boldsymbol{\eta}_1+\mathbf{R}_0\boldsymbol{\eta}_1=\mathbf{0} \tag{13}$$

$$\begin{aligned}\mathbf{M}_0 D_0^2\boldsymbol{\eta}_2+\mathbf{R}_0\boldsymbol{\eta}_2=&-2\mathbf{M}_0 D_0 D_1\boldsymbol{\eta}_1-\mathbf{M}_1(\boldsymbol{\eta}_1)D_0^2\boldsymbol{\eta}_1-\mathbf{K}_2(\boldsymbol{\eta}_1,\boldsymbol{\eta}_1)\boldsymbol{\eta}_0\\&-\mathbf{K}_1(\boldsymbol{\eta}_1)\boldsymbol{\eta}_1+\mathbf{P}_{12}D_1\boldsymbol{\eta}_1+\mathbf{P}_2(\boldsymbol{\eta}_1,D_0\boldsymbol{\eta}_1)\end{aligned} \tag{14}$$

$$\begin{aligned}\mathbf{M}_0 D_0^2\boldsymbol{\eta}_3+\mathbf{R}_0\boldsymbol{\eta}_3=&-2\mathbf{M}_0 D_0 D_2\boldsymbol{\eta}_1-2\mathbf{M}_0 D_0 D_1\boldsymbol{\eta}_2-\mathbf{M}_0 D_1^2\boldsymbol{\eta}_1\\&-2\mathbf{M}_1(\boldsymbol{\eta}_1)D_0 D_1\boldsymbol{\eta}_1-\mathbf{M}_1(\boldsymbol{\eta}_1)D_0^2\boldsymbol{\eta}_2-\mathbf{M}_1(\boldsymbol{\eta}_2)D_0^2\boldsymbol{\eta}_1\\&-\mathbf{M}_2(\boldsymbol{\eta}_1,\boldsymbol{\eta}_1)D_0^2\boldsymbol{\eta}_1-\mathbf{K}_1(\boldsymbol{\eta}_1)\boldsymbol{\eta}_2-\mathbf{K}_1(\boldsymbol{\eta}_2)\boldsymbol{\eta}_1-\mathbf{K}_2(\boldsymbol{\eta}_1,\boldsymbol{\eta}_1)\boldsymbol{\eta}_1\\&-2\mathbf{K}_2(\boldsymbol{\eta}_1,\boldsymbol{\eta}_2)\boldsymbol{\eta}_0-\mathbf{K}_3(\boldsymbol{\eta}_1,\boldsymbol{\eta}_1,\boldsymbol{\eta}_1)\boldsymbol{\eta}_0+\mathbf{P}_{12}D_1\boldsymbol{\eta}_2+\mathbf{P}_{12}D_2\boldsymbol{\eta}_1\\&+2\mathbf{P}_{21}(\boldsymbol{\eta}_1,D_0\boldsymbol{\eta}_2+D_1\boldsymbol{\eta}_1)+2\mathbf{P}_{21}(\boldsymbol{\eta}_2,D_0\boldsymbol{\eta}_1)+\mathbf{P}_3(\boldsymbol{\eta}_1,D_0\boldsymbol{\eta}_1)\\&+\mathbf{P}_{211}(\boldsymbol{\eta}_1,\boldsymbol{\eta}_2)+\mathbf{P}_{22}(D_0\boldsymbol{\eta}_1,D_0\boldsymbol{\eta}_2+D_1\boldsymbol{\eta}_1)\end{aligned} \tag{15}$$

The expressions of each matrix and vector in the equations are detailed in the appendix.

The solution is solved from equation (12) as follows

$$\boldsymbol{\eta}_0=\mathbf{K}_0^{-1}\mathbf{P}_0 \tag{16}$$

When the matrix $\mathbf{M}_0$ is nonsingular, equation (13) can be rewritten as follows

$$D_0^2\boldsymbol{\eta}_1+\mathbf{S}_0\boldsymbol{\eta}_1=\mathbf{0} \tag{17}$$

where $\mathbf{S}_0=\mathbf{M}_0^{-1}\mathbf{R}_0$ is a matrix function with respect to $\boldsymbol{\eta}_0$. Solving equation (17) yields the first-order approximate solution as

$$\boldsymbol{\eta}_1 = A_1(T_1,T_2)e^{\mathrm{i}\omega_1 T_0}\mathbf{p}_1 + \overline{A_1}(T_1,T_2)e^{-\mathrm{i}\omega_1 T_0}\mathbf{p}_1 + A_2(T_1,T_2)e^{\mathrm{i}\omega_2 T_0}\mathbf{p}_2 + \overline{A_2}(T_1,T_2)e^{-\mathrm{i}\omega_2 T_0}\mathbf{p}_2 + A_3(T_1,T_2)e^{\mathrm{i}\omega_3 T_0}\mathbf{p}_3 + \overline{A_3}(T_1,T_2)e^{-\mathrm{i}\omega_3 T_0}\mathbf{p}_3 \tag{18}$$

where $\overline{A_j}$ denotes the conjugate of A_j, ω_j and $\mathbf{p}_j$ represent the natural frequency and mode shape of the equation around $\boldsymbol{\eta}_0$, respectively, and the following eigenvalue equation is satisfied as follows

$$\omega_j^2\mathbf{p}_j = \mathbf{S}_0\mathbf{p}_j (j=1,2,3) \tag{19}$$

The parameters $d_1 = 3$ and $\kappa_2 = \kappa_3 = 1$ are set, and the effects of parameters b_1 and κ_1 on the system frequency are analyzed to determine the possibility of internal resonance occurring. It can be seen from figure 3 that internal resonance cases including 2:1, 3:1 and 4:1 occur on each curve respectively, and the upper right corner of the figure and the area between each curve are all non-resonance areas. In the actual antenna system design, it is usually necessary to avoid the range of the resonance region. Once the parameters are near the boundary of the resonance curve, the system will occur internal resonance, which may destroy the antenna system. The detailed analysis of the case of 3:1 internal resonance of the system will be given following figure 3.

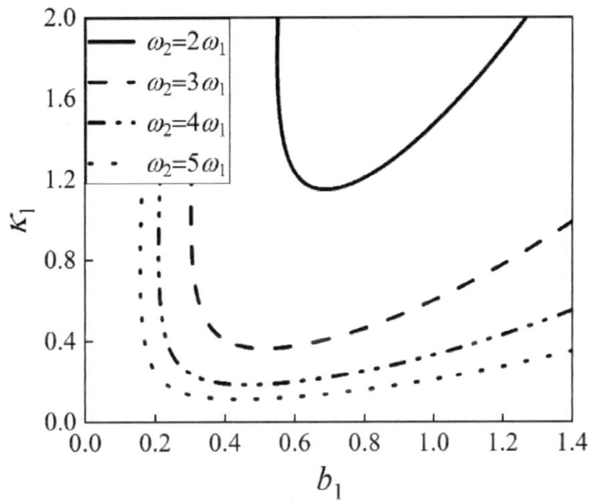

Figure 3. Condition diagram of complete internal resonance for a 3-DOF T-beam system.

The 3:1 internal resonance occurs when the derived system frequency satisfies $\omega_2 \approx 3\omega_1$. Suppose the relationship is as follows

$$\omega_2 = 3\omega_1 + \varepsilon\sigma \tag{20}$$

By inserting equations (16) and (18) into equation (14) and removing non-resonant terms, the governing equation simplifies to:

$$D_1 A_j\left(T_1,T_2\right) = 0, (j=1,2,3) \tag{21}$$

Then the second-order approximate solution of the equation is given as follows

$$\boldsymbol{\eta}_2 = \boldsymbol{\Delta}(\omega)[-\mathbf{M}_1(\boldsymbol{\eta}_1)D_0^2\boldsymbol{\eta}_1 - \mathbf{K}_2(\boldsymbol{\eta}_1,\boldsymbol{\eta}_1)\boldsymbol{\eta}_0 - \mathbf{K}_1(\boldsymbol{\eta}_1)\boldsymbol{\eta}_1 + \mathbf{P}_2(\boldsymbol{\eta}_1, D_0\boldsymbol{\eta}_1)] \tag{22}$$

where $\boldsymbol{\Delta}(\omega) = \left(\mathbf{R}_0 - \mathbf{M}_0\omega^2\right)^{-1}$ is the matrix matching each frequency in the harmonic term at the end of expression (22). Substituting the expressions for $\boldsymbol{\eta}_0$, $\boldsymbol{\eta}_1$ and $\boldsymbol{\eta}_2$ into equation (15), there exists a unique solution $\boldsymbol{\eta}_3$ to the equation when the secular term is orthogonal to $\mathbf{u}_j\left(j=1,2,3\right)$, where $\mathbf{u}_j$ is the solution of the following equation as

$$\omega_j^2\mathbf{M}_0^T\mathbf{u}_j = R_0^T\mathbf{u}_j (j=1,2,3) \tag{23}$$

Thus, the following solvability condition can be obtained as

$$\left(\mathbf{u}_1^T\boldsymbol{\alpha}_1\right)A_1' + \left(\mathbf{u}_1^T\boldsymbol{\alpha}_{10}\right)A_1^2\overline{A}_1 + \left(\mathbf{u}_1^T\boldsymbol{\alpha}_{12}\right)A_1A_2\overline{A}_2 + \left(\mathbf{u}_1^T\boldsymbol{\alpha}_{13}\right)A_1A_3\overline{A}_3 + \left(\mathbf{u}_1^T\boldsymbol{\alpha}_{14}\right)A_2\overline{A}_1^2e^{\mathrm{i}\sigma T_2} = 0 \tag{24}$$

$$\left(\mathbf{u}_2^T\boldsymbol{\alpha}_2\right)A_2' + \left(\mathbf{u}_2^T\boldsymbol{\alpha}_{20}\right)A_2^2\bar{A}_2 + \left(\mathbf{u}_2^T\boldsymbol{\alpha}_{21}\right)A_2A_1\bar{A}_1 + \left(\mathbf{u}_2^T\boldsymbol{\alpha}_{23}\right)A_2A_3\bar{A}_3 + \left(\mathbf{u}_2^T\boldsymbol{\alpha}_{24}\right)A_1^3e^{-\mathrm{i}\sigma T_2} = 0 \quad (25)$$

$$\left(\mathbf{u}_3^T\boldsymbol{\alpha}_3\right)A_3' + \left(\mathbf{u}_3^T\boldsymbol{\alpha}_{30}\right)A_3^2\bar{A}_3 + \left(\mathbf{u}_3^T\boldsymbol{\alpha}_{31}\right)A_3A_1\bar{A}_1 + \left(\mathbf{u}_3^T\boldsymbol{\alpha}_{32}\right)A_3A_2\bar{A}_2 = 0 \quad (26)$$

The expressions of $\boldsymbol{\alpha}_j$, $\boldsymbol{\alpha}_{j0}$ and $\boldsymbol{\alpha}_{sr}$ in the formula are given in the appendix.

Setting $A_j(T_2) = a_j(T_2)e^{(i\beta_j(T_2))/2}(j=1,2,3)$, substituting it into equations (24) to (26), and separating the real and imaginary parts, one can obtain the following equations

$$4\dot{a}_1 + a_1^3v_{101} + a_1a_2^2v_{121} + a_1a_3^2v_{131} + a_1^2a_2\left(v_{141}\cos\gamma - v_{142}\sin\gamma\right) = 0 \quad (27)$$

$$4a_1\dot{\beta}_1 + a_1^3v_{102} + a_1a_2^2v_{122} + a_1a_3^2v_{132} + a_1^2a_2\left(v_{141}\sin\gamma + v_{142}\cos\gamma\right) = 0 \quad (28)$$

$$4\dot{a}_2 + a_2^3v_{201} + a_2a_1^2v_{211} + a_2a_3^2v_{231} + a_1^3\left(v_{241}\cos\gamma + v_{242}\sin\gamma\right) = 0 \quad (29)$$

$$4a_2\dot{\beta}_2 + a_2^3v_{202} + a_2a_1^2v_{212} + a_2a_3^2v_{232} + a_1^3\left(-v_{241}\sin\gamma + v_{242}\cos\gamma\right) = 0 \quad (30)$$

$$4\dot{a}_3 + a_3^3v_{301} + a_3a_1^2v_{311} + a_3a_2^2v_{321} = 0 \quad (31)$$

$$4a_3\dot{\beta}_3 + a_3^3v_{302} + a_3a_1^2v_{312} + a_3a_2^2v_{322} = 0 \quad (32)$$

where $\gamma(T_2) = \beta_2(T_2) - 3\beta_1(T_2) + \sigma T_2$. The approximate solution of the system vibration is given as follows

$$\begin{aligned}\boldsymbol{\eta} = \boldsymbol{\eta}_0 &+ \varepsilon\left[\sum_{j=1}^{3}a_j(t)\cos\left(\omega_jt + \beta_j(t)\right)\mathbf{p}_j\right] \\ &+ \frac{1}{2}\varepsilon^2\{\sum_{j=1}^{3}\sum_{r=1}^{3}\Delta\left(\omega_j+\omega_r\right)a_ja_r\cos\left(\left(\omega_j+\omega_r\right)t + \beta_j(t) + \beta_r(t)\right)z_{1jr} \\ &+ \sum_{j=1}^{3}\sum_{r=1}^{3}\Delta\left(\omega_j-\omega_r\right)a_ja_r\cos\left(\left(\omega_j-\omega_r\right)t + \beta_j(t) - \beta_r(t)\right)z_{2jr}\}\end{aligned} \quad (33)$$

The expressions for z_{1jr} and z_{2jr} are given as

$$\begin{cases} z_{1jr} = \omega_r^2\mathbf{M}_1(\mathbf{p}_j)\mathbf{p}_r - \omega_j\omega_r\mathbf{P}_{22}(\mathbf{p}_j,\mathbf{p}_r) - \mathbf{K}_2(\mathbf{p}_j,\mathbf{p}_r)\boldsymbol{\eta}_0 - \mathbf{K}_1(\mathbf{p}_j)\mathbf{p}_r \\ z_{2jr} = \omega_r^2\mathbf{M}_1(\mathbf{p}_j)\mathbf{p}_r + \omega_j\omega_r\mathbf{P}_{22}(\mathbf{p}_j,\mathbf{p}_r) - \mathbf{K}_2(\mathbf{p}_j,\mathbf{p}_r)\boldsymbol{\eta}_0 - \mathbf{K}_1(\mathbf{p}_j)\mathbf{p}_r \end{cases} \quad (34)$$

By combining equations (28) and (30), we obtain the following governing equation:

$$\begin{aligned}4a_1a_2\dot{\gamma} = a_1a_2^3\left(3v_{122} - v_{202}\right) + a_2a_1^3\left(3v_{102} - v_{212}\right) + a_1a_2a_3^2\left(3v_{132} - v_{232}\right) \\ +3a_1^2a_2^2\left(v_{141}\sin\gamma + v_{142}\cos\gamma\right) - a_1^4\left(-v_{241}\sin\gamma + v_{242}\cos\gamma\right) + a_1a_2\sigma\end{aligned} \quad (35)$$

The periodic solutions describing nonlinear vibrational modes of the system emerge as equilibrium states in the reduced-order dynamics governed by equations (27), (29), (31), (32) and (35).At the same time, the stability of the nonlinear mode solution is consistent with the stability of the steady-state solution of the appeal equation system. Setting $\dot{\gamma} = 0$ in equation (35) yields the following equation as

$$\begin{aligned}v_{242}h^4 - \left(3v_{102} - v_{212} + \beta_1\left(3v_{132} - v_{232}\right)\right)h^3 - 3v_{142}h^2 \\ -\left[\left(3v_{122} - v_{202}\right) + \beta_2\left(3v_{132} - v_{232}\right) + \frac{\sigma}{a_2^2}\right]h = 0\end{aligned} \quad (36)$$

where the expressions for h, β_1 and β_2 are given as

$$h = \frac{a_1}{a_2}, \beta_1 = -\frac{v_{311}}{v_{301}}, \beta_2 = -\frac{v_{321}}{v_{301}} \quad (37)$$

The system parameters are set as $d_1 = 3$, $b_1 = 0.426$, $\kappa_1 = 0.386$, $\kappa_2 = \kappa_3 = 1$ and $L = 10$ such that the frequency satisfies $\omega_2 \approx 3\omega_1$ and the system undergoes 3:1 internal resonance. The variation curve of h with σ / a_2^2 for 3:1 internal resonance is shown in figure 4. As shown in figure 4, the system has one stable and one unstable coupled mode solution when $\sigma / a_2^2 < -39.12$, two unstable coupled mode solutions when $-39.12 < \sigma / a_2^2 < -24.43$, three unstable and one stable coupled mode solutions when $-24.43 < \sigma / a_2^2 < -21.98$, and one stable and one unstable coupled mode solution when $\sigma / a_2^2 > -21.98$.

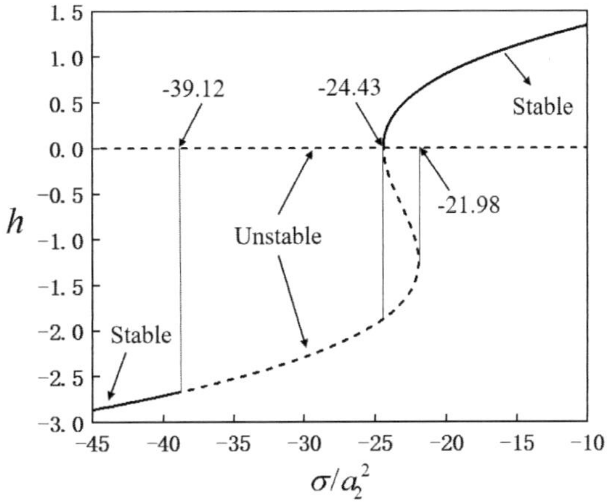

Figure 4. The variation curve of h with σ / a_2^2 in 3-DOF T-shaped antenna system with 3:1 internal resonance.

4. Numerical analysis

In this section, the numerical solution of the 3-DOF T-shaped antenna system Eq. (10) solved via the Runge–Kutta method is presented to validate the analytic solution. For the 3:1 internal resonance, the system parameters $d_1 = 3$, $b_1 = 0.426$, $\kappa_1 = 0.386$, $\kappa_2 = \kappa_3 = 1$ and $L = 10$ are set, and the system frequency satisfies $\omega_2 \approx 3\omega_1$. Taking the initial state as $[a_{10}, a_{20}, a_{30}, \beta_{10}, \beta_{20}, \beta_{30}]^T = [0, 0.1, 0, 0, 0, 0]^T$, figure 5 demonstrates excellent agreement between numerical simulations and analytical predictions for the 3:1 internally resonant system, validating our perturbation approach. As shown in figure 5, the numerical result and the analytical result are agree very well with each other, indicating the validity and high precision of the analytical solution analysis.

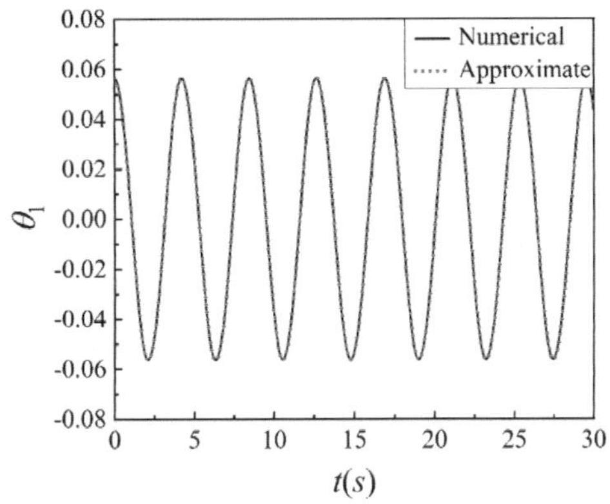

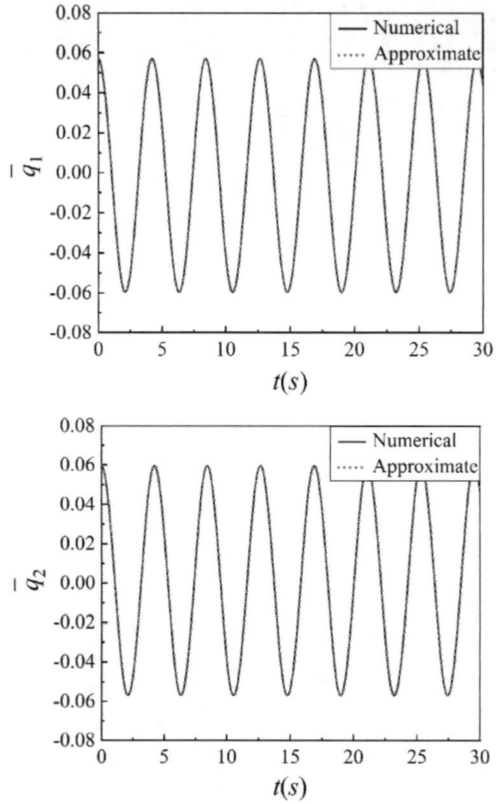

Figure 5. Comparison of numerical and analytical solution results of 3:1 internal resonance.

5. Conclusions

The study focuses on internal resonance phenomena in a three-DOF T-configuration antenna with hybrid rigid-flexible dynamics. In the 3:1 internal resonance analysis, the approximate analytical solutions of the high-dimensional rigid-flexible coupling system can be obtained by using the matrix form multi-scale method, and the critical conditions for periodic and aperiodic motions are given. Examination of the frequency response characteristics indicates potential multimodal coupling mechanisms during 3:1 internal resonance conditions. Finally, the numerical and analytical solutions are simulated to verify the validity and high accuracy of the analytical analysis.

A key advance in this work is the development of a coupled rigid-flexible dynamic model, enabling precise analysis of antenna system nonlinearities. The analytical solution and the possibility of the internal resonance are given under appropriate structural, which provides an important reference for the parameters design of antenna structures. The proposed analytical framework enables comprehensive investigation of nonlinear dynamics in deployable antenna structures, offering valuable insights for stability-oriented design optimization.

6. References

[1] Santiago-Prowald J, Baier H 2019 Advances in deployable structures and surfaces for large apertures in space *Ceas Space Journal* **5** 89-115

[2] Xing Z, Zheng G 2014 Deploying process modeling and attitude control of a satellite with a large deployable antenna *Chinese Journal of Aeronautics* **27** 299-312

[3] Sun Y, Zhang W, Yao M H 2017 Multi-Pulse Chaotic Dynamics of Circular Mesh Antenna with 1:2 Internal Resonance *International Journal of Applied Mechanics* **9**

[4] Fraux V, Lawton M, Reveles J R and You Z 2013 Novel large deployable antenna backing structure concepts for foldable reflectors *Ceas Space Journal* **5** 195-201

[5] Siriguleng B, Zhang W, Liu T and Liu YZ 2020 Vibration modal experiments and modal

interactions of a large space deployable antenna with carbon fiber material and ring-truss structure *Engineering Structures* **207**

[6] Salehian A, Inman D J 2008 Dynamic analysis of a lattice structure by homogenization: Experimental validation *Journal of Sound and Vibration* **316** 180-197

[7] Liu F, Wang L, Jin D and Wen H 2019 Equivalent continuum modeling of beam-like truss structures with flexible joints *Acta Mechanica Sinica* **35** 1067-1078

[8] Chen J, Zhang W, Zhang Y F 2020 Equivalent continuum model and nonlinear breathing vibrations of rotating circular truss antenna subjected to thermal excitation *Thin-Walled Structures* **157**

[9] Liu T, Zhang W, Zheng Y and Zhang YF 2021 Andronov-Hopf bifurcations, Pomeau-Manneville intermittent chaos and nonlinear vibrations of large deployable space antenna subjected to thermal load and radial pre-stretched membranes with 1:3 internal resonance *Chaos, Solitons & Fractals* **144**

[10] Zhang W, Xi A, Siriguleng B, et al 2019 An equivalent cylindrical shell model of vibration analysis based on simplified repeating unit cell for ring truss structure *Journal of Sound and Vibration* **459**

[11] Wu Q L, Zhang W, Dowell E H 2018 Detecting multi-pulse chaotic dynamics of high-dimensional non-autonomous nonlinear system for circular mesh antenna *International Journal of Non-Linear Mechanics* **102** 25-40

[12] Niu M, Su J, Zhang Z and Hua H 2019 In-plane vibration of a circular ring with arbitrary concentrated elements by an analytical method *Archive of Applied Mechanics* **89** 2215-2228

[13] Lee S, Yan Q 2015 Exact static analysis of in-plane curved timoshenko beams with strong nonlinear boundary conditions *Mathematical Problems in Engineering* **2015** 1-12

[14] Hu W, Wang F, Cao D, Chen J and Feng J 2019 Quantitative validation of the analytical mode shapes of a beam-like structure with a Z-shaped configuration *Journal of Mechanical Science and Technology* **33** 2059-2065

[15] Gao X, Jin D, Chen T 2018 Analytical and experimental investigations of a space antenna system of four DOFs with internal resonances *Communications in Nonlinear Science and Numerical Simulation* **63** 380-403

[16] Gao X, Jin D, Chen T 2018 Nonlinear analysis and experimental investigation of a rigid-flexible antenna system *Meccanica* **53** 33-48

[17] Gao X, Jin D, Hu H 2017 Internal resonances and their bifurcations of a rigid-flexible space antenna *International Journal of Non-Linear Mechanics* **94** 160-173

[18] Lee S Y, Yan Q Z 2015 An analytical solution for out-of-plane deflection of a curved Timoshenko beam with strong nonlinear boundary conditions *Acta Mechanica* **226** 3679-3694

[19] Zhou Y, Nyberg T R, Xiong G, Zhou H and Li S 2016 Precise deflection analysis of laminated piezoelectric curved beam *Journal of Intelligent Material Systems and Structures* **27** 2179-2198

[20] Rajasekaran S 2014 Analysis of curved beams using a new differential transformation based curved beam element *Meccanica* **49** 863-886

[21] Zhang W, Wu R Q and Behdinan K 2019 Nonlinear dynamic analysis near resonance of a beam-ring structure for modeling circular truss antenna under time-dependent thermal excitation *Aerospace Science and Technology* **86** 296-311

[22] Wu R Q, Zhang W, Chen J E, Feng J and Hu W 2023 Nonlinear vibration and stability analysis of a flexible beam-ring structure with one-to-one internal resonance *Applied Mathematical Modelling* **119** 316-337

[23] Wu R Q, Zhang W and Behdinan K 2019 Vibration Frequency Analysis of Beam - Ring Structure for Circular Deployable Truss Antenna *International Journal of Structural Stability and Dynamics* **19**

[24] Wu J, Zhao Z H and Ren G X 2012 Dynamic Analysis of Space Structure Deployment with Transient Thermal Load *Advanced Materials Research* **479-481** 803-807

[25] Liu J Y and Lu H 2007 Thermal effect on the deformation of a flexible beam with large kinematical driven overall motions *European Journal of Mechanics - A/Solids* **26** 137-151

[26] Liu L, Cao D, Huang H, Shao C and Xu Y 2017 Thermal-structural analysis for an attitude maneuvering flexible spacecraft under solar radiation *International Journal of Mechanical Sciences* **126** 161-170

Acknowledgement
This work was supported by the National Natural Science Foundation of China (Grant No. 12002151).

Appendix

The details of $\mathbf{M}$, $\mathbf{K}$ and $\mathbf{P}$ in (10) read

$$\mathbf{M}(\boldsymbol{\eta})=\begin{bmatrix} \frac{A_{xx}}{L}\left(\bar{q}_1^2+\bar{q}_2^2\right)+\frac{2b_1A_x}{L}\left(\bar{q}_1-\bar{q}_2\right)+\frac{2}{3}+2b_1^2+\frac{7}{12}d_1b_1^2 & \frac{B_x}{L^2} & \frac{B_x}{L^2} \\ \frac{B_x}{L^2} & \frac{A_{xx}}{L} & 0 \\ \frac{B_x}{L^2} & 0 & \frac{A_{xx}}{L} \end{bmatrix} \tag{A.1}$$

$$\mathbf{K}=\begin{bmatrix} \kappa_1 & 0 & 0 \\ 0 & \kappa_2 & 0 \\ 0 & 0 & \kappa_3 \end{bmatrix} \tag{A.2}$$

$$\mathbf{P}(\boldsymbol{\eta},\dot{\boldsymbol{\eta}})=\begin{bmatrix} -2\dot{\theta}_1\left(\frac{A_{xx}}{L}\bar{q}_1\dot{\bar{q}}_1+\frac{A_x}{L}b_1\dot{\bar{q}}_1+\frac{A_{xx}}{L}\bar{q}_2\dot{\bar{q}}_2-\frac{A_x}{L}b_1\dot{\bar{q}}_2\right) \\ \dot{\theta}_1^2\left(\frac{A_x}{L}b_1+\frac{A_{xx}}{L}\bar{q}_1\right) \\ -\dot{\theta}_1^2\left(\frac{A_x}{L}b_1-\frac{A_{xx}}{L}\bar{q}_2\right) \end{bmatrix} \tag{A.3}$$

$$A_x=\int_0^L\varphi(x)dx, A_{xx}=\int_0^L\varphi^2(x)dx, B_x=\int_0^L x\varphi(x)dx \tag{A.4}$$

The details of expressions for each matrix and vector in (12)-(15) read

$$\mathbf{M}_0=\mathbf{M}(\boldsymbol{\eta}_0),\mathbf{K}_0=\mathbf{K}(\boldsymbol{\eta}_0),\mathbf{P}_0=\mathbf{P}(\boldsymbol{\eta}_0,D_0\boldsymbol{\eta}_0) \tag{A.5}$$

$$\mathbf{M}_1(x)=\left[\frac{\partial\mathbf{M}}{\partial\eta_i}x_i\right],\mathbf{K}_1(x)=\left[\frac{\partial\mathbf{K}}{\partial\eta_i}x_i\right],\mathbf{P}_{11}=\left[P_{11ij}\right]=\left[\frac{\partial\mathbf{P}_i}{\partial\eta_j}\right],\mathbf{P}_{12}=\left[P_{12ij}\right]=\left[\frac{\partial\mathbf{P}_i}{\partial\dot{\eta}_j}\right] \tag{A.6}$$

$$\mathbf{M}_2(x,y)=\left[\frac{1}{2}\frac{\partial^2\mathbf{M}}{\partial\eta_i\partial\eta_j}x_iy_j\right],\mathbf{K}_2(x,y)=\left[\frac{1}{2}\frac{\partial^2\mathbf{K}}{\partial\eta_i\partial\eta_j}x_iy_j\right] \tag{A.7}$$

$$\begin{cases}
\mathbf{P}_2(x,y)=\{P_{2k}(x,y)\}=\left\{\dfrac{1}{2}\left(\dfrac{\partial\mathbf{P}_k}{\partial\eta_i}\mathbf{x}_i+\dfrac{\partial\mathbf{P}_k}{\partial\dot{\eta}_j}\mathbf{y}_j\right)^2\right\}\\
\mathbf{P}_{21}(x,y)=\{P_{21k}(x,y)\}=\left\{\dfrac{1}{2}\dfrac{\partial^2\mathbf{P}_k}{\partial\eta_i\partial\dot{\eta}_j}\mathbf{x}_i\mathbf{y}_j\right\}\\
\mathbf{P}_{211}(x,y)=\{P_{211k}(x,y)\}=\left\{\dfrac{1}{2}\dfrac{\partial^2\mathbf{P}_k}{\partial\eta_i\partial\eta_j}\mathbf{x}_i\mathbf{y}_j\right\}\\
\mathbf{P}_{22}(x,y)=\{P_{22k}(x,y)\}=\left\{\dfrac{1}{2}\dfrac{\partial^2\mathbf{P}_k}{\partial\dot{\eta}_i\partial\dot{\eta}_j}\mathbf{x}_i\mathbf{y}_j\right\}\\
\mathbf{P}_3(x,y)=\{P_{3k}(x,y)\}=\left\{\dfrac{1}{6}\left(\dfrac{\partial\mathbf{P}_k}{\partial\eta_i}\mathbf{x}_i+\dfrac{\partial\mathbf{P}_k}{\partial\dot{\eta}_j}\mathbf{y}_j\right)^2\right\}\\
\mathbf{K}_3(x,y,z)=\left[\dfrac{1}{6}\dfrac{\partial^3\mathbf{K}}{\partial\eta_i\partial\eta_j\partial\eta_k}\mathbf{x}_i\mathbf{y}_j\mathbf{z}_k\right]\\
\mathbf{R}_0=\mathbf{K}_0-\mathbf{P}_{11}-\mathbf{P}_{12}D_0+\mathbf{Q}_0,\mathbf{Q}_0=\left[\left\{\dfrac{\partial\mathbf{K}}{\partial\eta_1}\boldsymbol{\eta}_0\right\}\left\{\dfrac{\partial\mathbf{K}}{\partial\eta_2}\boldsymbol{\eta}_0\right\}\left\{\dfrac{\partial\mathbf{K}}{\partial\eta_3}\boldsymbol{\eta}_0\right\}\right]_{\boldsymbol{\eta}_0}
\end{cases}\tag{A.8}$$

The details in (24)-(26) read

$$\begin{cases}
\mathbf{v}_j=-2i\omega_j\mathbf{M}_0\mathbf{p}_j\\
\mathbf{v}_{j0}=4\omega_j^2\mathbf{M}_1(\mathbf{p}_j)\mathbf{z}_{j1}+\omega_j^2\mathbf{M}_1(\mathbf{z}_{j1})\mathbf{p}_j+\omega_j^2\mathbf{M}_1(\mathbf{z}_{j3})\mathbf{p}_j\\
\qquad+3\omega_j^2\mathbf{M}_2(\mathbf{p}_j,\mathbf{p}_j)\mathbf{p}_j+4i\omega_j\mathbf{P}_{21}(\mathbf{p}_j,\mathbf{z}_{j1})\\
\qquad+2i\omega_j(\mathbf{P}_{21}(\mathbf{z}_{j3},\mathbf{p}_j)-\mathbf{P}_{21}(\mathbf{z}_{j1},\mathbf{p}_j))\\
\mathbf{v}_{jk}=\omega_j^2\mathbf{M}_1(\mathbf{z}_{k3})\mathbf{p}_j+2\omega_j^2\mathbf{M}_2(\mathbf{p}_k,\mathbf{p}_k)\mathbf{p}_j\\
j,k=1,2,3
\end{cases}\tag{A.9}$$

$$\begin{cases}
\mathbf{z}_{j1}=\left[\mathbf{R}_0-4\omega_j^2\mathbf{M}_0\right]^{-1}\left[\omega_j^2\mathbf{M}_1(\mathbf{p}_j)\mathbf{p}_j+i\omega_j\mathbf{P}_2(\mathbf{p}_j,\mathbf{p}_j)\right]\\
\mathbf{z}_{j3}=2\mathbf{R}_0^{-1}\left[\omega_j^2\mathbf{M}_1(\mathbf{p}_j)\mathbf{p}_j\right]\\
\mathbf{z}_{k3}=2\mathbf{R}_0^{-1}\left[\omega_k^2\mathbf{M}_1(\mathbf{p}_k)\mathbf{p}_k\right]
\end{cases}\tag{A.10}$$

Journal of Physics: Conference Series **3041** (2025) 012007 doi:10.1088/1742-6596/3041/1/012007

Freejet tests on a variable geometry supersonic inlet driven by shape memory alloys

Yuhui Wang, Junhua Zhang, Zhaoyang Tian and Lei Shi*

National Key Laboratory of Solid Rocket Propulsion,
Northwestern Polytechnical University, Xi'an 710072, PR China

*E-mail: shilei@nwpu.edu.cn

Abstract: Conventional variable geometry supersonic inlets have defects of high redundant mass, large size and poor control effect. A variable geometry inlet with the SMAs as the control element is experimentally studied in a high-speed wind tunnel. The control scheme for inlet shapes is proposed, and the operating characteristics of the variable geometry inlet under typical conditions have been examined. The deformation of the throat and forebody can ensure stable operation of the inlet in a wide flight Mach number domain.

1. Introduction

As a core component of hypersonic air-breathing aircrafts, the ramjet inlet operates in a large range of speed and airspace[1]. Variable geometry techniques have been proposed to ensure that inlets operate in complex conditions[2]. The shock system will be adjusted to improve the performance of the inlet by changing aerodynamic shapes. Axisymmetric inlets have been widely used. However, they have the defects of high aerodynamic resistance and easy to be stuck. Thus, researchers focus on the two-dimensional, variable-geometry inlet with a wide speed range of operation. Two-dimensional inlets with variable geometry by adjusting several components, such as cowl plate and throat plate[3-5]. Also, control devices will increase the weight and size of inlets. It is important to improve the performance of the inlet by adopting new control techniques.

Shape memory alloys (SMAs) are smart materials that widely used in smart devices because of the shape memory effect, high energy density and driving strain. They were first used in variable geometry inlets by NASA[6]. However, practical applications of SMAs on variable geometry inlets are still few at present. Therefore, experiments will be carried out on a variable geometry inlet driven by SMAs. The model and control scheme of the variable geometry inlet will be checked.

2. Control schemes and experimental models

2.1. Control schemes

The theoretical structure and control scheme of the variable geometry inlet proposed in this paper are shown in Figure 1. The operational performance of the inlet can be adjusted by changing the internal compression ratio and the structure of the forebody shock waves. The control scheme of the inlet in the wide Mach domain is as follows:

1) At Mach 2.5 to 3.5, the deformation device is not operating.

2) At Mach 4.0, heated SMAs drive the deformation device in the throat to increase the internal compression ratio.

3) At Mach 5.0 to 5.5, based on the configuration at Mach 4.0, the SMAs are heated to drive the deformation of the forebody. The secondary flow circulation cavity in the forebody will be opened to move the shock wave out of the cowl.

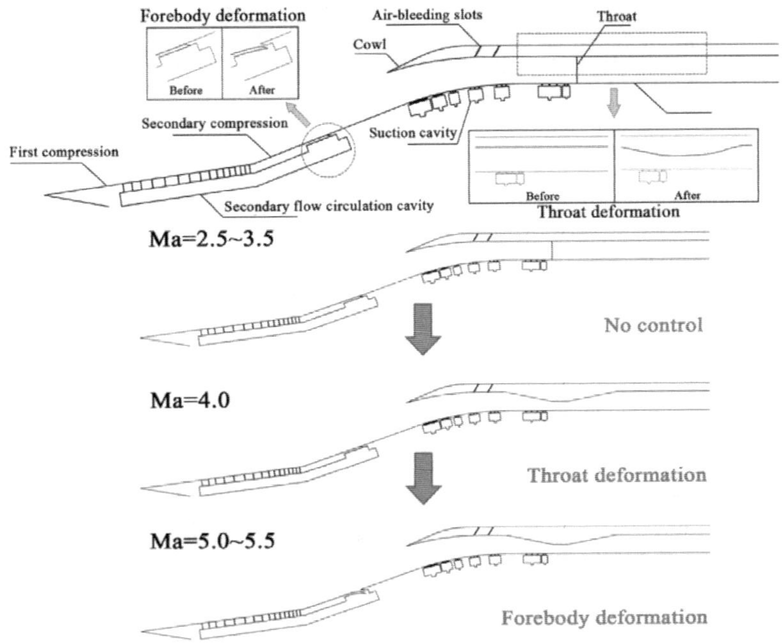

Figure 1. The theoretical structure and control scheme of the variable geometry inlet

2.2. Experimental models

The experimental model is shown in Figure 2. The main components contain the forebody, isolator and flow measurement section, in which pressure monitoring points and comb pitots are set up. Figure 3 shows the control devices for the deformation of the forebody and throat optimized based on the experimental conditions. SMAs are used instead of motors as control elements to reduce the size and weight of the variable geometry parts.

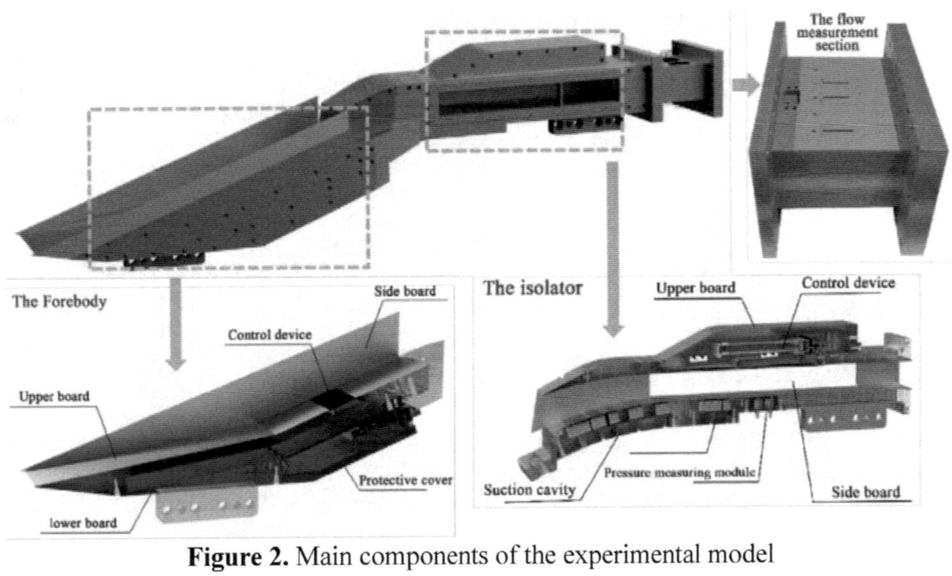

Figure 2. Main components of the experimental model

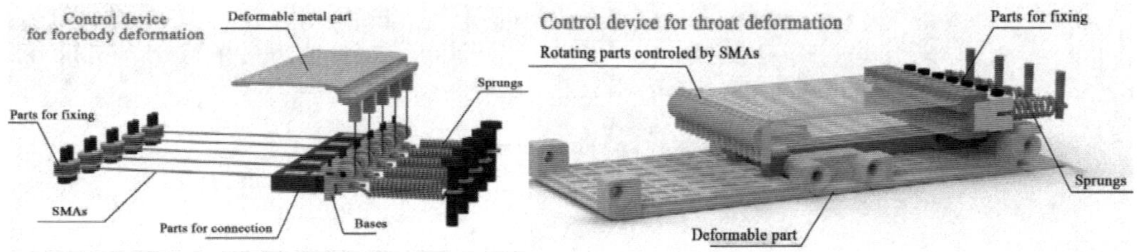

Figure 3. The structure of the deformation control device after optimization

The supersonic wind tunnel FD-30 is deployed for freejet test. The experimental model is implemented in the test section of FD-30 as shown in Figure 4, in which a wedge-shaped plug is located at the exit for simulating the combustion pressure. The theoretical parameters for stable operation of the inlet under typical conditions are shown in Table 1. The total pressure recovery coefficient σ and the flow coefficient Φ at the throat are obtained from the total pressure of incoming flows P_0, the static pressure of incoming flows P_∞, the static pressure at the wall of the isolator P and the total pitot pressure P_t[7].

Figure 4. The experimental model in the wind tunnel FD-30

Table 1. Theoretical parameters of the two-dimensional variable geometry inlet

Mach number	P_0 (MPa)	P_∞ (Pa)	σ at the throat	Φ
3.0	0.2	5445	0.8262	0.6110
4.0	0.5	3293	0.7662	0.8110
5.0	1.2	2268	0.6946	0.9338

3. Results and discussions

3.1. Freejet test at Mach 3.0

The inlet can operate normally at Mach 3.0 without deformation control. The performance of the inlet is discussed with backpressure. The backpressure ratio can be expressed as $PR = P(P_\infty)^{-1}$.

Figure 5 shows the variations of the static pressure distribution along the ramp wall and schlieren figures of the isolator. X1~X10 are pressure monitoring points, while X1 is near the throat and X10 is at the outlet of the isolator. It can be seen that the flow in the inlet is stable at t = 2s. The flow in the suction cavity is maximized gradually under the effect of the increased backpressure, which reflects the leading edge of shock trains moving towards the throat. This leads to an increase in the pressure of monitoring points. The total pressure recovery coefficient at the throat is 0.7980 and the flow coefficient is 0.6170. The deviations from the theoretical values are 3.41% and 0.98%, respectively.

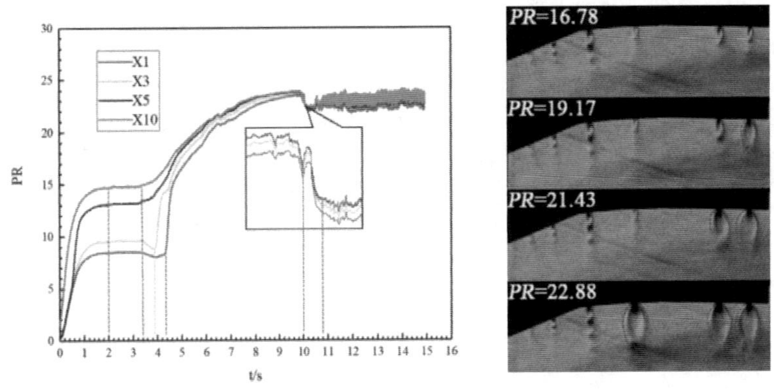

Figure 5. Static pressure distribution and schlieren figures of the isolator at Mach 3.0

3.2. Freejet test at Mach 4.0

The compression capacity of the inlet needs to be enhanced as the flight Mach number increases. The throat height can be changed by controlling the deformation device at Mach 4.0. Figure 6 shows the static pressure distribution along the ramp wall and schlieren figures of isolator during the throat deformation. The throat becomes a compression-to-expansion structure during the deformation, resulting in a reflected shock wave and two shock systems both propagating upstream. Thus, the pressure near the throat increases dramatically. It can be seen that the pressure of X1 and X2 decreases slightly from t = 5.6s to t = 5.9s because of the faster propagation of the reflected shock wave. The walls downstream of the throat are mainly affected by the expansion fan. The expansion fan is gradually enhanced in the deformation, which eventually stabilizes with the shock wave at X3 and X5, respectively. Therefore, the pressure of X3 and X5 stabilized at t = 7.7s after a slight change. The deviations of the total pressure recovery coefficient and flow coefficient at the throat from the theoretical values are 2.86% and 0.26%, respectively.

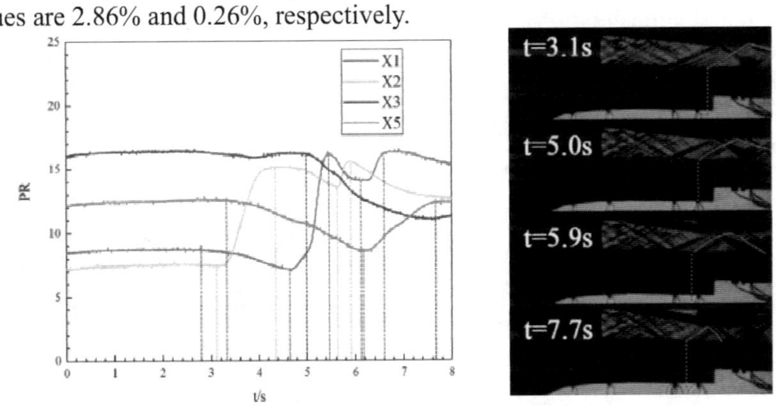

Figure 6. Static pressure distribution and schlieren figures during throat deformation at Mach 4.0

Table 2 shows the parameters in the isolator after deformation at Mach 4.0. It can be seen that the pressure in the isolator after deformation is significantly increased especially at the throat. The total pressure recovery coefficient and flow coefficient at the outlet of the isolator decreased by only 4.44% and 0.80%, respectively. As a result, the stability of the inlet is improved.

Table 2. Parameters in the isolator after deformation at Mach 4.0

	Pressure-lift ratio at the throat	Pressure-lift ratio at the outlet	σ at the outlet	Φ
Before deformation	-	-	0.6578	0.8154
After deformation	1.79	1.01	0.6286	0.8089

Journal of Physics: Conference Series **3041** (2025) 012007 doi:10.1088/1742-6596/3041/1/012007

3.3. Freejet test at Mach 5.0

The forebody shock wave will enter the cowl lip because of the small angle itself when the Mach number rises to 5.0. Thus, it is necessary to deform the forebody. Figure 7 shows the static pressure distribution along the ramp wall when the throat is deformed first. Similar to Mach 4.0, the pressure firstly decreases and then increases under the upstream propagation of the two shock systems. At the end of the deformation, the pressure stabilizes under the effect of the expansion fan.

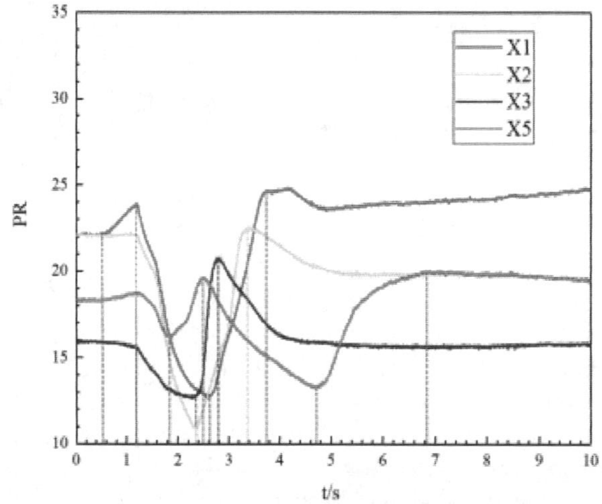

Figure 7. Static pressure distribution when throat is deformed first at Mach 5.0

Figure 8 shows the static pressure distribution along the ramp wall and schlieren figures during the forebody deformation. The forebody deformation increases the thickness of the boundary layer in the isolator. The shock waves propagate upstream at a low speed resulting in a slight decrease of the static pressure. Meanwhile, the opening of the secondary flow circulation cavity increases the angle of the forebody shock. The forebody shock moved upstream beyond the cowl lip where it meets the cowl shock. It can be seen that shock waves are enhanced outside of the cowl. Compared with theoretical values, the total pressure recovery coefficient of the throat is reduced by 4.53%, and the flow coefficient is reduced by 1.25%.

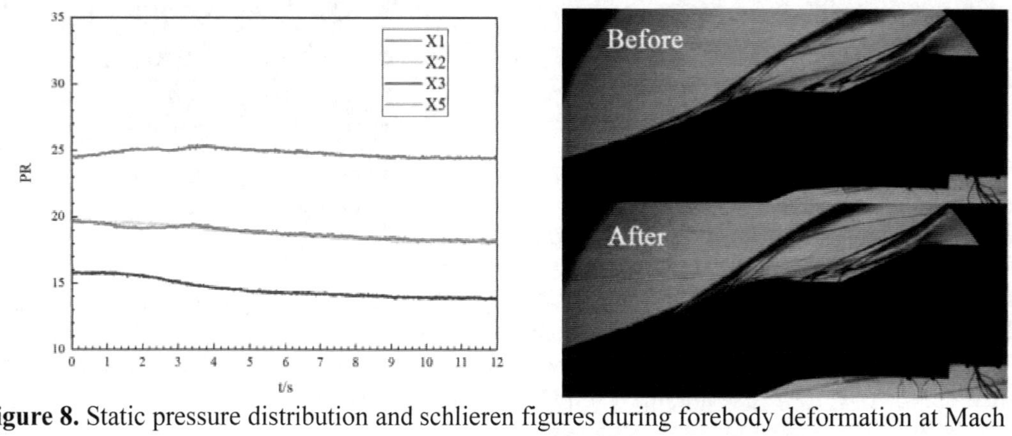

Figure 8. Static pressure distribution and schlieren figures during forebody deformation at Mach 5.0

The parameters in the isolator after deformation at Mach 5.0 is shown in Table 3. Because of the increase of the intensity of the shock wave and the pressure inside the isolator after throat deformation, the total pressure recovery coefficient and flow coefficient at the outlet decrease slightly. After the forebody deformation, the intensity of the shock wave decreases. The total pressure recovery coefficient at the outlet increases to 0.5060, which is only 0.69% lower than that before the deformation. And the flow coefficient decreased by 3.70%.

Table 3. Parameters in the isolator after deformation at Mach 5.0

	Pressure-lift ratio at the throat	Pressure-lift ratio at the outlet	σ at the outlet	Φ
Before deformation	-	-	0.5095	0.9575
After throat deformation	1.09	1.18	0.4872	0.9449
After forebody deformation	-	-	0.5060	0.9221

4. Conclusions

In this paper, a variable geometry inlet driven by the shape memory alloys is designed. The control scheme and performance of the inlet under typical conditions in the wide Mach domain are tested at Mach 3.0, 4.0, and 5.0. The results show that the structure of the shock system can be adjusted by reasonably changing the shape of the throat and the forebody, which ensures the stable operation of the inlet under different Mach numbers. Meanwhile the performance of the inlet is not significantly reduced. The total pressure recovery coefficients at the inlet throat are 0.7980, 0.7443 and 0.6631 under three typical conditions, while the flow coefficients are 0.6170, 0.8089 and 0.9221, respectively. The deviations from the theoretical model are within 5%.

5. References

[1] Fry R S. "A century of ramjet propulsion technology evolution," *J. Propul. Power.* 20(1), 27-58(2004).

[2] Dalle D, Torrez S, Driscoll J. "Performance analysis of variable-geometry scramjet inlets using a low-order model," *47th AIAA/ASME/SAE/ASEE joint propulsion conference & exhibit.* p.5756(2011).

[3] Lawrence H, Kenneth R, Edward R, David W, Earl A. "Hyper-X flight engine ground testing for X-43 flight risk reduction," *10th AIAA/NAL-NASDA-ISAS International Space Planes and Hypersonic Systems and Technologies Conference.* 2001.

[4] Falempin F, Goldfeld M. "Design and experimental evaluation of a Mach 2 Mach 8 inlet," *AIXA/NAL-NASDA-ISAS 10th International Space Planes and Hypersonic Systems and Technologies conference, AIAA Paper.* p.1980(2001).

[5] Falempin F, Wendung E, Goldfeld M, Starov A. "Experimental investigation of starting process for a variable geometry air inlet operating from mach 2 to mach 8," *42nd AIAA/ASME/SAE/ASEE Joint Propulsion Conference & Exhibit.* p.4513(2006).

[6] Dale M P, James P D, Edward V W. "Design and Test of a SMA Powered Adaptive Aircraft Inlet Internal Wall," *43rd AIAA/ASME/ASCE/AHS/ASC Structures, Structural Dynamics, and Materials Conference.* p.1356(2002).

[7] Pan P Y. "Aerodynamic Performance Research and Wind Tunnel Test of Two-Dimensional and Variable Geometry TBCC Inlet," Nanjing University of Aeronautics and Astronautics. 2021.

Acknowledgments

This research work is supported by the National Natural Science Foundation of China through grant 52272405, and the Key Research and Development Program of Shaanxi Province through grant 2022ZY2-JCYJ-01-02 and 2023ZY1-JCYJ-01-01.

Numerical study of ignition energy influence on flame acceleration and detonation transition in supersonic flow

Dejiao Luo[1], Zhehao Wei[2], Siqi Zhang[1], Lili Zhang[2], Mingchao Huang[1], Wandong Zhao[1,*], Xiong Yang[1,]**

[1]College of Aerospace Science and Engineering, National University of Defense Technology, Changsha, 410073, China
[2]Basic Education College, National University of Defense Technology, Changsha, 410073, China

Wandong Zhao (zhao.wandong@qq.com), Xiong Yang (yx12321@126.com)

Abstract. Deflagration-to-detonation transition (DDT) is crucial for detonation engines, particularly in supersonic flows. This study explores ignition energy's effect on FA and DDT. Using a detailed chemical reaction mechanism (12 species, 27 steps), coupled with reactive Navier-Stokes equations, high-resolution simulations were performed within an adaptive mesh refinement to study flame acceleration (FA) and DDT mechanism. Results show that a curved leading shock wave (LSW) develops into Mach stem structure. Upon reflection from a solid boundary, these structures produce a small explosive point that subsequently evolves into detonation wave. With a constant incoming Mach number, high ignition energy condition (which do not directly cause detonation) results in dominant initial flame acceleration but require the longest time for detonation transition, leading to extended run-up time (TDDT) and distance (LDDT) to detonation. Conversely, moderate ignition energy facilitates the rapid formation of the LSW, resulting in significantly shorter TDDT. Therefore, it is not necessarily true that higher ignition energy results in faster TDDT and LDDT in supersonic flows. This phenomenon can be attributed to that high-energy ignition causes excessive flame expansion and intense shocks, increasing blockage ratio and hindering rapid LSW formation. This research offers valuable insights for optimizing FA and detonation initiation in air-breathing propulsion systems.

1. Introduction

Flame acceleration and the transition from deflagration to detonation (DDT) are phenomena of critical concern across safety engineering, fire protection systems, and in the design of detonation-based propulsion devices. For several decades, these processes have garnered considerable attention from both industrial engineers and academic researchers [1]. In particular, as detonation-based engines emerge as a promising frontier in propulsion technology, understanding and harnessing flame acceleration and DDT have become an especially vital and complex challenge for the development of pulse detonation engine (PDE) and rotating detonation engine (RDE) [2].

To grasp the intricate physical phenomena and potential transition mechanisms associated with flame acceleration and DDT, considerable efforts have been expended by researchers and engineers over the years. The majority of experimental, numerical, and theoretical investigations have traditionally centered on static and homogeneous mixtures to simplify the analysis. However, in real-world applications, such as those encountered in safety engineering industries and detonation combustion-based ramjet engines, the combustible mixtures are typically characterized by non-uniform conditions, including gradients of temperature, velocity, pressure, and concentration. These gradients introduce additional layers of complexity to the combustion processes, significantly influencing flame behavior and the likelihood of DDT occurring. Therefore, it is crucial for ongoing research to address these heterogeneous conditions to better understand and predict flame acceleration and DDT in practical systems. This understanding is vital for enhancing the design and operational safety of industrial facilities and advanced propulsion systems, ensuring reliable performance while minimizing risks associated with unintended detonations [3]. As such, Shamsadin et al. [4] numerically studied the effect of the concentration gradient of hydrogen mixtures on deflagration and found that lateral concentration gradients can either increase or decrease flame acceleration, depending on the average hydrogen concentration. Ettner et al. [5] and Vollmer et al. [6] also conducted numerical simulations of flame acceleration and DDT under conditions of concentration gradients. Furthermore, Boeck et al. [7] carried out experimental studies on these phenomena. Their results showed that many mixtures undergo turbulent deflagration combustion following the leading shock wave (LSW), and mixtures with higher average concentrations or lower gradients lead to multi-headed detonation cell structures. Frolov et al. [8] conducted numerical simulations of the multiphysics working processes of air-breathing PDE, including charging, mixing, flame acceleration, and deflagration processes. The study showed that under inhomogeneous mixing conditions, the distribution of the mixture's mass within the combustion chamber is extremely non-uniform.

Indeed, the understanding of flame acceleration and DDT in inhomogeneous mixtures under stationary conditions has been significantly advanced by previous studies. However, when it comes to supersonic flow states, especially those pertinent to ramjet PDE, the consistency of the DDT mechanism is less understood. The dynamics of DDT in supersonic flows introduce unique challenges that differ from subsonic or stationary cases due to the rapid movement of the mixture and the interaction between shock waves and combustion. The run-up time to detonation (TDDT) and run-up distance to detonation (LDDT) are critical parameters for PDE performance and efficiency, and their dependency on ignition energy is a topic of ongoing research. It is hypothesized that increasing ignition energy could potentially reduce TDDT and LDDT, thereby facilitating quicker and more reliable detonation initiation. However, this relationship is not straightforward and may be influenced by other factors such as the initial state of the mixture, the geometry of the combustion chamber, and the specific characteristics of the fuel used.

On the basis of the aforementioned reasons, this study considers three ignition energy scenarios, including 0.1 MPa, 0.5 MPa, and 1.0 MPa in a slightly supersonic flow state. Notably, the three ignition energy levels do not lead to direct detonation but rather aim to explore the threshold and mechanisms leading up to DDT. The Navier-Stokes equations are solved within the framework of Adaptive Mesh Refinement (AMR) technology [9], and the combustion process is modeled via a detailed chemical reaction mechanism (11 species, 27 steps). Such a study is indispensable for advancing the design and operational capabilities of detonation-based propulsion systems.

2. Physical configuration and numerical model

Figure 1 provides a schematic illustration of the geometry and boundary conditions for the ramjet PDE. The PDE configuration consists of two parts: the inlet isolation section (dimensions $L_i \times Ly$=0.18 m × 0.02 m) and the combustion chamber (dimensions $L_c \times Ly$=0.8 m × 0.02 m). Simultaneously, ten pairs of symmetric solid obstacles are placed within the combustion chamber, with geometric parameters of $d \times h$=0.003 m × 0.002 m, resulting in a lower blockage ratio (Br) of 0.2 [10]. The obstacles are positioned within the range x=0-0.4 m of the combustion chamber, with an obstacle spacing set at S=0.04 m. As noted in previous literature, larger obstacle spacings facilitate the formation of Mach stems, which are conducive to the onset of detonation [11].

The study posits the operational parameters of the ramjet PDE at an altitude of 1 km, characterized by a static temperature of T = 281.7 K and a different ignition energy [10]. Ignition conditions are established at three levels: 0.1 MPa, 0.5 MPa, and 1.0 MPa, designated as cases 1, 2, and 3, respectively. Table 1 delineates the ignition energies, ignition temperatures, and AMR refinement criteria for each scenario. Notably, our previous research [10] has validated the mesh resolution for studying flame acceleration and DDT, thus the simulation employs a minimum mesh size of $dx = dy$ = 2.5E-5 m with the AMR approach.

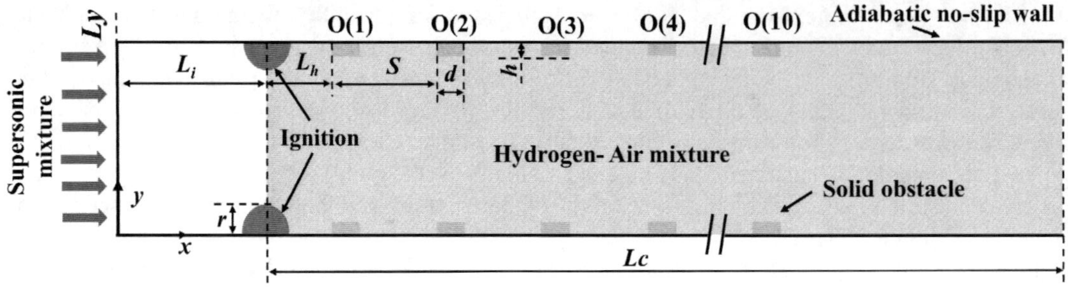

Figure 1. Schematic of the combustion configuration and boundary conditions for a ramjet PDE under supersonic conditions.

As depicted in figure 1, the computational setup features two semi-circular hot spots to initiate combustion within the supersonic flow mixture inside a combustion chamber pre-filled with hydrogen-air premixed mixture. For all conditions, the flow is simulated over $t = 0.2$ ms, establishing non-uniform fields of temperature, pressure, and velocity within the chamber. Ignition commences post intake valve closure, at $t = 0.21$ ms. Throughout the flame acceleration phase, the inlet valve remains closed, and the left boundary of the domain applies a no-slip adiabatic wall condition. Solid boundaries throughout the domain are subject to adiabatic no-slip conditions, while the right boundary adopts a free outflow condition.

Table 1. Detailed parameters employed for flame acceleration and DDT in supersonic flow mixtures.

Operating Case	Research Objective	Ignition Pressure (MPa)	Ignition Temperature (K)	Mach Number	Refinement Levels and Factors	Minimum Grid Size in *dx and dy* (m)
1	Influence of	0.1	2500	1.27	L4-1(2，2，2)	2.5E-5
2	Ignition	0.5	2500	1.27	L4-1(2，2，2)	2.5E-5
3	Energy	1.0	2500	1.27	L4-1(2，2，2)	2.5E-5

The study employs unsteady reactive two-dimensional Navier-Stokes equations to solve for multi-component flow, flame acceleration, DDT, as well as detonation propagation processes. The governing equations are as follows:

$$\frac{\partial Q}{\partial t}+\frac{\partial (F_x-G_x)}{\partial x}+\frac{\partial (F_y-G_y)}{\partial y}=S_{chem}, \tag{1}$$

where Q represents the vector of state parameters; F_x and F_y represent the convective fluxes; and S_{chem} represents the source terms. These symbols are subject to the following constraints,

$$U=\begin{bmatrix}\rho\\ \rho u\\ \rho v\\ \rho E\\ \rho_1\\ \vdots\\ \rho_{N_{sp}-1}\end{bmatrix},\ F_x=\begin{bmatrix}\rho_i u\\ \rho u^2+p\\ \rho uv\\ u(\rho E+p)\\ \rho_1 Y_1 u\\ \vdots\\ \rho_{N_{sp}-1}Y_{N_{sp}-1}u\end{bmatrix},\ F_y=\begin{bmatrix}\rho_i v\\ \rho uv\\ \rho v^2+p\\ v(\rho E+p)\\ \rho_1 Y_1 v\\ \vdots\\ \rho_{N_{sp}-1}Y_{N_{sp}-1}v\end{bmatrix}. \tag{2}$$

The detailed governing equations and parameter variables can be referred to Ref. [12]. This study conducts numerical simulations of flame acceleration and DDT using AMROC [9]. The time terms are integrated using a semi-implicit generalized Runge-Kutta scheme for chemical kinetics at CFL=0.4, with dynamic time stepping [13]. A level-set method for virtual fluid is employed to handle multiple solid obstacles [14].

Furthermore, the study employs a detailed chemical reaction mechanism, comprising 11 species and 27 steps as proposed by Burke et al. [15], to accurately simulate DDT reactions across a broad spectrum of pressures and temperatures. This mechanism has been widely adopted in the literature for modeling flame acceleration and DDT processes [16-18], lending credibility and robustness to the simulation outcomes. The use of this comprehensive reaction mechanism ensures that the complex interactions between chemical kinetics and fluid dynamics are adequately captured, providing a reliable foundation for understanding the intricate phenomena associated with DDT in ramjet PDEs under varying ignition conditions.

3. Results and discussion

The initial stage of flame acceleration and DDT for mixtures ignited under operating cases 1, 2, and 3 are illustrated in figure 2, with columns from left to right corresponding to cases 1, 2, and 3, respectively. As highlighted by circle C on the right, the high ignition energy from the hot spots causes the flame to rapidly expand, generating a shock wave. Conversely, as seen in circle A, case 1 results in only a limited combustion area. In case 3, the supersonic incoming flow deflects the flame front, while the separated flames merge more swiftly than in case 1, thereby increasing the initial flame velocity. The flame acceleration rate in case 2 falls between that of cases 1 and 3, as indicated in circle B, where the flame expands and also generates a shock wave. The early evolution of the flame front clearly differs from the flame acceleration observed in static mixtures, as described by Zhao et al. [19], which is characterized by a slower flame speed and a curved front. Consequently, ignition energy plays a critical role in influencing both early flame acceleration and the subsequent formation of the LSW.

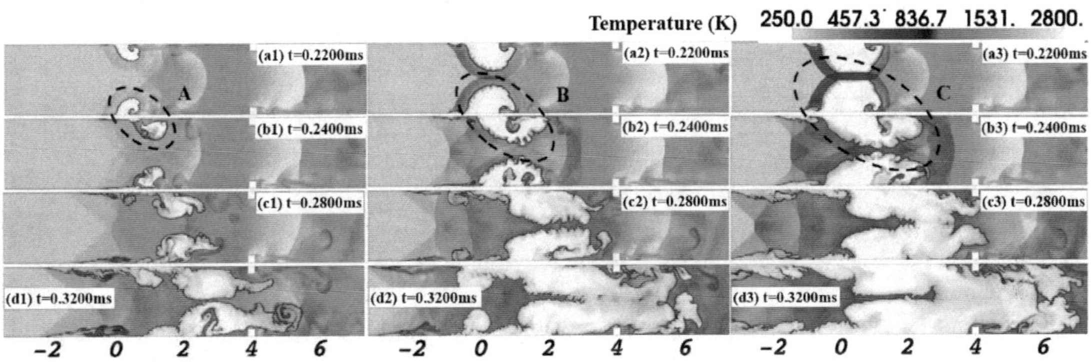

Figure 2. Temperature contour evolution throughout the initial flame acceleration process in: (Left) case 1, (Middle) case 2, and (Right) case 3 after mixture ignition.

The instantaneous snapshots of flame acceleration and DDT in the later stages for cases 1, 2, and 3 are depicted in figure 3. In case 1, the interaction of multiple compression waves leads to the formation of a LSW, as seen in figure 3(b1). This LSW subsequently results in Mach stems propagating from the upper and lower walls between t = 0.4400 ms and t = 0.4600 ms. Upon these intense Mach stems impacting the solid walls and reflecting off them, a localized explosion point emerges in the corners at t = 0.4750 ms (figure 3(f1)), which then penetrates the solid boundary and propagates both upstream and downstream within the combustion chamber.

Concerning operating case 2, an intense LSW is observed earlier at t = 0.4275 ms. The impact of the Mach stem on the upper solid surface triggers a localized explosion in the corners before t = 0.4550 ms (figure 3(f2)). However, in case 3, despite the higher ignition energy, no LSW is observed from t = 0.4000 ms to t = 0.4800 ms, indicating a different macroscopic behavior associated with LSW and Mach stem development compared to the other two cases. Given the described DDT process, it appears that the mechanism involving stronger Mach stem reflection in the corners is similar across all three cases. However, in case 3 with its high ignition energy, both TDDT and LDDT are greater than those observed in the lower-energy ignitions of cases 1 and 2. Therefore, this suggests that higher ignition energy does not necessarily lead to shorter TDDT and LDDT, highlighting the complex relationship between ignition conditions and DDT characteristics.

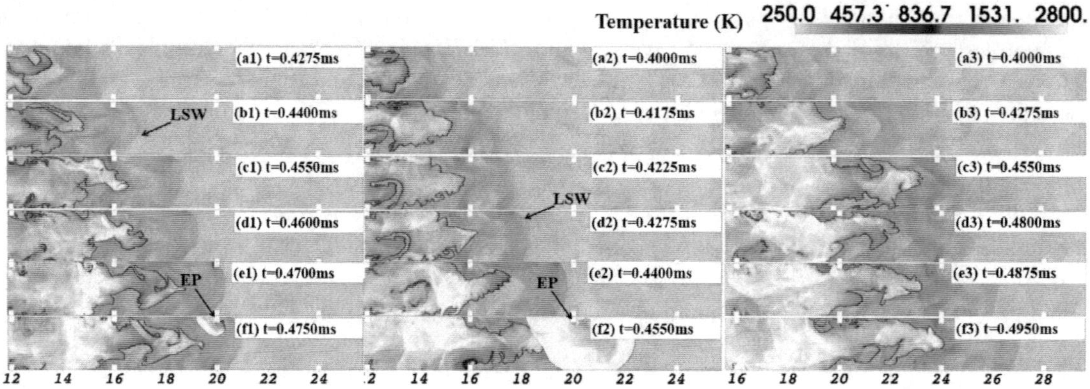

Figure 3. DDT evolution in (Left) case 1, (Middle) case 2, and (Right) case 3.

To elucidate the DDT mechanism illustrated in figure 3, figure 4 offers a magnified view of the transient temperature, pressure, and OH radical contours in a localized area near the upper solid wall following the Mach stem reflection in operating case 1. The images reveal that after the Mach stem reflects, a small explosion point emerges on the solid surface, as highlighted in circles A, B, and C, subsequently transitioning into the detonation combustion phase. This observation underscores that even with relatively low burning rates within the combustion chamber, the reflection of the Mach stem is potent enough to initiate detonation. Thus, it is evident that the intense reflected Mach stem plays a crucial role in facilitating the occurrence of DDT.

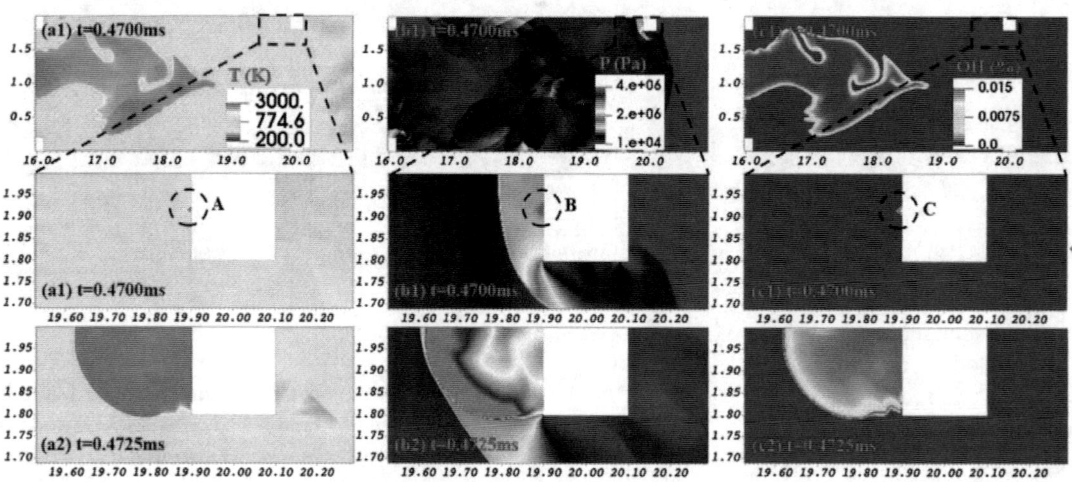

Figure 4. The instantaneous temperature, pressure, and OH radical distribution during the DDT in case 1.

Figure 5 further presents the axial positions of the flame fronts and the corresponding flame front propagation speeds (U_f) for operating cases 1, 2, and 3. As indicated by the flame front positions, points A, B, and C represent the onset of detonation, which can be evidenced by the gradient changes in the propagation speeds. In the initial stage, although flame acceleration is predominant under high ignition energy conditions, TDDT is longer. Under the intermediate ignition energy condition of case 2, the TDDT decreases from 0.5100 ms to 0.4450 ms. This results in a 21% reduction in the TDDT for case 2 compared to case 3. The LDDT in cases 1 and 2 are identical, representing a 31.25% shortening relative to case 3. Consequently, moderate ignition energy appears advantageous for initiating DDT with the shortest TDDT and LDDT in supersonic mixtures. The underlying reason may be that excessively high ignition energy causes the initial flame to expand more aggressively, generating a stronger shock wave with higher pressure. This intense expansion can lead to greater blockage within the supersonic flow, which in turn hinders the rapid formation of the LSW necessary for efficient DDT.

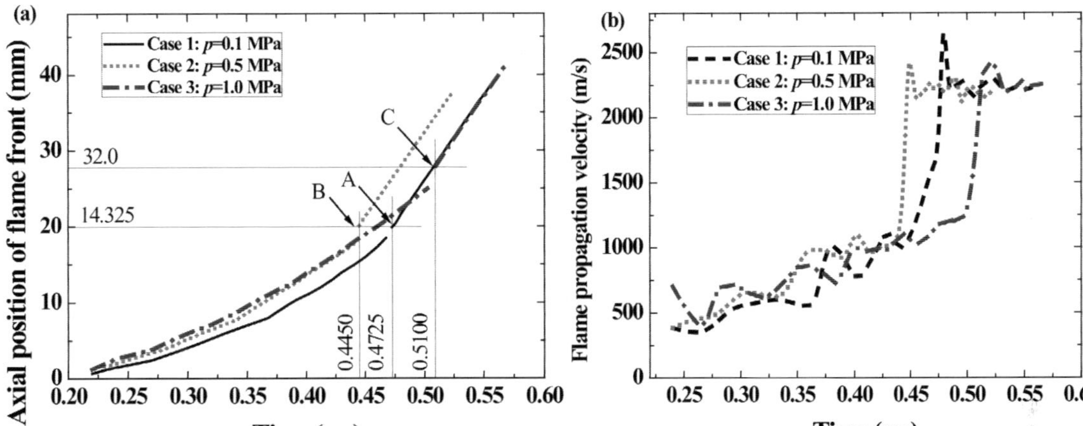

Figure 5. The (a) axial positions of the flame fronts and the corresponding (b) flame front speeds over time under different ignition energy conditions.

To substantiate this point, figure 6 presents pressure recordings along the centreline of the combustion chamber during the interval from 0.3975 ms to 0.4375 ms for both the moderate "soft" ignition condition (case 2) and the high-energy ignition condition (case 3). Notably, as illustrated in figure 6(a2), case 3 initially generates a higher pressure value due to the strong shock wave produced by the intense ignition energy, which propagates through the inlet section. Under the moderate ignition energy condition, the flame accelerates more gradually, leading to the formation of a single, well-defined shock wave at t = 0.4325 ms, with peak pressure reaching 17.7 bar. In contrast, the high-energy ignition in case 3 results in multiple compression waves. This suggests that while high ignition energy creates an immediate and intense pressure environment, and it can also lead to complex wave interactions that may impede the efficient formation of a LSW. Therefore, the rapid formation of the LSW in case 2 facilitates the swift occurrence of DDT, highlighting the advantage of moderate ignition energy in promoting a more controlled and effective transition to detonation in supersonic mixtures.

In summary, in supersonic mixtures, the ignition energy of a hot spot significantly impacts flame acceleration and DDT. Under high ignition energy conditions, while initial flame propagation is more vigorous, it paradoxically results in longer TDDT and LDDT due to the delayed formation of a strong LSW. The efficient occurrence of DDT in these mixtures primarily depends on the timely appearance of the LSW. Consequently, hot spots with intermediate ignition energy are more effective for triggering DDT, balancing rapid flame development with prompt LSW establishment.

4. Conclusion

This study utilized a high mesh resolution and adaptive mesh refinement (AMR) to numerically simulate flame acceleration and deflagration-to-detonation transition (DDT) in supersonic combustible mixtures under varying ignition energy conditions. The findings reveal that ignition energy significantly affects the time and distance required for detonation onset. For high ignition energy scenarios (which do not result in direct detonation initiation), more time is needed to develop an intense leading shock wave (LSW) despite an increased flame propagation speed during the initial acceleration phase. This results in extended TDDT and LDDT. The DDT process is critically dependent on the rapid formation of the LSW within the supersonic mixture. Hot spots with moderate ignition energy are found to be more effective in triggering DDT. This effectiveness may stem from the fact that high ignition energy causes the initial flame to expand rapidly, generating a strong shock wave that can create significant blockage in the supersonic flow. Such blockage adversely affects the swift formation of the LSW, thereby delaying the DDT process. However, the current study is limited to comparing only three cases for triggering DDT. The influence of a broader range of ignition energies, as well as other influencing factors on DDT in supersonic mixtures, should be investigated in greater detail in future work.

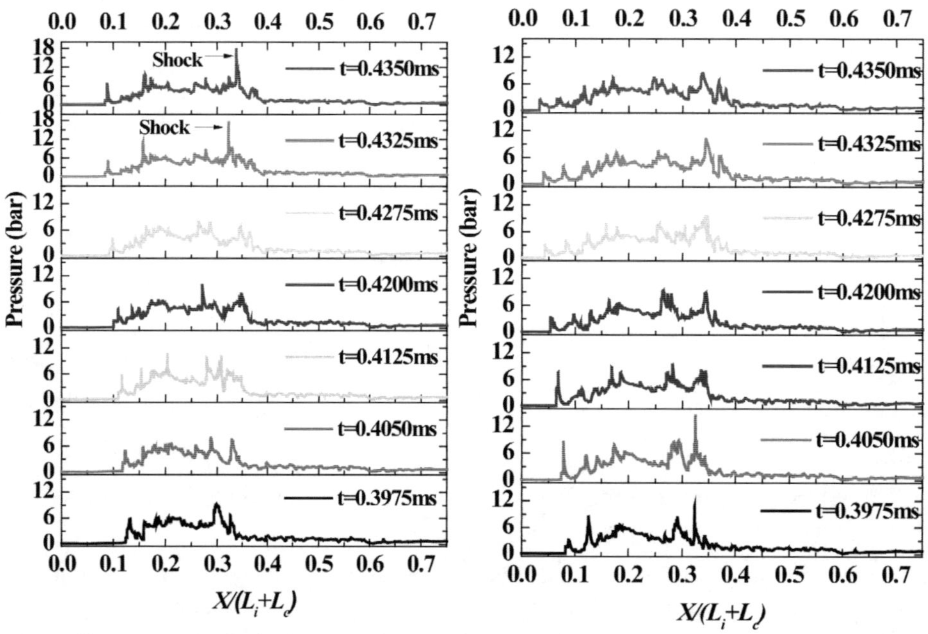

Figure 6. Pressure recordings along the centreline of the combustion chamber during flame acceleration and DDT for (left) case 2 and (right) case 3.

References

[1] Oran ES, Chamberlain G, Pekalski A. Mechanisms and occurrence of detonations in vapor cloud explosions 2020 *Prog. Energy Combust.* **77** 100804

[2] Roy GD, Frolov SM, Borisov AA, Netzer DW. Pulse detonation propulsion: challenges, current status, and future perspective 2004 *Prog. Energy Combust.* **30**(6) 545-672

[3] Zhao W, Deiterding R, Liang J, Cai X, Wang X. Detonation simulations in supersonic flow under circumstances of injection and mixing 2023 *Proc. Combust Inst.* **39**(3) 2895-903

[4] Shamsadin Saeid MH, Khadem J, Emami S. Numerical investigation of the mechanism behind the deflagration to detonation transition in homogeneous and inhomogeneous mixtures of H2-air in an obstructed channel 2021 *Int. J. Hydrog. Energy* **46**(41) 21657-71

[5] Ettner F, Vollmer KG, Sattelmayer T. Mach reflection in detonations propagating through a gas with a concentration gradient 2013 *Shock Waves* **23**(3):201-6

[6] Vollmer K, Ettner F, Sattelmayer T. Deflagration-To-detonation Transition in Hydrogen-Air Mixtures with a Concentration Gradient 2012 *Combust Sci. Technol.* **184**(10-11) 1903-5

[7] Boeck LR, Hasslberger J, Sattelmayer T. Flame Acceleration in Hydrogen/Air Mixtures with Concentration Gradients 2014 *Combust Sci. Technol.* **186**(10-11) 1650-61

[8] Frolov, S. M, Ivanov, V. S, Zangiev, A. E. Thrust characteristics of an airbreathing pulse detonation engine in flight at mach numbers of 0.4 to 5.0 2016 *Russ. J. Phys. Chem. B* **10** 272-83

[9] Deiterding R, editor Block-structured adaptive mesh refinement-theory, implementation and application 2011 *Esaim: Proceedings* (EDP Sciences) **34** p 97-150

[10] Zhao W, Deiterding R, Liang J, Wang X, Cai X, Duell J. Adaptive simulations of flame acceleration and detonation transition in subsonic and supersonic mixtures 2023 *Aerosp. Sci. Technol.* **136** 108205
[11] Gamezo VN, Ogawa T, Oran ES. Flame acceleration and DDT in channels with obstacles: Effect of obstacle spacing 2008 *Combust. Flame* **155** 302-15
[12] Zhao W, Deiterding R, Cai X, Liang J, Yang X, Sun M. Flame acceleration and detonation transition in premixed and inhomogeneous supersonic flows 2024 *J. Fluid Mech.* **1001** A36
[13] Kaps P and Rentrop P Generalized Runge-Kutta methods of order four with stepsize control for stiff ordinary differential equations 1979 *Numer. Math.* **33**(1) 55-68
[14] Deiterding R A parallel adaptive method for simulating shock-induced combustion with detailed chemical kinetics in complex domains 2009 *Comput. Struct.* **87**(11–12) 769-83
[15] Burke M, Chaos M, Ju Y, Dryer F, Klippenstein S. Comprehensive H2/O2 kinetic model for high-pressure combustion 2012 *Int. J. Chem. Kinet.* **44** 444-74
[16] Haghdoost MR, Edgington-Mitchell D, Nadolski M, Klein R, Oberleithner K. Dynamic evolution of a transient supersonic trailing jet induced by a strong incident shock wave 2020 *Phys. Rev. Fluids* **5**(7) 073401
[17] Li T, Wang X, Xu B, Kong F. An efficient approach to achieve flame acceleration and transition to detonation 2021 *Phys. Fluids* **33**(5) 056103
[18] Fan J and Xiao H Flame acceleration and transition to detonation in non-uniform hydrogen-air mixtures in an obstructed channel with different obstacle arrangements 2022 *Fire Saf. J.* **133** 103660
[19] Zhao W, Liang J, Deiterding R, Cai X, Wang X Flame-turbulence interactions during flame acceleration using solid and fluid obstacles 2022 *Phys. Fluids* **34**(10) 106106

Acknowledgments
This work was supported by the National Natural Science Foundation of China (No. 91741205) and the National College Students' Innovative Entrepreneurial Training Plan Program (S202490002110). Wandong Zhao also thanks the funding supported by the Natural Science Foundation of Hunan Province (No. 2024JJ6454) and the Youth Independent Innovation Science Foundation of National University of Defense Technology (ZK24-34).

Study on the influence of canard setting angle and longitudinal position on the sonic boom of supersonic business jet

Lintao Shao[1], Dawei Bie[1], Jinting Xuan[1], Menglong Ding[1,*]

[1] Tianmushan Laboratory, Hangzhou 311115, China

[*] corresponding author's e-mail: ddmmll@qq.com

Abstract: The aerodynamic layout design for reducing sonic boom intensity was a crucial technology in the development of supersonic business jets. The canard-wing configuration served as a strategic approach to achieving a low sonic boom for supersonic business jets. In this study, the near-field and far-field sonic boom calculation results of the model, as provided by SBPW2, were compared to verify the accuracy of the adopted sonic boom prediction method. A low-sonic-boom configuration featuring a canard wing was proposed for a supersonic business jet, and a comparative analysis was conducted to examine the effects of varying canard wing setting angles and longitudinal positions on the sonic boom results. The canard wing with a positive setting angle effectively generated multiple weak shock waves, thereby reducing the perceived level in decibels on the ground. Specifically, when the canard wing's setting angle was 3°, the perceived level in decibels reached a minimum of 84.60 PLdB. Additionally, positioning the canard wing further aft enhanced shock wave interference with the wing's shock wave system, further mitigating the sonic boom. When the longitudinal position was 14 meters, the ground-level perceived loudness was reduced to a minimum of 84.42 PLdB.

1. Introduction

Supersonic business jets represent a key development direction for the future of civil aviation. Currently, major aviation powers are actively engaged in in creating and advancing a new era of eco-friendly supersonic business jets. Sonic boom refers to the disturbance waves, primarily shock waves and expansion waves, created in the near field during the flight of a supersonic business jet. These waves evolve into intense sound waves in the far field, impacting the ground after propagating through the atmosphere[1, 2]. Sonic boom has been one of the key factors that has limited the long-term and continuous operation of early-stage supersonic business jets. For example, during the Concorde's cruise phase, the ground-level sonic boom loudness could reach as high as 108 PLdB, which was a major factor in its commercial failure[3]. The "N + X" generation supersonic business jet plan released by NASA sets strict limits on sonic boom as a key performance indicator, requiring the sonic boom level to not exceed 70 PLdB. The United States, Japan, Europe, and Russia have proposed various schemes for the low-sonic-boom aerodynamic layout of supersonic business jets, such as JAXA's S4[4] and Lockheed Martin's QSTA[5], with perceived level in decibels of sonic boom at the ground reduced to between 80 and 85 PLdB. NASA is also conducting a low-sonic-boom flight verification project and has successfully developed the X-59 QueSST technology demonstration aircraft[6].

Precise forecasting of sonic boom is crucial for both sonic boom assessment and creation of low-sonic-boom designs during the development of supersonic business jets. Based on the formation, propagation, and evolution of sonic boom, prediction methods are typically divided into two steps: forecasting the near-field sonic boom signals and the prediction of far-field propagation. High-precision near-field sonic boom signals are commonly computed using the CFD method. For far-field sonic boom propagation, the calculation method using the augmented Burgers equation has found widespread application[7]. This method is particularly advantageous as it accounts for the dispersion and dissipation effects of the atmosphere on the waveform during far-field propagation, allowing for an accurate description of waveform features, such as the rise time of the sonic boom.

Low-sonic-boom design refers to methods and techniques aimed at minimizing perceived level in decibels at the ground while preserving the aircraft's aerodynamic performance. In the SSBD project[8], NASA first demonstrated that modifying the aerodynamic shape of an aircraft can effectively alter the ground sonic boom waveform and reduce the peak overpressure, providing a solid foundation for the optimization of sonic boom shaping. The canard-wing layout is a strategic approach for achieving low sonic boom in supersonic business jets[9]. It facilitates the formation of a near-field pressure signal with multiple shock waves and prevents the shock waves produced by the wing from merging forward[10]. In this paper, a low-sonic-boom supersonic business jet layout featuring a canard wing was proposed. The CFD method and the augmented Burgers equation were used to calculate the near-field and far-field sonic-boom signals, respectively. Furthermore, the influence of canard-wing layouts with different setting angles and longitudinal positions on the wave-system distribution and sonic-boom calculation results were analyzed. The results obtained offer a reference point for designing the aerodynamic layout of low-sonic-boom supersonic business jets.

2. Near-field and far-field analysis methods in sonic boom prediction

2.1. Near - field CFD numerical simulation method

In the calculation of refined near-field sonic-boom signals, the CFD numerical simulation method must be adopted to accurately predict the wave system and its spatial distribution, enabling a highly reliable forecast of the near-field sonic boom. For CFD calculations of near-field sonic boom, a conical grid is employed. Specifically, based on the characteristics of supersonic flow, a conical solution domain was generated, with grid lines arranged according to the Mach angle to capture the wave system. Additionally, the grid in the cylindrical area around the aircraft was refined in the boundary layer. The RANS equations are used as the flow control equations, and the k-ω SST turbulence model is applied for the solution. The JWB low-sonic-boom standard model from the Second Sonic Boom Prediction Workshop (SBPW2) was used to verify the accuracy of the CFD method. The computational grid is shown in Figure 1, with a total of 28.4 million grids. The near-field sonic-boom signal at a distance of 0.85 times the body length from the fuselage, directly below the trajectory, was extracted and compared with results from different researchers[11] of SBPW2, as shown in Figure 2. As can be observed from the figure, the CFD results presented in this paper show strong agreement with the reference values, demonstrating that the near-field calculation results are highly reliable.

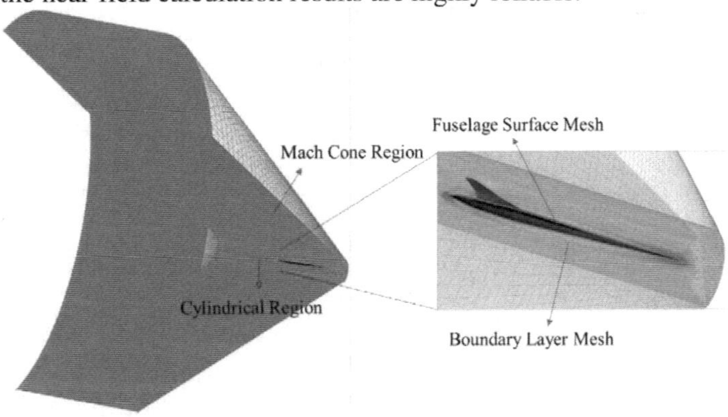

Figure 1. The JWB low-sonic-boom benchmark mesh used for CFD validation in this study.

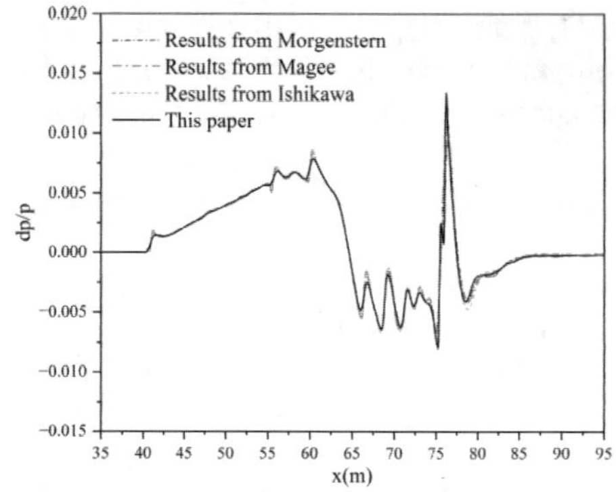

Figure 2. Comparison between the sonic boom predicted by the CFD simulation in this study and the results from researchers at the SBPW2.

2.2. Far-field augmented Burgers equation prediction method

The far-field propagation of the sonic boom is simulated by solving the augmented Burgers equation[12]. The non-dimensionalized form of the augmented Burgers equation is expressed as follows:

$$\frac{\partial P}{\partial \sigma}=P\frac{\partial P}{\partial \tau}+\frac{1}{\Gamma}\cdot\frac{\partial^2 P}{\partial \tau^2}+\frac{1}{2\rho_0 c_0}\cdot\frac{\partial \rho_0 c_0}{\partial \tau}P-\frac{1}{2S}\cdot\frac{\partial S}{\partial \sigma}p+\sum_{v}\frac{C_v\frac{\partial^2 P}{\partial \tau^2}}{1+\theta_v\frac{\partial}{\partial \tau}} \quad (1)$$

In Equation (1), the five terms on the right-hand side, in sequence, are the nonlinear effect term, the thermo-viscous absorption term, the atmospheric stratification effect term, the geometric spreading effect term, and the molecular relaxation term. Consequently, this equation can simulate the propagation of sound waves in a stratified and lossy real-world atmosphere. The operator-splitting method, as proposed by Cleveland, is used to solve this equation[12].

After obtaining the ground waveform through the solution of the augmented Burgers equation, the Mark IV method developed by Stevens[13] is employed as the evaluation criterion for sonic-boom intensity to calculate the perceived level in decibels (PLdB) when the sonic boom propagates to the ground. The near-field sonic-boom signal provided by Ishikawa[14] at SBPW2 was used as the input for the far-field sonic-boom propagation program employed in this study. A comparison between the calculation results of this study and those obtained from NASA's far-field sonic-boom prediction program, sBoom, is presented in Figure 3. The two waveforms were found to be nearly identical, with a difference of only 0.16 PLdB in the perceived sound pressure level, indicating that the far-field propagation program is highly reliable.

In summary, the precision of the sonic boom prediction method adopted in the study was confirmed by comparing it with the standard examples provided by the SBPW2.

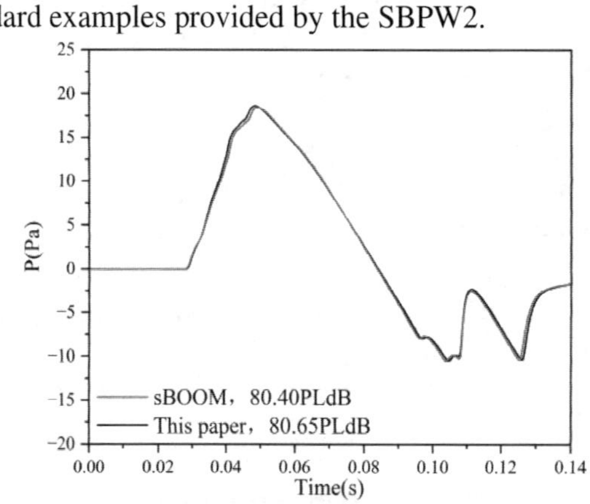

Figure 3. Comparison of the far-field sonic boom prediction results from this study with the results from sBoom.

3. Research

3.1. Basic model of a supersonic business jet with canard wings

A medium-sized supersonic business jet is chosen as the research subject. The aircraft meets the following specifications: a cruise weight of 60 tons, a cruise Mach number of 1.6, and a cruise altitude of 17 km. The proposed low-sonic-boom layout is shown in Figure 4. The aircraft's fuselage length is 65 m, wingspan is 24.5 m, wing area is 400 m², canard-wing area is 10 m², the setting angle is 3°, and the canard wing is positioned 15 meters longitudinally from the aircraft's nose. It exhibits the common characteristics of typical low-sonic-boom supersonic business jets, such as an oblong nose, a highly swept and dihedral arrow-shaped wing, a slender fuselage, and a cruciform tail.

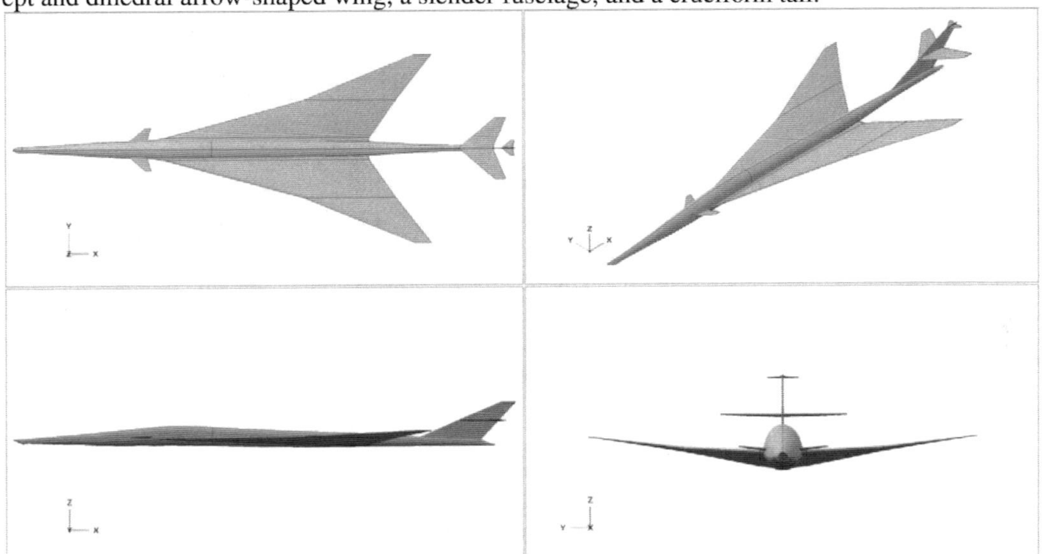

Figure 4. Different views of the canard configuration layout

Near-field CFD calculations were performed for the supersonic business jet configuration with canard wings. Through grid independence analysis, the grid quantity is set to 12 million, The conical grid and the grid on the fuselage surface are shown in Figure 5. The boundary layer of the fuselage is set to 30 layers, the initial boundary layer thickness is 3 mm, and the growth rate is set to 1.2.

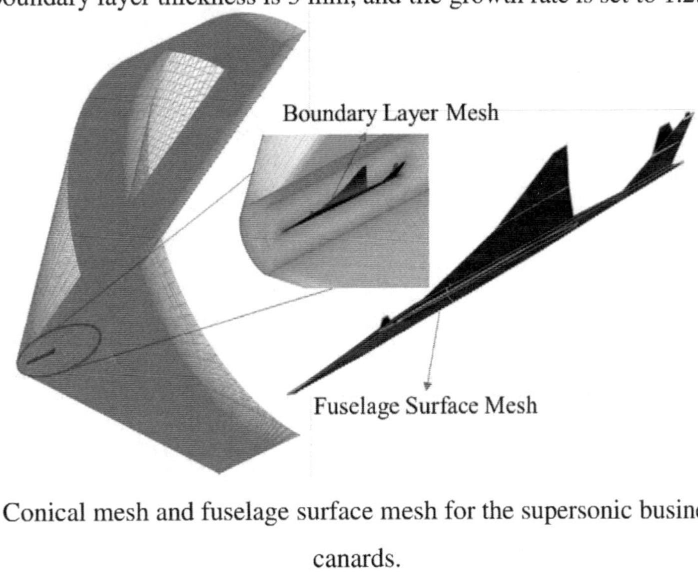

Figure 5. Conical mesh and fuselage surface mesh for the supersonic business jet with canards.

3.2. Influence of canard setting angle variation on sonic boom calculation results

Based on the initial configuration, parametric adjustments were made to the setting angle of the canard wing. Different canard-wing setting angle configurations (-3°, 0°, 3°, 6°) are shown in Figure 6. The impact of changes in the setting angle on the sonic-boom results was studied, and the calculation results for the configuration without a canard wing were included for comparison. The near-field calculation results for the different canard-wing setting angle configurations, as well as the configuration without a canard wing, are shown in Figure 7. The far-field and ground sonic-boom calculation results are presented in Figure 8. The near-field overpressure plot near the canard wing is shown in Figure 9, where the black dashed-line box highlights the wave-system plot corresponding to the canard-wing area.

As observed from the pressure signals, the presence of the canard wing and variations in the setting-angle parameter significantly affected the distribution of the wave pattern in the vicinity of the canard wing. Comparing configurations with and without a canard wing, a positive canard-wing setting angle caused the third shock wave in the near field to appear earlier, effectively breaking the subsequent strong shock wave under the wing into multiple weaker shock waves. In the far-field signal, this results in a longer rise time for the shock wave, thereby reducing the sonic-boom intensity. When the canard-wing setting angle was -3°, the presence of the canard wing strengthens the expansion wave beneath it, which led to an increase in the shock wave under the wing. Compared with the configuration without a canard wing, this results in a higher shock-wave peak, causing the ground-level sonic-boom intensity to increase. As the canard-wing setting angle changed from 0° to 6°, the pressure wave under the canard wing strengthend, while the pressure-wave peak under the wing gradually decreased. When the canard-wing setting angle was 3°, a balance is achieved between the pressure wave under the canard wing and the pressure wave under the wing, resulting in the lowest calculated perceived level in decibels at ground level of 84.60 PLdB.

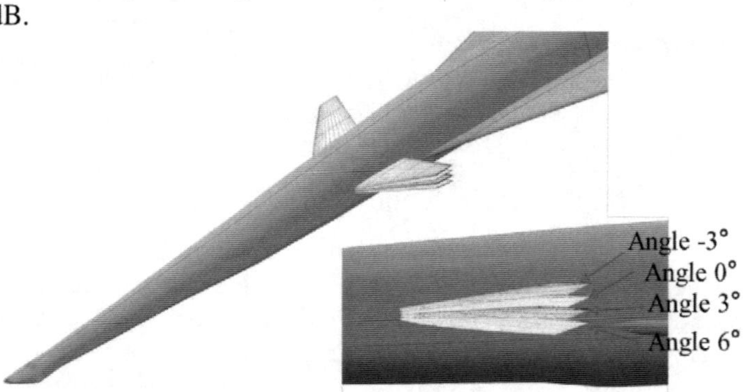

Figure 6. Comparison of canard configurations with varying setting angles

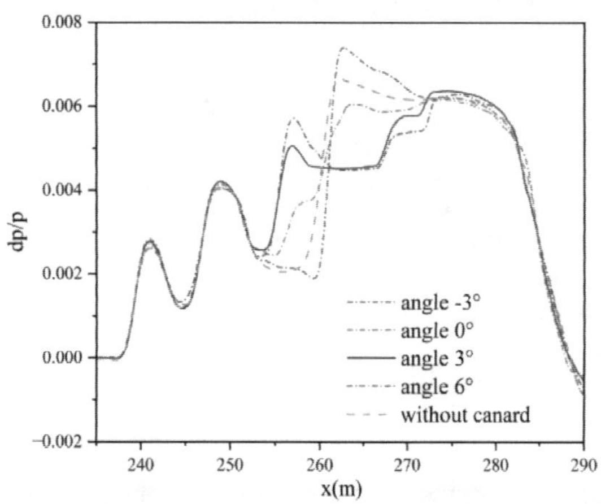

Figure 7. Near-field calculation results for different canard wing setting angle configurations near the aircraft nose

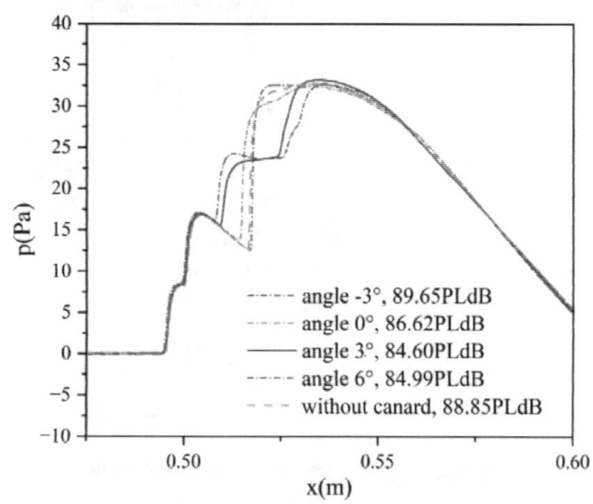

Figure 8. Far-field and PLdB results for different canard wing setting angle configurations near the aircraft nose

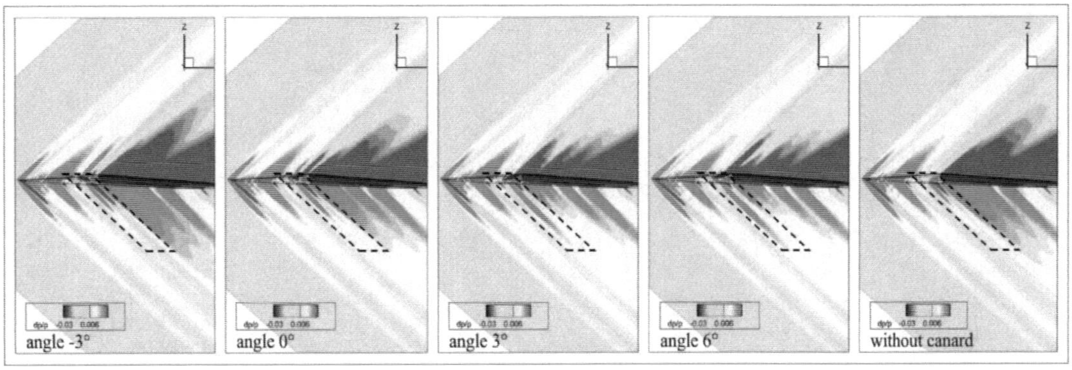

Figure 9. Near-field overpressure contour maps for different canard wing setting angle configurations

3.3. Influence of longitudinal position change of the canard wing on sonic boom calculation results

Based on the initial configuration, parametric adjustments were made to the position of the canard wingtip along the longitudinal axis with respect to the aircraft nose. The impact of adjusting the longitudinal position of the canard wing (ranging from 11m to 15m, as shown in Figure 10) on the sonic boom results was analyzed, with comparisons made to the configuration without a canard wing. The near-field calculation results for different canard-wing longitudinal-position configurations are shown in Figure 11. The far-field and ground sonic-boom calculation results were presented in Figure 12. The near-field overpressure plot near the canard wing is shown in Figure 13, where the black dashed-line box highlights the wave-system plot corresponding to the canard-wing area.

As seen from the pressure signals, the placement of the canard wing affected the distribution of the wave system in both the near- and far-fields. Moving the canard wing forward caused the shock wave produced by the canard wing to appear earlier. However, if the canard wing was positioned too far from the main wing's leading edge, the shock waves generated by both the canard wing and the main wing did not interfere constructively, reducing the potential for sonic boom mitigation. As the canard wing's longitudinal position increased, the shock wave from the canard wing moved backward, with its peak becoming smaller. The wave systems of the canard and main wings began to interfere with each other, smoothing the overall wave system and effectively reducing the sonic boom. When the longitudinal position of the canard wing reached 15m, it was too close to the main wing, causing shock-wave peaks to merge. As a result, compared to the configuration with the canard wing at 14m, the sonic boom intensity slightly increased. The findings indicated that when the longitudinal position of the canard wing was 14m, the ground perceived level in decibels reached its lowest value of 84.42 PLdB.

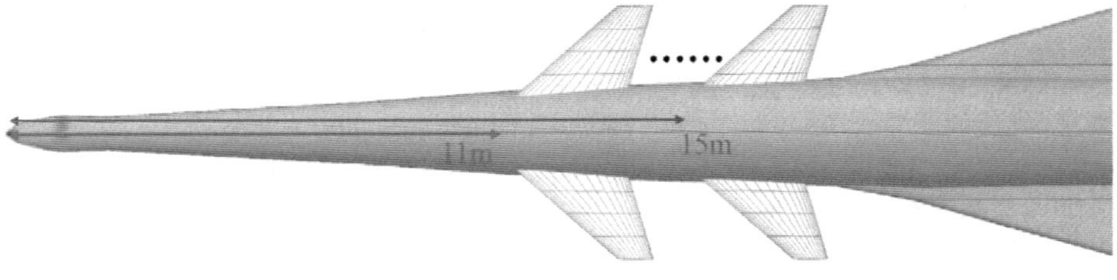

Figure 10. Comparison of canard configurations at different longitudinal positions

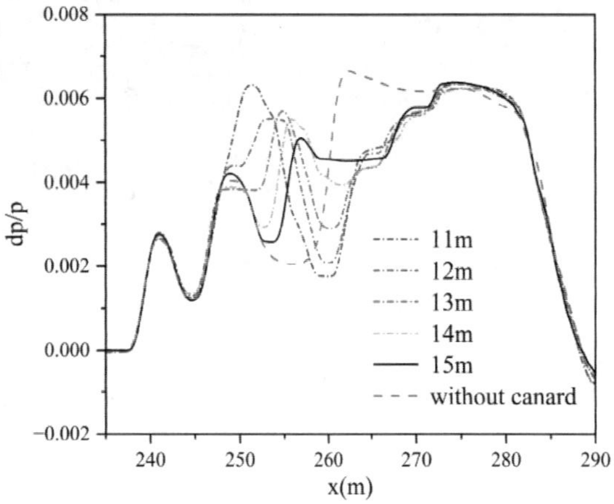

Figure 11. Near-field calculation results for canard configurations with different longitudinal positions near the aircraft nose

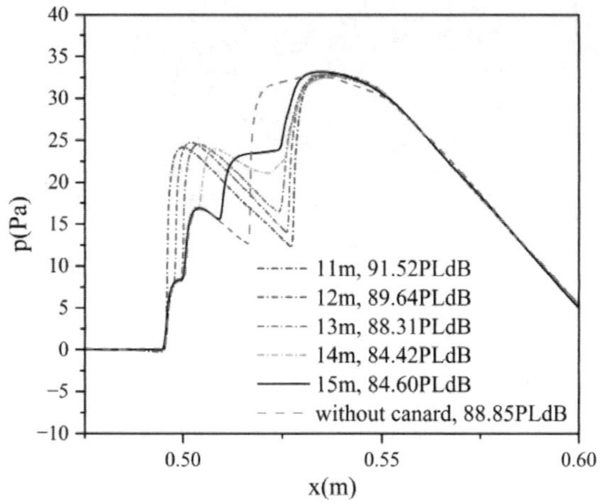

Figure 12. Far-field and PLdB results for canard configurations with different longitudinal positions near the aircraft nose

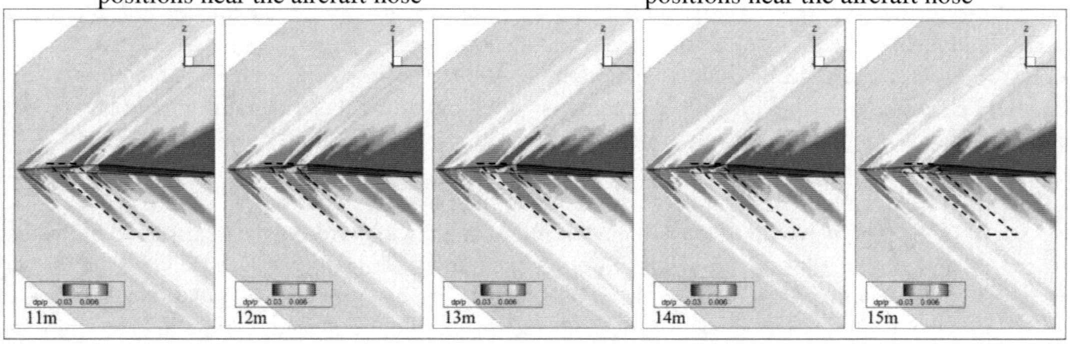

Figure 13. Near-field overpressure contour maps for canard configurations at different longitudinal positions

4. Conclusion

In this study, the accuracy of the calculation methods was confirmed by comparing them with standard models from SBPW2. A low-sonic-boom configuration for a supersonic business jet featuring a canard wing and a cruise weight of 60 tons was proposed. The effects of the canard-wing setting angle and longitudinal position on the sonic boom were analyzed. The primary conclusions are summarized:

(1) The near-field CFD method and the far-field augmented Burgers method developed in this study were compared with the standard models provided by SBPW2. The results demonstrated that both the near-field and far-field predictions exhibited high accuracy, making them suitable for the precise prediction and calculation of sonic booms in supersonic business jets.

(2) The effects of different canard-wing setting angles on the sonic boom were compared. A negative canard-wing setting angle increased the ground-level sonic-boom intensity. However, when the setting angle ranged from 0° to 6°, the canard-wing configuration, compared to the baseline configuration without a canard wing, effectively split the shock wave beneath the wing into multiple weaker waves, thereby reducing the sonic-boom intensity on the ground. The lowest perceived level in decibels, 84.60 PLdB, was achieved when the canard-wing setting angle was 3°, where a balance between the shock waves under the canard wing and the wing was reached.

(3) The effects of different longitudinal positions of the canard wing on the sonic boom were analyzed. As the longitudinal position of the duck wings moved further away from the aircraft nose, the shock wave generated by the canard wing shifted rearward. This resulted in beneficial interference with the wing's shock wave system, leading to a decrease in the shock wave's peak value and a reduction in sonic-boom intensity. However, if the canard wing was positioned too close to the wing, the shock-wave peaks

merged. The lowest perceived level of 84.42 PLdB occurred when the canard wing was placed at 14 meters.

References

[1] Alonso J J, Colonno M R. Multidisciplinary Optimization with Applications to Sonic-Boom Minimization[J]. *Annual Review of Fluid Mechanics*, 2012, 44(1): 505-526.
[2] Plotkin K J. State of the art of sonic boom modeling[J]. *The Journal of the Acoustical Society of America*, 2002, 111(1): 530-536.
[3] Chudoba B, Coleman G, and Oza A, et al. What price supersonic speed? A design anatomy of supersonic transportation Part 1[J]. *Aeronautical Journal*, 2008, 112(1129): 141-151.
[4] Liebhardt B, Lütjens K, and Ueno A, et al. JAXA's S4 Supersonic Low-Boom Airliner - A Collaborative Study on Aircraft Design, Sonic Boom Simulation, and Market Prospects[C]. *AIAA Aviation 2020 Forum. Virtual Event: American Institute of Aeronautics and Astronautics*, 2020.
[5] Buonanno M, Chai S, and Marconi F, et al. Overview of Sonic Boom Reduction Efforts on the Lockheed Martin N+2 Supersonic Validations Program[C]. *32nd AIAA Applied Aerodynamics Conference. American Institute of Aeronautics and Astronautics*, 2014.
[6] Richwine D, Brandon J. Quiet SuperSonic Technology (QueSST) Aircraft Preliminary Design Status and Low-Boom Flight Demonstration (LBFD) Project Update[C]. *AIAA SciTech Forum & Exposition (SciTech 2018)*. 2018.
[7] Rallabhandi S K. Advanced sonic boom prediction using the augmented Burgers equation[J]. *Journal of Aircraft*, 2011, 48(4): 1245-1253.
[8] Pawlowski J, Graham D, Boccadoro C, et al. Origins and overview of the shaped sonic boom demonstration program[C], *43rd AIAA Aerospace Sciences Meeting and Exhibit*. 2005: 5.
[9] Kishi, Yuki, Risato Yashiro, and Masahiro Kanazaki. Low-Boom Design for Supersonic Transport with Canard and Forward-Swept Wings Using Equivalent Area Design Method[J]. *Aerospace*, 10.8 (2023): 717.
[10] Sun Y, Smith H. Review and prospect of supersonic business jet design[J]. *Progress in Aerospace Sciences*, 2017, 90: 12-38.
[11] Park M A, Nemec M. Nearfield summary and statistical analysis of the second AIAA sonic boom prediction workshop[J]. *Journal of Aircraft*, 2019, 56(3): 851-875.
[12] Cleveland R O. Propagation of sonic booms through a real, stratified atmosphere[D]. *The University of Texas at Austin*, 1995.
[13] Stevens S S. Perceived level of noise by Mark VII and decibels (E)[J]. *The Journal of the Acoustical Society of America*, 1972, 51(2B): 575-601.
[14] Ishikawa H, Makino Y, Ueno A, Kasuga Y. Near-field Pressure Signature Prediction by JAXA[R]. https://lbpw-ftp.larc.nasa.gov/sbpw2/workshop/sbpw2-talks-nearfield/05-sbpw2-jaxa-ishikawa-makino-ueno.pdf

Acknowledgments

This work was supported by the Reward Funds for Research Project of Tianmushan Laboratory (TK-2024-D-008).

Journal of Physics: Conference Series **3041** (2025) 012010 doi:10.1088/1742-6596/3041/1/012010

Numerical simulation of powder fuel blending characteristics in a supersonic partially covered cavity-based combustor

Wenxue Han[1], Chunbo Hu[1], Chao Li[2]*, Rong Lei[1], Wei Cui[1], Jiaxin Dong[1]

[1]National Key Laboratory of Solid Rocket Propulsion, Northwestern Polytechnical University, Xi'an, Shaanxi 710072, People's Republic of China

[2]Unmanned System Research Institute, Northwestern Polytechnical University, Xi'an, Shaanxi 710072, People's Republic of China

*Corresponding author: lichao@nwpu.edu.cn

Abstract. A numerical simulation study of the powder mixing characteristics in a supersonic partially covered cavity combustor was carried out to explore the combustion organization scheme of powder-fueled scramjets. The study analysed powder mixing patterns within combustion chambers featuring various baffle lengths, cavity dimensions, and nozzle quantities. It was found that compared with the cavity chamber structure, the partially covered cavity chamber structure can double the particle retention time, and the particle retention time increases and then decreases with the increase of the baffle length. The particle retention time reaches the maximum in the baffle length of about 35 mm, and the degree of powder blending can achieve as high as 0.8 under optimal conditions. An increase in the length of the cavity negatively impacts the retention time of particles in the combustor. As the powder is injected into the combustor cyclically, a higher number of spray holes leads to a decrease in particle retention time.

1. Introduction

Scramjets have the advantage of simple structure and can realize high Mach number flight [1-3]. Researchers have conducted a more in-depth study on combustion organization and other key technologies of scramjets fueled by hydrocarbons [4,5]. Compared with hydrocarbon fuels, boron, and aluminium powders have the advantages of high calorific value, no dissociation effect under high temperatures, safe and convenient fuel storage, and so on [6]. As a new concept of powder-fueled scramjets, which can more easily realize high Mach number (Ma>7) flight, it has become a hot research issue in recent years [7,8]. At this stage, the powder enters into the supersonic combustor by gas-solid two-phase flow under the coercion of fluidized gas to go through the processes of evaporation, oxide film rupture, surface non-homogeneous reaction, gas-phase reaction, surface condensation, etc[9,10]. These complex processes result in a prolonged ignition delay; however, the fast gas flow within the supersonic combustor significantly shortens the fuel retention time, often to the millisecond level [11]. Therefore, achieving full mixing of powdered fuels in a supersonic combustor is a crucial aspect of carrying out combustion organization. Powder-fueled scramjets represent a new class of engines still in the early stages of development, and there is a notable lack of experimental data regarding their mixing and combustion technologies. It remains uncertain whether the mixing enhancements observed in hydrocarbon fuel scramjets apply to powder scramjets, highlighting the urgent need for further research.

Currently, researchers have mainly investigated the mixing of powder particles in a supersonic combustor using numerical simulation. Ding[12] et al. used numerical simulation to reveal the flow mechanism of hydrogen/metal powder fuel in a supersonic gas stream and found that the depth of penetration and mixing efficiency were greater when the fuel was injected through double holes. Liu [13] et al. studied the turbulent boundary layer of a jet entering a supersonic flow using PIV and NPLS techniques and obtained the detailed structures of horseshoe vortices, separation zones, and collision shocks. Luo[14] et al. used numerical simulation to study the effect of powder fuel blending characteristics under different injection modes in a combustor with a cavity structure. They found that the particles are concentrated in the cavity under the injection scheme within the cavity. Some of the particles enter the mainstream region under the effect of reflux. The regions with better mixing performance are the shear layer near the bottom wall of the cavity and the expansion section. Che[15] investigated the influence laws of injection parameters and mixing enhancement methods on parameters such as penetration depth and diffusion angle of transverse powder jets in supersonic flow. The study revealed that employing tower gate and porous injection methods can significantly increase the depth of the powder jet. Conversely, while cavity and physical slope mixing enhancement methods reduce the jet depth, they also generate vortices and surge waves, which facilitate the mixing of incoming air and fuel.

From the above research, it can be found that the current stage of research on powder fuel mixing in supersonic combustors has a relatively obvious characteristic, focusing on the enhanced mixing structure in the combustor of the hydrocarbon fuel scramjets to carry out the relevant laws of the study, which involves the cavity and transverse jets and so on[16,17]. However, there remains a significant gap in the comprehensive investigation of mixing and combustion dynamics in powder-fueled engines, particularly those characterized by long ignition delays and limited flowability of powdered fuel. This paper aims to analyze the particle mixing characteristics within a supersonic combustor featuring a partially covered cavity. The objective is to identify effective mixing structures and injection schemes that facilitate improved mixing of powder fuel in these combustors. The findings from this study will provide valuable insights and guidance for optimizing combustion organization in powder-fueled scramjets.

2. Numerical methodology

2.1. Gas phase governing equation

The gas-phase equations in a supersonic gas flow use a set of N-S control equations containing the continuity, energy, and momentum equations, supplemented by a set of gas equations of state to form a closure, as shown in the following equation:

$$\frac{\partial Q}{\partial t}+\frac{\partial E}{\partial x}+\frac{\partial F}{\partial y}+\frac{\partial G}{\partial z}=\frac{\partial E_v}{\partial x}+\frac{\partial F_v}{\partial x}+\frac{\partial G_v}{\partial x}+S \tag{1}$$

where Q represents the conserved variable, E , F , G represent the convective terms in the x, y, and z directions, and E_v , F_v , G_v represent the viscous terms in the x, y, and z directions.

Powder particles enter the combustion chamber of a scramjet engine under the coercion of fluidized gas and are mainly mixed with oxygen in the shear layer inside the combustion chamber. To calculate the flow mixing in the shear layer more accurately, this paper adopts the compressibility-modified shear stress transport (SST) two-equation turbulence model for simulation. Menter developed this model, and it has been widely used in the numerical simulation of supersonic flow [18,19]. The principle is to consider the flow with shear at a low Reynolds number at the near wall and to neglect the turbulent viscosity at the far wall for the calculation.

2.2. Particle phase governing equation

Considering that the volume concentration of the particles is much less than 10^{-3} in the calculations within the supersonic combustion chamber in this paper, the random orbital model in the particle orbital model is chosen. The following equation can characterize the dynamics of the particles:

$$\frac{dX_p}{dt} = V_p \tag{2}$$

$$\frac{dV_p}{dt} = F_p + F \tag{3}$$

Where V_p and X_p denote the velocity and positional distributions of the particle phases, respectively, F_p is the drag force received by the particle population per unit mass, and F is the other forms of force on the particles, including Stefan flow, pressure gradient forces, and so on.

where the drag force received by the particle population per unit mass can be defined as:

$$F_p = \frac{3}{4}\frac{C_D \rho}{\rho_p d_p}(V - V_p)\left|V - V_p\right| \tag{4}$$

C_D is the resistance coefficient, which can be calculated by the following equation [20].

$$C_D = \begin{cases} \frac{24}{Re_p}\left(1 + \frac{1}{6} Re_p^{\frac{2}{3}}\right), & Re_p < 1000 \\ 0.44, & Re_p > 1000 \end{cases} \tag{5}$$

Re_p in the above equation represents the Reynolds number of the particles and is defined as follows:

$$Re_p = \frac{\rho d_p \left|V - V_p\right|}{\mu} \tag{6}$$

2.3. Particle phase governing equation

To verify the rationality and accuracy of the numerical calculation model, we carried out an experimental study on the mixing law of powder fuel in the supersonic gas flow. The inlet conditions of the combustion chamber are shown in Table 1.

Table 1. Entry parameters for experimental section

Incoming flow parameters	numerical value
Total pressure, MPa	1
Total temperature, K	1150
Mach number	1.85

During the experimental process, supersonic gas with a Mach number of 1.8 enters the combustor through the isolation section. The powdered fuel is supplied to the combustor via a pipeline system, entering behind the tower door structure. As the powdered fuel mixes with the supersonic gas flow, pressure measurement points are established on the side wall of the combustion chamber to monitor changes in pressure within the combustor.

Figure 1 illustrates the calculated wall pressure distribution within the combustion chamber under the specified conditions. The mixing structure and powder injection holes in the combustor lead to a sudden pressure change in a localized area. Overall, the numerical results effectively capture the pressure change trends in the test section and align well with the data obtained from the experimental measurement points. Consequently, the gas-solid two-phase numerical calculation model we developed for the supersonic combustor demonstrates a degree of rationality and accuracy, providing a reliable representation of the flow dynamics of powder particles within the combustor.

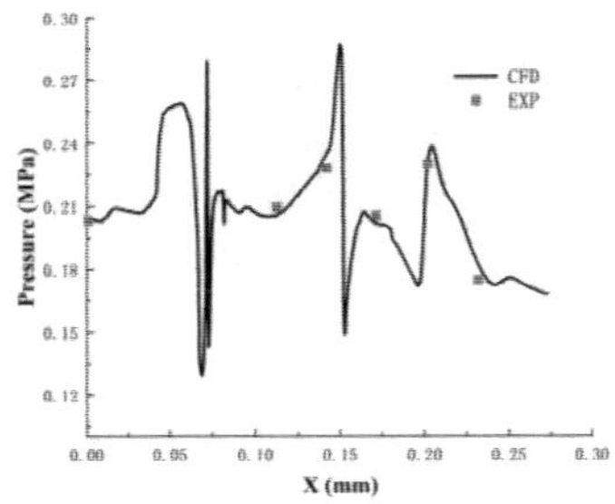

Figure 1. Comparison of numerical calculation results and experimental data

3. Numerical calculation model and evaluation index of blending performance

3.1. Numerical model

The present study focuses on a partially covered cavity structure within a supersonic combustor, which primarily consisting of a cavity and a baffle plate. In future analyses, the length of the baffle plate will be referred to as Lc, while L will denote the length of the cavity. D will represent the depth of the cavity. Incorporating the partially covered cavity structure into the circular combustor, the powder spray holes are positioned in the covered area at an angle of 60° to the wall of the combustion chamber. This arrangement aims to increase the retention time of the particles in the combustion chamber and enhance their mixing effectiveness with the air. The numerical computational model developed in this study includes two main components: an isolation section and an expanded combustion chamber. The inlet diameter of the circular isolation section is 64 mm, while the combustion chamber features a unilateral expansion angle of 1°.

A direct-connection experimental study simulating the exact flight conditions of H=25 km and Ma=6 conducted, as detailed in the literature[21]. This study provided inlet flow parameters for the isolation section, which have been utilized as boundary conditions in numerical simulations related to the scramjets[22]. As a result, the incoming flow parameters at the entrance of the isolation section in the numerical calculations presented in this paper align with those measured in the direct-connection experiments. Specifically, the Mach number at the entrance of the isolation section is 2.52, and the total incoming temperature is 1600 K. Figure 2 illustrates the structural mesh diagram for the partially covered cavity configuration within the combustion chamber. In this mesh design, the influence of the attached boundary layer on the flow is taken into account, resulting in a refined mesh in the wall area of the combustion chamber to ensure accuracy in the simulations.

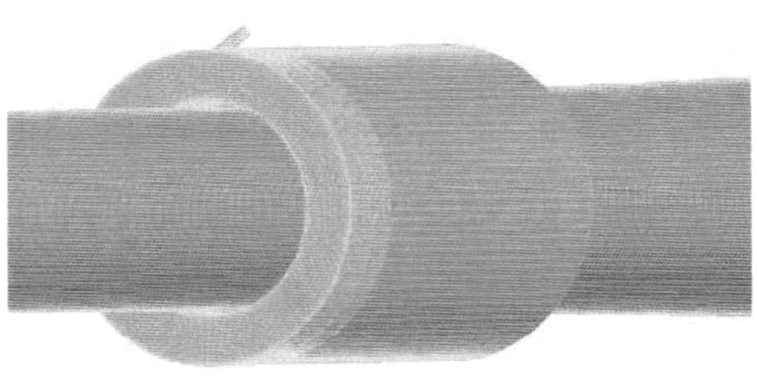

Figure 2. Mesh diagram of partially covered cavity structure

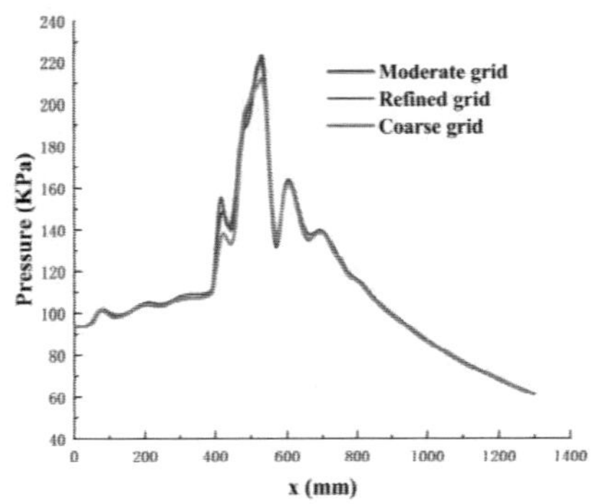

Figure 3. Calculation results of combustor model with different number of meshes

Figure 3 shows the comparison of the pressure distribution on the central axis, and it can be observed that the calculation results of 2.3 million and 3 million grids are the same. Grid calculation results are the same. Under the premise of guaranteeing the calculation accuracy and considering the time cost, this paper finally selects the moderate grid (2.3 million) for calculation.

Journal of Physics: Conference Series **3041** (2025) 012010 doi:10.1088/1742-6596/3041/1/012010

3.2. Blending performance evaluation index

Compared to hydrocarbon fuels, boron powder fuels exhibit a longer ignition delay and a more complex combustion process. However, the high-speed airflow in the supersonic combustor often causes the particles to exit the combustion chamber before ignition can occur. To address these challenges, we have selected particle retention time and the mixing degree between the powder and oxygen as the key evaluation criteria for assessing mixing performance.

Particle retention time is defined as the duration from the moment a particle enters the combustor until it exits the combustion chamber. The magnitude of this value directly influences the mixing, heating, and heat release of the particles within the combustion chamber. The degree of blending is a quantitative index to evaluate the uniformity of the concentration distribution of each component in the blending process. In this study, the mixing degree between boron particles and oxygen is mainly involved, so the concept of particle mixing degree is introduced to evaluate the mixing effect of boron particles and oxygen in the supersonic combustion chamber, which is defined as shown in Equation 7[23].

$$\beta_p = 1 - \frac{1}{1 + \lg\left[1 + \frac{\rho v Y_{O_2} n_p v_p m_p \eta}{\left(\rho v Y_{O_2} - n_p v_p m_p \eta\right)^2}\right]} \tag{7}$$

In the formula, ρ and v indicates the air density and velocity, respectively; n_p, v_p and m_p indicate the number density of particles, particle velocity, and particle mass, respectively; η is the mass flow rate of oxygen and particles into the system.

It is considered that the powder inside the partially covered cavity structure enters the supersonic combustor by means of cyclonic injection. The variation of the parameters such as baffle length, dimensions of the cavity, and the number of powder injection holes will have an effect on the mixing of the powdered fuel with the pressurized air into the supersonic combustor, so the influence of the above factors will be carried out in the following, and the numerical calculations of the working conditions are shown in Table 2:

Table 2. Numerical calculation worksheet

Case	Baffle length (mm)	cavity length (mm)	Number of holes
1	0	150	1
2	25	150	1
3	35	150	1
4	45	150	1
5	25	120	1
6	25	180	1
7	25	150	2
8	25	150	3

4. Numerical calculation results and discussion

4.1. Study of the baffle length

The powder is introduced into the supersonic combustor through a cyclonic flow at the partially covered cavity structure, where the baffle plays a crucial role in shaping the trajectory distribution of the particles. One key benefit of the baffle is that it prevents direct contact between the particles and the supersonic mainstream, promoting increased collisions between the particles and the baffle itself. This creates a buffer zone, enhancing the heating of the powdered fuel. Additionally, the combination of the baffle structure and the powder cyclone injection scheme extends the retention time of the particles, thus establishing favourable conditions for their combustion.

Figure 4 illustrates the particle trajectory distribution within the supersonic combustion chamber for varying baffle lengths. It is observed that in the cavity-based combustor without a baffle, a limited zone of particle aggregation forms at the leading edge of the cavity chamber, while most particles traverse the shear layer and exit the chamber with the main flow. In contrast, in the partially covered cavity-based combustor, a greater concentration of particles accumulates in the covered region. Additionally, there is an increase in the number of particle trajectories rotating axially around the combustion chamber in the uncovered area, leading to an overall longer particle retention time in the partially covered cavity-based design.

The particle retention times in the above combustor configurations were quantified, and the results are shown in figure 5. It is evident that the particle retention time in the partially covered cavity combustor surpasses that of the cavity structure. The particle retention time in the combustion chamber configuration with Lc=35mm reaches the maximum value, and the particle retention time in the structure with Lc=25mm differs from the maximum value by about 4ms but is higher than that in the combustion chamber structure with Lc=45mm.

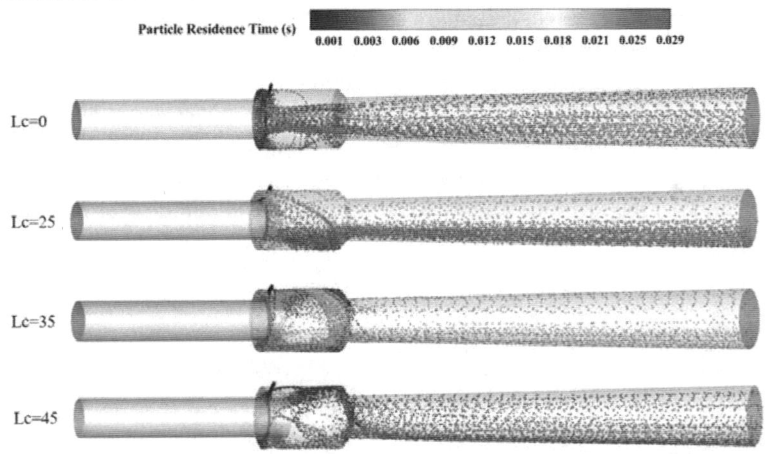

Figure 4. Comparison of particle trajectories in combustor with different baffle lengths

Figure 7 shows the local flow diagram at the central section of the combustor with the nozzle position. It is observed that a symmetrical return region is formed in the structure with Lc=0mm, with the return center located near the leading edge of the cavity. This return region creates favourable conditions for fuel ignition and combustion, contributing to flame stabilization. When the baffle length is set to Lc=25mm, a significant return area is evident in the uncovered region, alongside the appearance of partially formed return vortices in the covered area. As the baffle length increases to 35mm, the return flow in the uncovered area remains relatively unchanged; however, a more pronounced return vortex structure emerges in the covered area. This return vortex is induced by the flow dynamics in the uncovered area, resulting in the formation of two return vortex structures in the bottom section of the covered area. At Lc=45mm, the increased baffle length diminishes the influence of the return flow structure in the uncovered area on the airflow within the covered area, leading to the transformation of the original double vortex structure into a larger return vortex. The impact of the return zone on boron powder fuel is illustrated by its ability to control the trajectory of the particles as they enter the partially covered cavity region, while the double return vortex structure enhances the retention time of the powder particles in the covered region at the Lc=35mm configuration.

Figure 6 compares the mixing degree within the combustor at various baffle lengths. It reveals that the mixing degree in the region of the partially covered cavity is lower than that in the expansion section of the combustion chamber overall. Notably, the introduction of the baffle results in a slight reduction in the mixing degree of the powder particles within the mixing region. However, the mixing degree in the expansion section remains relatively unchanged, reaching approximately 0.8.

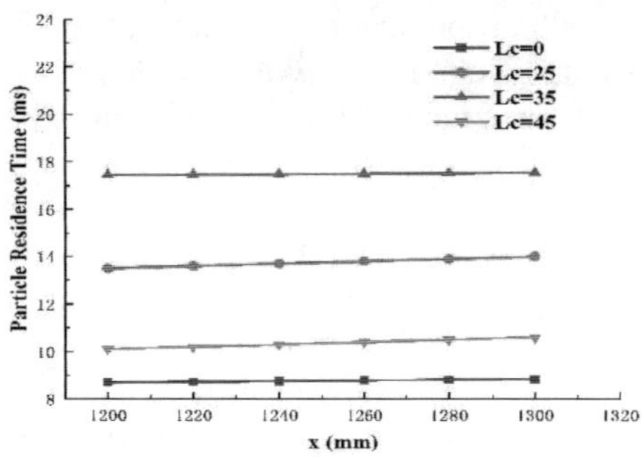

Figure 5. Particle retention time in the combustor for different calculation conditions

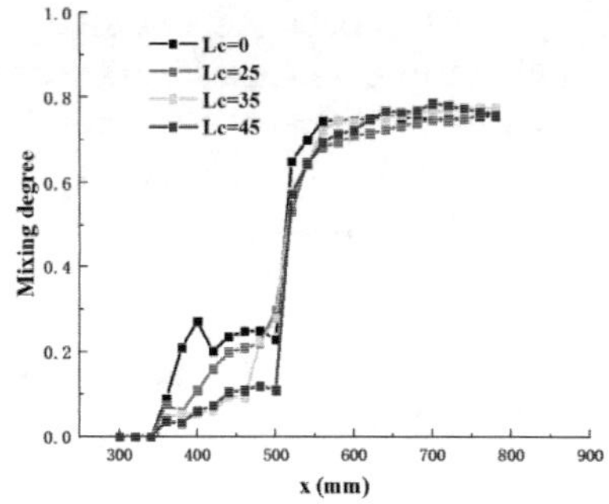

Figure 6. Flow diagram of partially covered cavity structure area under different operating conditions

This phenomenon can be attributed to the increased retention time of particles in the partially covered cavity structure due to the baffle. As a result, the concentration of particles in the covered area increases, leading to poorer mixing with the supersonic mainstream in that region. In contrast, beyond the expansion section, particles mix more effectively with the supersonic mainstream, resulting in a rapid increase in the overall mixing degree. The analysis indicates that the mixing effect of particles and air is enhanced in the combustion chamber with the partially covered cavity structure. Moreover, the higher particle concentration in the covered area facilitates the ignition and combustion of powder fuel within the supersonic combustor.

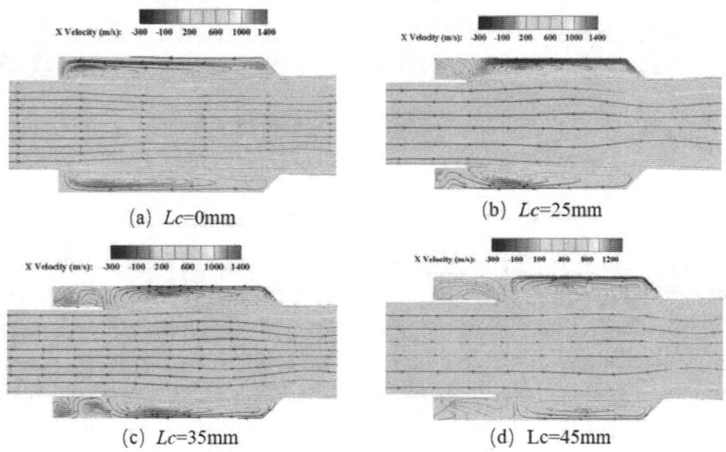

(a) *Lc*=0mm (b) *Lc*=25mm

(c) *Lc*=35mm (d) Lc=45mm

Figure 7. Flow diagram of partially covered cavity structure area under different conditions

4.2. Study of the dimension of the cavity

In this section, we investigated the influence of cavity length on the trajectory of powder particles in the combustion chamber with a partially covered cavity structure. Based on the previous findings, a baffle length of 25 mm was selected for this study. Figure 8 displays the corresponding cloud diagrams. The results show that as the cavity length increases, the concentration of particles in the covered area slightly diminishes, and there is a corresponding decrease in the trajectories of particles rotating along the wall near the inner surface of the uncovered area. An examination of the retention time of boron particles within the three combustors reveals that in the L=120 mm chamber, the trajectories of particles with longer retention times are most pronounced. In contrast, as the concave chamber length increases, the particle retention time decreases, with the L=180 mm chamber exhibiting the shortest retention time.

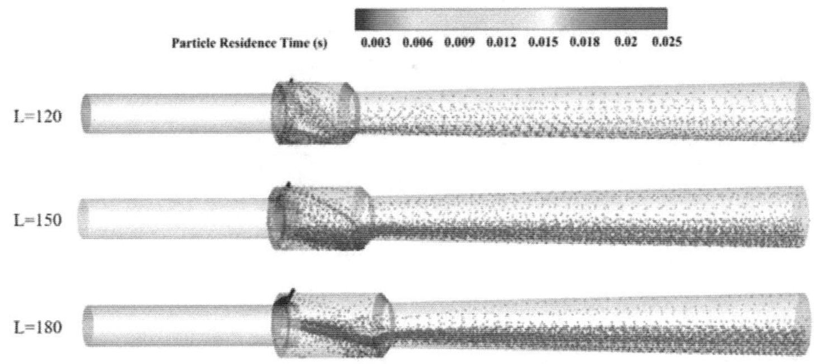

Figure 8. Comparison of particle trajectories in combustor with different cavity lengths

This phenomenon can be explained by the presence of a cavity shear layer between the reflux region inside the concave chamber and the main flow of the external supersonic velocity. As shown in figure 9, this shear layer exhibits different characteristics depending on the length-to-depth ratio of the cavity. When the cavity length is 180 mm, the shear layer is positioned closer to the bottom wall of the cavity. This configuration enhances the mass exchange between the interior and exterior of the cavity, accelerating the ejection of powder particles from the low-velocity region and thereby reducing the particle retention time in the combustion chamber. In the combustion chamber configuration with a concave chamber length of L=120 mm, the shear layer is positioned far from the bottom wall and is attached to the back wall of the concave chamber. This configuration inhibits mass exchange between the interior of the concave chamber and the external environment. As a result, boron particles within the low-velocity reflux region of the L=120 mm configuration experience a longer retention time, leading to a relatively high concentration of particles in the concave chamber.

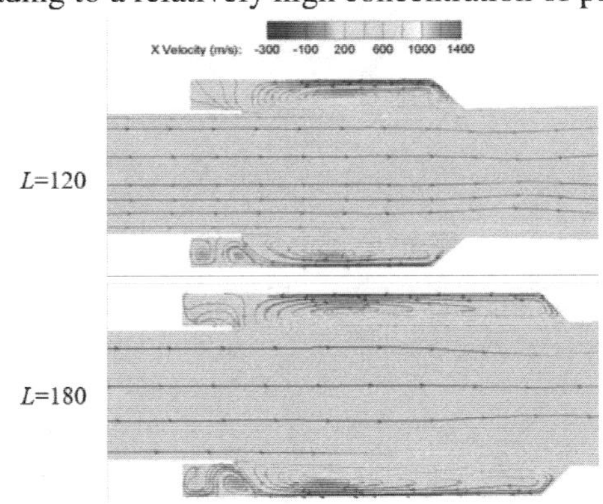

Figure 9. Particle retention time in the combustor for different calculation conditions

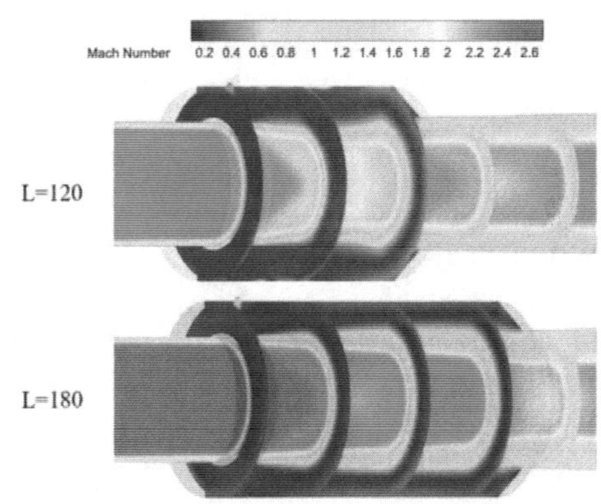

Figure 10. Comparison of Mach numbers in combustor of different configurations

Figure 10 illustrates the Mach number distribution within combustors featuring different cavity lengths. In the configuration with a cavity length of 180 mm, the penetration depth of the high-velocity gas flow in the return region of the cavity structure is greater than that observed in the configuration with a cavity length of 120 mm. Consequently, more particles are exposed to the high-velocity airflow in the chamber with a cavity length of 180 mm, leading to a decrease in particle retention time. This observation is consistent with the results discussed above.

4.3. Characterization of powder mixing in partially covered cavity structures

In this section, we investigate the behavior of powder particles entering the combustor through cyclonic flow in the partially covered cavity, considering the impact of the number of powder injection holes on the particle trajectory. Figure 12 compares the particle trajectories in the combustion chamber with varying numbers of powder orifices. The results show that in the single-hole injection mode, the distribution of particles in the partially covered cavity structure is more concentrated. However, as the number of powder orifices increases, the particles become more concentrated in the covered area, while the distribution of particles in the uncovered area decreases.

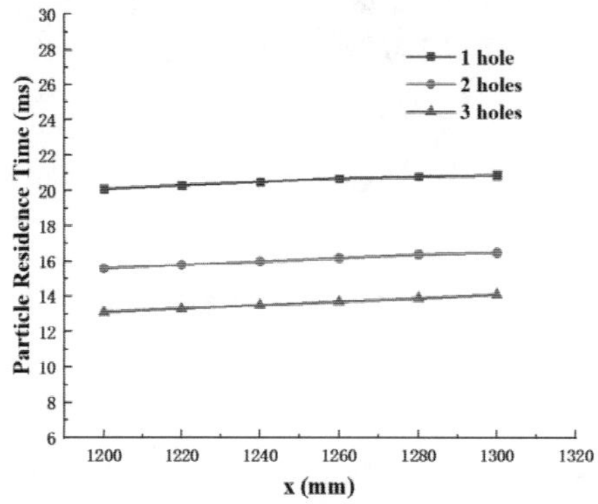

Figure 11. Comparison of particle retention time across different numbers of orifices

The observed phenomenon is attributed to the increase in powder injection holes, which enhances mass exchange between the covered and uncovered areas of the partially covered cavity structure. This results in more particles coming into direct contact with the supersonic mainstream and leaving the combustion chamber. Figure 11 presents a statistical comparison of the retention times under these conditions, demonstrating that the increase in spray holes coincides with a decrease in particle retention time, confirming the conclusions drawn from the previous analysis.

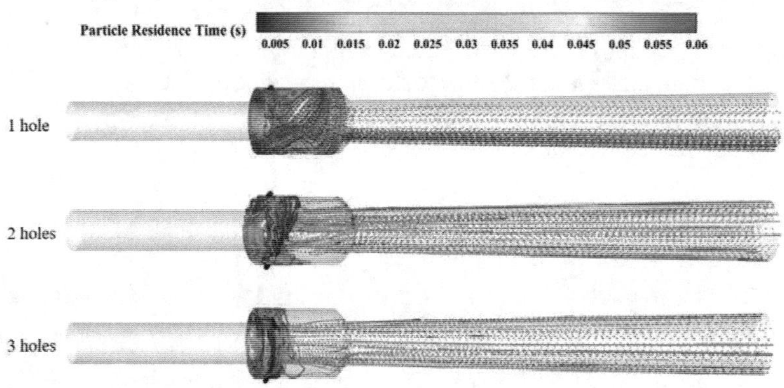

Figure 12. Particle trajectories in the combustion chamber with varying numbers of orifices

5. Conclusions

To explore the combustion organization scheme for powder-fueled scramjets, this study examines the mixing characteristics of powder particles in a partially covered cavity combustor. The findings indicate that:

The partially covered cavity structure created by the addition of the baffle significantly enhances the retention time of particles in the supersonic combustion chamber. It establishes a high concentration area of powder within the covered region, which is beneficial for the ignition and combustion of the particles. It is also found that the particle retention time increases and then decreases with the increase of the baffle length, and the particle retention time reaches the maximum in the baffle length of about 35 mm, and the particle mixing degree can reach about 0.8 in the combustion chamber of this configuration.

Increasing the length of the cavity in the partially covered cavity structure does not contribute to an increase in particle retention time within the combustor. In the scheme where the powder in the partially covered cavity is injected into the combustor by means of cyclonic injection, the increase of the powder injection holes instead results in a reduction of the particle retention time.

6. References

[1] Zhang C, Wang Y L and Ye L 2020 Summary of the technological development of overseas hypersonics in the past ten years Tactical Missile Technology 6 81-86
[2] Curran E T. 2012 Scramjet engines: The first forty years Journal of Propulsion & Power 17 1138-48
[3] Liu S X, Liu S J and Li Y 2023 Current developments in foreign hypersonic vehicles and defense systems Air & Space Defence 6 39-51
[4] Smart M 2007 Scramjets The Aeronautical Journal 111 605-619
[5] Wu X J and Wei Z.J 2023 Analysis of the characteristics of scramjet mode and ramjet mode of axisymmetric dual-combustion ramjet Acta Astronaut 203 125-134
[6] Zhao X, Xia Z X, Ma L K and Li C L 2021 Research progress on solid-fueled scramjet Chinese J. Aeronaut 35 398-415
[7] Hu C B, Li C and Sun H J 2017 A summary of powder-fueled ramjet Journal of Solid Rocket Technology 40 269-276
[8] Lv Z, Xia Z X, Liu B and Liu Y C 2015 Experimental and numerical investigation of a solid-fuel rocket scramjet combustor Journal of Propulsion and Power
[9] Huang Y, Risha G A, Yang V and Yetter R A 2005 Analysis of nano-aluminum particle dust cloud combustion in different oxidizer environments Aerospace Sciences Meeting & Exhibit 3013-3026
[10] Feng Y C, Xia Z X, Huan L Y and Yan X T 2016 Experimental investigation on the combustion characteristics of aluminum in air Acta Astronaut 129 1-7
[11] Ma L K, Li C L and Xia Z X 2021 Experimental investigation of solid rocket scramjet combustor with cavity flameholder Journal of Propulsion Technology 42 319-326
[12] Ding H M, Zhuo C F and Chen X 2022 Numerical study on the transverse jet flow and mixing characteristics of hydrogen/metal powder fuel in powder fuel scramjet Fuel 326
[13] Liu Y, Sun M B and Yang Y N 2022 Turbulent boundary layer subjected to a sonic transverse jet in a supersonic flow Aerospace Science and Technology 104 106016
[14] Luo S B, Feng Y C and Song J W 2022 Powder fuel transport process and mixing characteristics in cavity-based supersonic combustor with different injection schemes Aerospace Science and Technology 128 107798
[15] Che Y H 2022 Research on blending characteristics of powder fuel in supersonic air flow Xi'an: Northwestern Polytechnical University 55-57.
[16] Feng Y B, Luo S B and Song J W 2023 Numerical investigation on the combustion characteristics of powder fuel under different regulation parameters Aerospace Science and Technology 142 108608
[17] Xiao W, Jin T, Luo K and Dai Q 2020 Eulerian–Lagrangian direct numerical simulation of preferential accumulation of inertial particles in a compressible turbulent boundary layer. Journal of Fluid Mechanics 903
[18] Ladeinde F 2010 Advanced computational-fluid-dynamics techniques for scramjet combustion simulation AIAA Journal 48 513-514
[19] Star J, Edwards J and Smart M 2005 Numerical simulation of transient scramjet combustion in a shock tunnel 43rd AIAA aerospace sciences meeting and exhibit 428
[20] Li C L, Xia Z X and Ma L K 2019 Experimental and numerical study of solid rocket scramjet combustor equipped with combined cavity and struct device Acta Astronaut 162 145-154
[21] Song X L, Wang H B and Sun M B 2019 Mixing and combustion characteristics in a cavity-based supersonic combustor with different injection schemes Acta Astronaut 159 584-592
[22] Feng Y B, Luo S B and Song J W 2023 Numerical investigation on the combustion characteristics of aluminium powder fuel in a supersonic cavity-based combustor Applied Thermal Engineering 221 119842
[23] Chen B B, Xia Z X 2013 Study on mixing process assessment method in solid ramjet. Journal of Solid Rocket Technology 36 731-735

Numerical investigation on establishment process of opposing jet in supersonic regimes

Liang Zhu[1,2], Xiang Tang[1], Jun Song[1], Hao-bi Bai[1]

1, Xi'an Modern Control Technology Research Institute, Xi'an 710065, PR China
2, National Key Laboratory of Land & Air Based Information Perception and Control, Xi'an 710065, PR China
zhuliang@njust.edu.cn

Abstract: As an innovative flow control strategy, opposing jets have proven effective in both drag reduction and thermal protection for high-speed aerospace systems. In the present paper, a developed CFD code was used to investigate the evolution process of the opposing jet in supersonic flows. At first, a classical experiment was selected for validating the current code in analysing the unsteady flow features containing shock interactions. The transitional dynamics governing the opposing jet's formation in high-speed flows are quantitatively characterized. The obtained findings demonstrated that the evolution process of opposing jet in supersonic regimes contains extremely complex flow features. Besides, the aerodynamic drag drops sharply at first and then fluctuates.

Keywords: opposing jet; unsteady flow; numerical investigation

1. Introduction

Nowadays, supersonic and hypersonic vehicles have attained worldwide attentions and lots of scholars have paid numerous efforts into this field [1]. As known, when a vehicle cruises at supersonic and hypersonic speed, aerodynamic drag and thermal environment are challenges in the engineering application. In the last century, opposing jet was introduced to achieve drag reduction for high-speed flight system, and both experimental measurements and numerical simulations were employed to analyze the flowfield patterns and thermal mitigation effects induced by counterflow jets in supersonic conditions by Hayashi et al. [2]. Actually, the flow features generated by opposing jet contains rich and complex flow structures. Always, there exist typical under-expanded jet structure including Mach disk, triple point, barrel shock and some other important flow features. Daso et al. [3] systematically examined how inflow Mach number and angle of attack affect flow characteristics, aerodynamic drag, and thermal protection performance in 2009. They found that the short penetration mode appears when the opposing jet mass flow rate exceeds a certain value. In 2010, Tamada et al. [4] numerically explored the influences of inflow Mach on heat flux reduction in supersonic and hypersonic flows. Chen et al. [5] utilized large eddy simulation (LES) approach to analyze the opposing jet flow mode and their findings showed that the drag reduction becomes bad as the size of opposing jet exit increases. In 2015, Gerdroodbary et al. [6] developed a porous opposing jet for thermal protection and investigated the associated flow field. In 2016, Li et al. [7] quantitatively assessed how opposing orifice parameters impact drag coefficient and heat flux reduction by computing 3D Navier-Stokes equations, and the results showed that square apertures exhibit enhanced drag and heat reduction characteristics. Zhang et al. [8] explored a pulsed opposing jet concept and they found that long penetration mode disappears even the opposing jet pressure is in low level.

As evidenced by previous studies, while the flow characteristics generated by opposing jets have received considerable research attention, the formation mechanism of the jet itself remains relatively unexplored. Thus, in the current investigation, we developed a code to analyze the evolution process of opposing jet in supersonic flows. And The organizational framework of this study is outlined below. Section 1 exhibits a brief introduction to opposing jet. Numerical approaches are listed in section 2 and section 3 describes the computation model. Section 4 gave a code validation. Results and discussion are given in section 5. At last, some findings are listed in section 6.

2. Governing equations

An in-house code was programmed to numerically solve the 2D unsteady axisymmetric NS equations and they are listed as below.

$$\frac{\partial}{\partial t}\iint_{\Omega} \boldsymbol{U}\mathrm{d}V + \oint_{\partial\Omega} \boldsymbol{F}_{\mathrm{c}} \cdot \boldsymbol{n}\mathrm{d}S - \oint_{\partial\Omega} \boldsymbol{F}_{\mathrm{v}} \cdot \boldsymbol{n}\mathrm{d}S = \iint_{\Omega} \boldsymbol{H}\mathrm{d}V \tag{1}$$

where t is time, $\boldsymbol{U}$, $\boldsymbol{F}_{\mathrm{c}}$ ***and*** $\boldsymbol{F}_{\mathrm{v}}$ are the conservative vector, convective flux vectors and viscous flux vectors, respectively. $\boldsymbol{H}$ is the axisymmetric term, which are presented as following.

$$\boldsymbol{U} = \begin{bmatrix} \rho \\ \rho u \\ \rho v \\ \rho E \end{bmatrix} \qquad \boldsymbol{H} = \frac{1}{y}\begin{bmatrix} -\rho v \\ -\rho uv + \tau_{xy} \\ -\rho v^2 + \tau_{yy} - \tau_{\theta\theta} \\ -(\rho E + p)v + u\tau_{xy} + v\tau_{yy} + q_y \end{bmatrix} \tag{2}$$

$$\boldsymbol{F}_{\mathrm{c}} = \begin{bmatrix} \rho u\boldsymbol{i} + \rho v\boldsymbol{j} \\ (\rho u^2 + p)\boldsymbol{i} + \rho uv\boldsymbol{j} \\ \rho vu\boldsymbol{i} + (\rho v^2 + p)\boldsymbol{j} \\ (\rho E + p)u\boldsymbol{i} + (\rho E + p)v\boldsymbol{j} \end{bmatrix} \qquad \boldsymbol{F}_{\mathrm{v}} = \begin{bmatrix} 0 \\ \tau_{xx}\boldsymbol{i} + \tau_{xy}\boldsymbol{j} \\ \tau_{yx}\boldsymbol{i} + \tau_{yy}\boldsymbol{j} \\ (u\tau_{xx} + v\tau_{xy} + q_x)\boldsymbol{i} + (u\tau_{yx} + v\tau_{yy} + q_y)\boldsymbol{j} \end{bmatrix} \tag{3}$$

Equation (2) and (3) provide the computational methods for obtaining the directional shear stress τ and heat flux q in x and y coordinates.

$$\tau_{xx} = \frac{2}{3}\mu(2\frac{\partial u}{\partial x} - \frac{\partial v}{\partial y} - \frac{v}{y}), \tau_{yy} = \frac{2}{3}\mu(2\frac{\partial v}{\partial y} - \frac{\partial u}{\partial x} - \frac{v}{y}) \tag{4}$$

$$\tau_{xy} = \mu(\frac{\partial u}{\partial y} + \frac{\partial v}{\partial x}), \tau_{\theta\theta} = 2\mu\frac{v}{y} - \frac{2}{3}\mu(\frac{\partial u}{\partial x} + \frac{\partial v}{\partial y} + \frac{v}{y}) \qquad q_x = \lambda\frac{\partial T}{\partial x}, q_y = \lambda\frac{\partial T}{\partial y} \tag{5}$$

Furthermore, as known, turbulence model is an important factor in CFD computation, especially for simulating flow separation, vortex and so on. Therefore, SST turbulence model is employed in the present simulation and it can be written in the following form.

$$\frac{\partial \rho k}{\partial t} + \frac{\partial(\rho u_j k)}{\partial x_j} = P_K - \beta^* \rho\omega k + \frac{\partial}{\partial x_j}\left[(\mu_L + \mu_T\sigma_k)\frac{\partial k}{\partial x_j}\right] \tag{6}$$

$$\frac{\partial \rho\omega}{\partial t} + \frac{\partial(\rho u_j \omega)}{\partial x_j} = \frac{\gamma}{\mu_T}P - \beta\rho\omega^2 + 2(1 - f_1)\frac{\rho\sigma_{\omega 2}}{\omega}\frac{\partial k}{\partial x_j}\frac{\partial \omega}{\partial x_j} + \frac{\partial}{\partial x_j}\left[(\mu_L + \sigma_\omega\mu_T)\frac{\partial \omega}{\partial x_j}\right] \tag{7}$$

Reference [9] provides a complete description of the implemented SST turbulence model.

3. Computational model

In this investigation, The blunt body diameter, namely, D=50mm. And the opposing jet hole diameter is 4mm. Besides, it is noted that the condition of the free flow is the same to case of Hayashi et al. [2] and flow Ma is 3.98 and the pressure ratio is set as 0.6 in the present investigation.

Journal of Physics: Conference Series **3041** (2025) 012011 doi:10.1088/1742-6596/3041/1/012011

4. Code validation

Before initiating numerical computations, code validation is essential. In this section, a classical experiment [10] is applied to validate the code. Fig. 1 gives the computed and experimental comparison images.

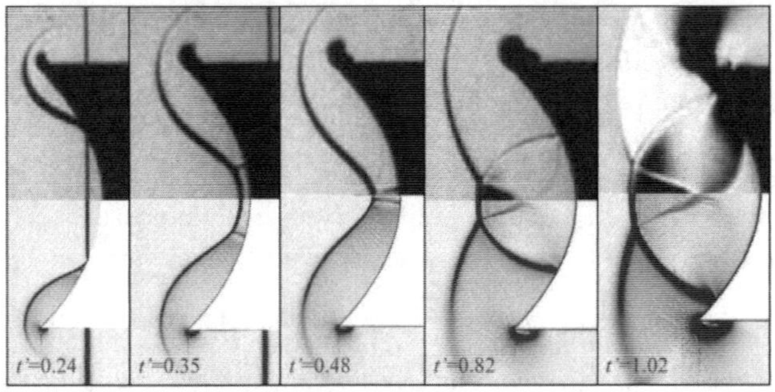

Figure 1. Comparison of computed and experimental results.

As displayed in Fig. 1, obviously, the numerical results can accurately predict the unsteady complex flow field including shock interactions. Thus, it is considered that the present code is reliable for computing the establishment evolution process of opposing jet in supersonic flows

5. Results and discussion

At first, Fig. 2 displays the Mach number distribution at key time intervals, illustrating the development process of opposing jets in supersonic flows.

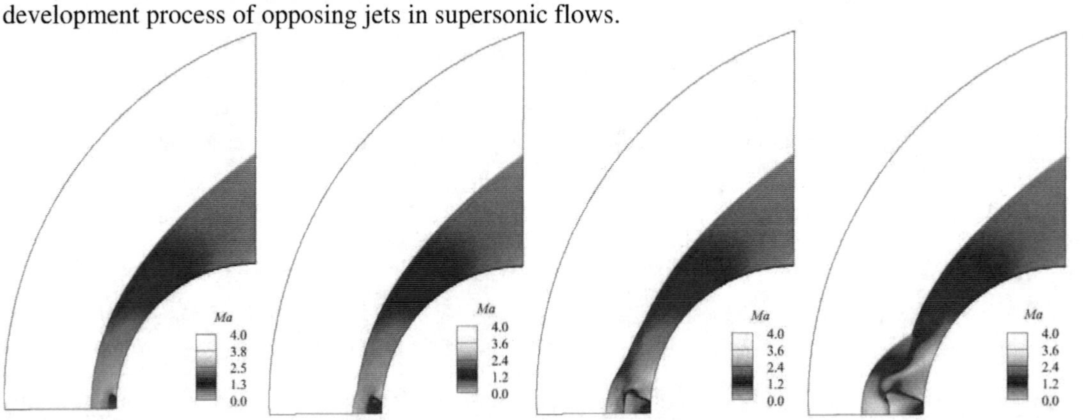

(a) t=5 us (b) t=10 us (c) t=25 us (d) t=50 us

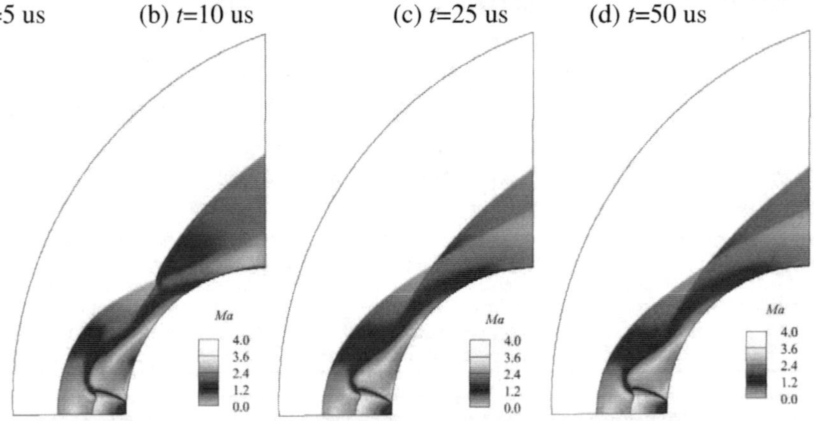

(e) t=100 us (f) t=200 us (g) t=500 us

Figure 2. Flow Mach contours at some typical moments

Obviously, it is seen that the flow structures experience remarkably changes in such short time. At first, when *t*=5us, the leading shock is observed and moves to the bow shock. Then when *t*=10us, the leading shock pushes the bow shock, and soon the shape of bow shock changes significantly at *t*=20us. Typical under-expanded jet structure can be clearly observed at *t*=50us. When *t* varies from 50us to 100us, the bow shock in front of the opposing jet changes remarkably and interacts severely with the reattachment shock. At *t*=200us, the numerical results show remarkable consistency with the flow structures captured by schlieren image in the reference [2]. Further, Fig. 3 illustrates the aerodynamic drag variations.

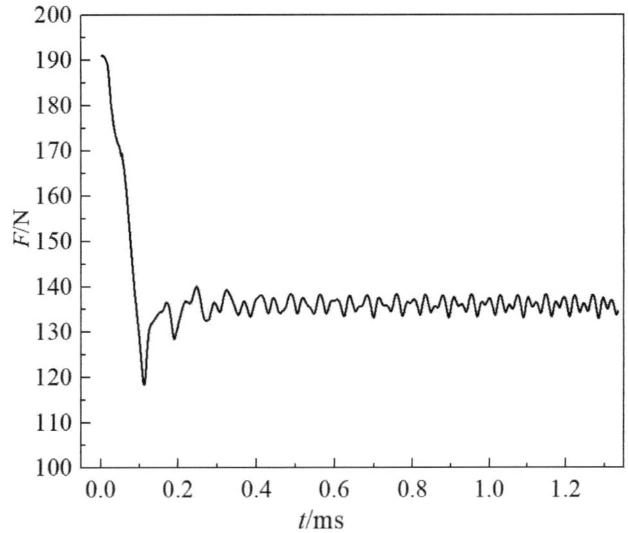

Figure 3. Aerodynamic drag variations

It is clear from Fig. 3 that the aerodynamic drag exhibits nonlinear variation characteristics during opposing jet development in supersonic regimes. At first. The aerodynamic drag on the blunt body decreases to a low value and soon increases to a relatively higher value. However, the aerodynamic drag fluctuates. Herein, a phenomenon should be noted. In the reference, Hayashi et al. presented a look-like steady flow schlieren image. Nevertheless, the aerodynamic drag variation illustrated in Fig.3 indicates the pressure along the blunt body changes with time. The phenomenon is explained as follows. Though the aerodynamic drag fluctuates, the main shock waves changes not significantly. As seen in Fig. 2, the shock waves at *t*=200us nearly seems no significant difference from that at *t*=500us. In fact, the pressure fluctuates resulting from the shear layer interaction while the main shock wave seems no change. Generally speaking, time-resolved measurements confirm the nonstationary nature of the opposing jet flowfield. And the opposing jet flow field establishment process is extremely complex.

6. Conclusion

Numerical investigation has been performed to investigate the establishment process of the opposing jet in supersonic regimes. Clear opposing jet development is presented and also the aerodynamic drag variation is shown. Herein, some meaningful findings are presented as follows.

(1) The developed code is accruable for modelling unsteady flow field containing severe shock interactions.

(2) The establishment process of the opposing jet contains rich and complex flow features and the whole establishment process belongs to strong transient process.

(3) The blunt body experiences an initial sharp reduction in aerodynamic drag followed by oscillatory behaviour.

Journal of Physics: Conference Series **3041** (2025) 012011 doi:10.1088/1742-6596/3041/1/012011

References

[1] Seleznev R K, Surzhikov S T, Shang J S. A review of the scramjet experimental data base[J]. Progress in Aerospace Sciences, 2019, 106:43-70.
[2] Hayashi K, Aso S, Tani Y. Numerical Study of Thermal Protection System by Opposing Jet[C]: 43rd AIAA Aerospace Sciences Meeting, 2005
[3] Daso E, Pritchett V, Wang T S, et al. The Dynamics of Shock Dispersion and Interactions in Supersonic Freestreams with Counterflowing Jets[J]. AIAA journal, 2009, 47(6):1313-1326.
[4] Tamada I, Aso S, Tani Y. Reducing Aerodynamic Heating by the Opposing Jet in Supersonic and Hypersonic Flows[J]. 48th AIAA Aerospace Sciences Meeting Including the New Horizons Forum and Aerospace Exposition, 2010
[5] Chen L W, Wang G L, Lu X Y. Numerical investigation of a jet from a blunt body opposing a supersonic flow[J]. Journal of Fluid Mechanics, 2011, 684:85-110
[6] Barzegar Gerdroodbary M, Ganji D D, Amini Y. Numerical study of shock wave interaction on transverse jets through multiport injector arrays in supersonic crossflow[J]. Acta Astronautica, 2015, 115:422-433
[7] Li S B, Wang Z G, Huang W, et al. Drag and Heat Reduction Performance for an Equal Polygon Opposing Jet[J]. Journal of Aerospace Engineering, 2017, 30(1):4016065
[8] Zhang R, Huang W, Yan L, et al. Drag and heat flux reduction induced by the pulsed counterflowing jet with different waveforms on a blunt body in supersonic flows[J]. Acta Astronautica, 2019, 160:635-645
[9] Guo J, Lin G, Bu X, et al. Parametric study on the heat transfer of a blunt body with counterflowing jets in hypersonic flows[J]. International Journal of Heat and Mass Transfer, 2018, 121:84-96
[10] Izumi K, Aso S, Nishida M. Experimental and computational studies focusing processes of shock waves reflected from parabolic reflectors[J]. Shock Waves, 1994, 3(3):213-222.

Fourier neural operator-based data-driven method for predicting structural dynamic response

Xuliang Luo[1], Laihao Yang[1,*], Changfeng Nan[2], Xuefeng Chen[1], Yu Sun[1]

[1] National Key Lab of Aerospace Power System and Plasma Technology, Xi'an Jiaotong University, Xi'an 710049, China
[2] AECC Commercial Aircraft Engine Co., Ltd, Shanghai 710049, China

[*] Corresponding author's e-mail: yanglaihao@xjtu.edu.cn

Abstract. Deep learning methods have made significant advancements in solving partial differential equations (PDEs), where physics-informed neural networks (PINN) are the most widely applied methods across various fields. However, for structural dynamic response prediction, employing PINN to solve the PDEs of structural dynamic models may result in poor accuracy and low training efficiency. To address these challenges, a novel method based on Fourier Neural Operator (FNO) for predicting structural dynamic responses is presented. Specifically, the FNO is harnessed to capture the main characteristics of the structural dynamic response in the frequency domain, achieving an accurate and efficient solution. Numerical experiments with a cantilever beam structure demonstrate that the proposed method can predict the structural response with high accuracy and low computational cost, effectively capturing the dynamic characteristics of the structure. The proposed method achieves a prediction error for both low-frequency and high-frequency responses that is one to two orders of magnitude lower than the PINN method. Additionally, its prediction time is six orders of magnitude faster than the Finite Element Method and four orders of magnitude faster than PINN.

1. Introduction

Accurate and efficient prediction of structural dynamic responses is crucial for the optimization design and structural health monitoring. Structural dynamics responses contain information about the inherent characteristics of the structure as well as its health status, playing a critical role in fields such as mechanical engineering, aerospace[1], and construction. The finite element method (FEM), with high-fidelity modeling capacity, is widely used in predicting structural dynamic responses. However, for large and complex structures such as aircraft engine blades[2], bridges, and high-speed gears, the FEM models with high fidelity can have tens of thousands of degrees of freedom (DOF) and may also exhibit nonlinear characteristics[2]. Solving these large-scale, high-fidelity models requires not only substantial computational resources but also considerable time. Therefore, it is crucial to explore cost-effective computational techniques to address these issues.

Deep learning methods have shown tremendous potential in solving partial differential equations (PDEs). Neural networks, by establishing a mapping between input data and the solutions of equations, possess powerful nonlinear modeling capabilities, which enable them to capture nonlinear relationships in complex datasets and solve equations efficiently, offering a novel approach to solving PDEs. Conventionally, the neural network learns to map inputs to outputs by training on large datasets. As the most recent and widely adopted advancement in this domain, the Physics Information Network (PINN) embeds physical equations into a loss function. This framework leverages automatic differentiation to determine the output function's partial derivatives across spatial and temporal dimensions. It has been validated and applied in various fields, such as fluid dynamics analysis[3] and seismic wave simulation[4]. In real-world use, however, PINN often encounters the vanishing gradient problem. This issue can lead solutions to forming around local minima, hindering optimal convergence[5]. The PINN is subject to spectrum bias[6], for dynamic equations involving high frequency solutions, which often makes it difficult for PINN to accurately approximate PDEs' solutions. In contrast to conventional numerical approaches, PINN relies on gradient descent optimization. This method can result in significantly slower computational speeds[5] compared to other techniques. Furthermore, PINN uses point-to-point physical constraints and requires a lot of data points for iterative optimization, which greatly increases the cost of training. Therefore, PINN can hardly predict the dynamic response accurately and efficiently.

Recent advancements have seen deep neural operators emerge as effective tools for solving partial differential equations (PDEs), demonstrating notable success in this domain[7]. Demonstrating exceptional efficacy, the Fourier Neural Operator (FNO) achieves notable precision in addressing parametric PDEs[8]. In contrast to the conventional approach, FNO does not depend on the grid, nor is it restricted to the solution of one PDE. It makes use of the fast Fourier transformation to parameterize the Fourier space, and transforms the time domain differential equation into the frequency domain. This makes the solution of the equations much simpler. Therefore, it is highly possible to use FNO to solve PDEs in structural dynamics models.

This paper presents a new data-driven approach to predict the dynamic response of structures based on FNO. Firstly, the excitation of the structure dynamic model is mapped to high dimensional Fourier space. Then, the solutions and the inherent properties of PDEs for structural dynamic models can be learned in the frequency domain, which enables achieving high-accuracy and low-time-cost predictions of structural dynamic responses. Ultimately, the effectiveness of the FNO-based data-driven method in predicting the low-frequency and high-frequency dynamic responses of structures was validated through numerical analysis using a cantilever beam structure.

2. Methodology

This section outlines the foundational principles and operational workflow of the Fourier Neural Operator (FNO). The methodology's ability to transform temporal data into spectral representations is detailed, highlighting how this property is harnessed to enable streamlined learning and accurate prediction within neural network architectures.

2.1. Fourier neural operator

The inspiration for FNO comes from the Green Function Method, which is commonly used to solve differential equations[9]. For any differential equation

$$L_b u = f \tag{1}$$

where, L_b represents the differential operator, which depends on the parameter b, while f denotes the external force. Solution $u(x)$ is expressed as an integral of Green's function $G_b(x,y)$, which is dependent on the parameter b. The solution $u(x)$ can be expressed as

$$u(x) = \int_D G_b(x,y) f(y) dy \tag{2}$$

where, $x, y \in D \subset \mathbb{R}^d$. In theory, if the Green's function can be determined, then the differential equation can be solved. However, the analytical solution of the Green's function can only be found in a limited number of instances. Given that neural networks have strong robust fitting capabilities, the kernel functions obtained through neural network parameterization could be employed to estimate Green's functions[9]. To elaborate, the integral covariate can be expressed in this way:

$$\left(K_\phi(b)v_i\right)(x) = \int_D \kappa_\phi\left(x, y, b(x), b(y)\right) v_i(y) dy \tag{3}$$

where, the kernel integral operator K_ϕ is parameterized by a neural network with learnable parameters $\phi \in \Theta_K$ and Θ_K is the finite-dimensional parameter space. We parameterize κ_ϕ as a neural network with trainable parameters ϕ, and κ_ϕ is used as the kernel function to approximate the Green's function. v_i is the latent representation generated by the Fourier layer.

According to Refs.[9-10], the Fourier layer in the FNO is described as an iterative structure, namely:

$$v_{i+1}(x) = \sigma\left(Wv_i(x) + \left(K_\phi(b)v_i\right)(x)\right), \quad i = 0,1,\ldots,L-1 \tag{4}$$

where, L represents the number of Fourier layers, W represents the linear transformation, and σ is the nonlinear activation function. As for the kernel integration, the FNO neural network structure replaces the kernel integral operator κ_ϕ with a Fourier-defined convolution operator, and the Fourier transform is used to efficiently calculate equation (4)[10].

κ_ϕ is selected as a translation-invariant kernel function, and the parameter b is removed, leading to $\kappa_\phi(x, y, a(x), a(y)) = \kappa_\phi(x-y)$. According to the convolution theorem, it can be known that the Fourier transform obtained after the convolution of a function is equivalent to the result of multiplying the Fourier transform of a single function. That is to say, the convolution operation within the time domain range is equivalent to the multiplication operation within the frequency domain range, i.e., $\mathrm{FT}\left(\int_D \kappa_\phi(x-y)v_i(y)dy\right) = \mathrm{FT}(\kappa_\phi)\cdot\mathrm{FT}(v_i)$, where FT indicates the Fourier transform of the function. Thus,

$$\left(K_\phi(b)v_i\right)(x) = \mathrm{FT}^{-1}\left(\mathrm{FT}(\kappa_\phi)\cdot\mathrm{FT}(v_i)\right)(x) \tag{5}$$

Directly parameterizing κ_ϕ in Fourier space yields the Fourier integral operator:

$$\left(K_\phi v_i\right)(x) = \mathrm{FT}^{-1}\left(R_\phi \cdot \mathrm{FT}(v_i)\right)(x) \tag{6}$$

where, R_ϕ is the Fourier transform of the periodic function kernel κ, which is parameterized by $\phi \in \Theta_K$. The function undergoes discretization via a trainable Fourier coefficient-based weight vector and is subsequently truncated to retain a finite set of Fourier modes.

Figure 1 shows the general structure of the FNO, including the Fourier layer's inner structure. First, the input $b(x)$ is processed through the lifting layer P to obtain $v_0(x)$. The input $b(x)$ may represent components such as initial conditions, forcing terms, or other parameters within a differential equation, and P is a fully connected network that projects the input into a high-dimensional feature space. Secondly, there are two ways to deal with $v_i(x)$ within the Fourier layer: first, the kernel is parameterized by Fourier transformation in Fourier space, and then the inverse Fourier transform is used to transform the original space, and the second is by applying a linear transformation W to $v_i(x)$. These are processed through the nonlinear activation function σ, yielding the output $v_{i+1}(x)$. Following successive iterations, the final $v_L(x)$ is output as $u(x)$ through the projection layer Q. The FNO combines nonlocal kernel integration operators with the activation function σ, which enables the FNO to approximate highly nonlinear operators.

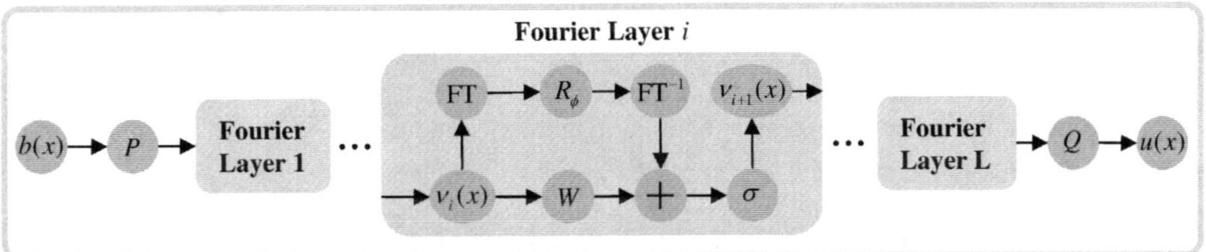

Figure 1. FNO architecture.

2.2. Frequency-domain solution of the structural dynamics model

The PDE of the structural dynamic model may be formulated as:

$$\mathbf{M}\frac{\partial^2\mathbf{u}(x,t)}{\partial^2 t}+\mathbf{C}\frac{\partial\mathbf{u}(x,t)}{\partial t}+\mathbf{K}\mathbf{u}(x,t)=\mathbf{F}(t),\ \ x\in\Omega, t\in[0,T] \tag{7}$$

where Ω is the domain of space, x and t are the space and time coordinates respectively. $\mathbf{u}(x,t)$, $\frac{\partial\mathbf{u}(x,t)}{\partial t}$, and $\frac{\partial^2\mathbf{u}(x,t)}{\partial^2 t}$ indicate the displacement, velocity, and acceleration, respectively; $\mathbf{M}$, $\mathbf{C}$, and $\mathbf{K}$ correspond to the mechanical structure's mass, damping, and stiffness matrices, respectively; $\mathbf{F}(t)$ represents the external force acting on the structure.

As a matter of fact, the Fourier layer of FNO can be transformed into a frequency domain differential equation. Applying a Fourier transform to the structural dynamic equation (7) yields:

$$\mathbf{M}\,\mathrm{FT}\left[\frac{d^2\mathbf{u}}{d^2 t}\right]+\mathbf{C}\,\mathrm{FT}\left[\frac{d\mathbf{u}}{dt}\right]+\mathbf{K}\,\mathrm{FT}[\mathbf{u}]=\mathrm{FT}[\mathbf{F}] \tag{8}$$

where, $\mathrm{FT}[\cdot]$ denotes the Fourier transform operator, characterized by its differential properties as follows:

$$\mathrm{FT}\left[\frac{d^n\mathbf{u}}{d^n t}\right]=(\mathbb{J}\omega)^n\,\mathrm{FT}[\mathbf{u}] \tag{9}$$

where $\mathbb{J}$ denotes the imaginary unit, and ω corresponds to the angular frequency. As shown in equation (9), the Fourier transform converts time-domain differential operations into frequency-domain algebraic counterparts. Inserting equation (9) into equation (8) yields:

$$-\omega^2\mathbf{M}\,\mathrm{FT}[\mathbf{u}]+\mathbb{J}\omega\mathbf{C}\,\mathrm{FT}[\mathbf{u}]+\mathbf{K}\,\mathrm{FT}[\mathbf{u}]=\mathrm{FT}[\mathbf{F}] \tag{10}$$

The equation above, when simplified, reduces to:

$$\mathrm{FT}[\mathbf{u}]=\frac{\mathrm{FT}[\mathbf{F}]}{-\omega^2\mathbf{M}+\mathbb{J}\omega\,\mathbf{C}+\mathbf{K}} \tag{11}$$

Structural vibration responses show continuous and complicated properties in the time domain, as illustrated in equation (11). Compared with the direct solution in the time domain, it is more convenient to solve the spectrum equation in the frequency domain. In traditional PINNs, it is very challenging to train the neural network to ensure that the output time-domain response aligns with the structural dynamics equations in the physical loss. The FNO embeds input data into a high-dimensional Fourier space, transforming continuous time-domain signals into sparse frequency-domain representations. In this paper, by means of the convolution theorem, we transform the learning of the differential equation into the frequency domain of the frequency domain, so that the sparse solution can be learned efficiently, and the difficulty of using FNO can be reduced effectively. Finally, the inverse Fourier transform reconstructs the time-domain data from its frequency-domain representation, enabling higher precision and accelerated convergence. This efficiency makes the FNO particularly adept at computing structural dynamic responses.

2.3. FNO predicts the structural dynamic response

In this section, we will elaborate on the overall design of FNO for predicting the structural dynamic response, aimed at predicting the response with high accuracy and low computational cost.

Prior research[11] has employed temporal data (e.g., displacement, velocity, or acceleration) as neural network inputs to predict corresponding dynamic response variables. In structural dynamic systems, excitation serves as the input, with the system's response as the output. It is apparent from equation (11) that the frequency domain solution of the structural dynamic equation is composed of excitation and system parameters. On the basis of all kinds of flexibility of FNO training methods, we propose an innovative approach to output the displacement response by means of excitation as input to FNO. Thus, the FNO directly translates the excitation input to the dynamic response output, which is in good agreement with the input/output relations of the actual vibration system. According to Figure 2, for one node j of the structure, we use the excitation $\mathbf{F}_j$ as the input to the FNO.

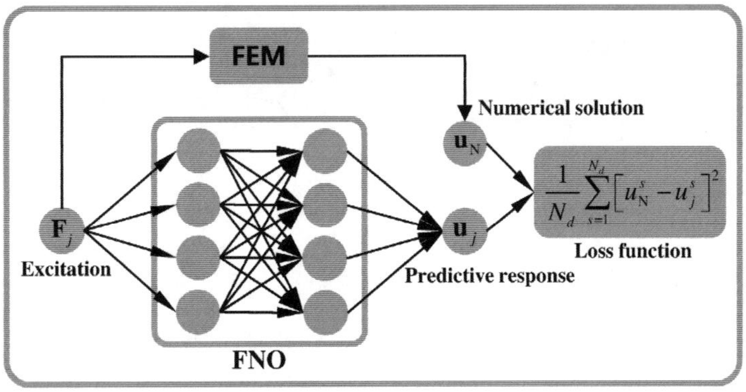

Figure 2. Flowchart of FNO predicting structural dynamic response

The excitation $\mathbf{F}_j$ passes through the lifting layer P and enters the Fourier layer for iteration to obtain $\mathbf{v}_i$. According to equation (11), the FNO learns the solution to the frequency-domain equation as:

$$\mathrm{F}\left[\mathbf{u}_j\right] = \frac{\mathbf{v}_i}{-\omega^2\mathbf{M} + \mathbb{J}\omega\,\mathbf{C} + \mathbf{K}} \tag{12}$$

Figure 2 illustrates that the loss function of the FNO is the residual between the predicted response and the numerical solution. The numerical solution $\mathbf{u}_\mathrm{N}$ of the structural vibration response is obtained through FEM, and the loss function is formed by combining it with the structural dynamic response $\mathbf{u}_j$ predicted by FNO. The loss function, which is represented by the structural dynamic response containing N_d data points, can be expressed as:

$$\mathrm{Loss} = \frac{1}{N_d}\sum_{s=1}^{N_d}\left[u_\mathrm{N}^s - u_j^s\right]^2 \tag{13}$$

The training paradigm of the FNO for solving equations in the frequency domain, as introduced in this paper, reduces the difficulty of predicting the structural dynamic response and significantly shortens the prediction time.

3. Numerical experiments

The traditional FEM method is time-consuming, whereas the deep learning method PINN suffers from low prediction accuracy and efficiency. Therefore, this paper proposes using FNO to quickly achieve high-accuracy structural response prediction. This section first develops a structural dynamics model to verify the effectiveness of the FNO in predicting responses under both low-frequency and high-frequency excitation. It also examines the FNO's ability to learn the structural dynamic properties through spectrum analysis of the response.

3.1. Structural dynamic model

To make it easier to generalize, the numerical experiment uses a generic structure, as illustrated in Figure 3. A cantilever beam, a foundational system in structural dynamics, is rigidly supported at one end while remaining unrestrained at the opposite extremity. A concentrated force $\mathbf{F}(t)$ is imposed at the free end's central point as the beam excitation.

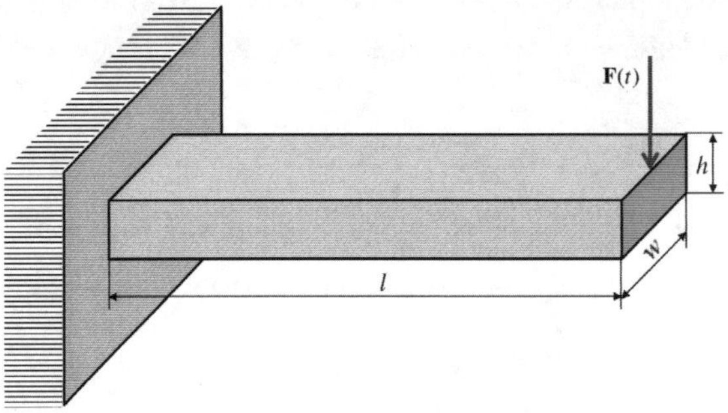

Figure 3. Cantilever beam structure.

The structure of the cantilever is modeled by ANSYS. The key physical and geometric properties are specified below: length $l=60\text{mm}$, width $w=15\text{mm}$, thickness $h=1.5\text{mm}$, elastic modulus $E=2\times10^{-5}E_{\text{s}}/\text{N}\cdot\text{mm}^{-3}$, Poisson's ratio $\nu=0.3$, and density $\rho=7.9\times10^{-6}\rho_{\text{s}}/\text{kg}\cdot\text{mm}^{-3}$. The structural model is meshed using 3D 20-node hexahedral solid element Solid186. To make the fitting of FNO more challenging and demonstrate its ability to predict the complex nonlinear responses of structural dynamics, according to Chen[12], neglecting stiffness-proportional damping will induce high-frequency oscillations. Therefore, Rayleigh damping is applied in this study with a lower stiffness-proportional damping to achieve a more oscillatory and complex nonlinear response curve, where Rayleigh damping is denoted as $\mathbf{C}=0.154\mathbf{M}+0.00078\mathbf{K}$.

3.2. Response Prediction

This section evaluates the accuracy of the FNO in predicting structural dynamic responses and compares it with the prediction errors of PINN.

In this section, low-frequency excitation $F_{\text{s1}}=10\sin(2\pi\cdot30t)$ at 30 Hz and high-frequency excitation $F_{\text{s2}}=10\sin(2\pi\cdot90t)$ at 90 Hz were applied as inputs to the structure. The response at the midpoint of the free end of the structure was then predicted. Because the structure vibration is induced by forced vibration, the frequency of the structural response is the same as that of the excitation force. To enhance waveform clarity across both low- and high-frequency regions, the predicted response duration was configured to 0.41 seconds. During the numerical experiments, the FNO is trained over 500 epochs, initialized with a learning rate of 0.001 that is reduced by a factor of 0.4 at 100-epoch intervals. The FNO is optimized using the Adam optimizer, configured with a weight parameter of 0.0001. Each Fourier layer has a width of 256.

Figure 4 (a) and Figure 5 (a) compare the predictions of the structural dynamic response by PINN and FNO. The results reveal that PINN exhibits significant discrepancies in predicting both low-frequency and high-frequency forced vibration responses when compared to the FEM-calculated responses. Its structural time-domain response deviates significantly from the FEM results, showing poor curve fitting and failing to make accurate predictions. Although PINN constrains the network's predictions using physical equations, the complexity of solving the structural dynamics model makes PINN optimization challenging, causing difficult convergence and resulting in low prediction accuracy. On the other hand, the FNO forecast for the structure time domain response is close to that of the FEM, and the response curve can be used to predict both the low frequency and high frequency response.

This study quantifies the accuracy of both approaches by computing the relative L_2 error between the model-predicted response and the FEM-generated solution. Table 1 highlights a significant disparity: the FNO achieves a relative L_2 error nearly two orders of magnitude lower than the PINN in modeling low-frequency forced vibration responses. In addition, the relative L_2 error of the FNO is approximately one order less than that of the PINN in the prediction of high frequency forced vibration response. Both the time domain response and the relative L_2 error result show that FNO is more accurate than PINN in predicting the response of the structure dynamic model, and it is able to predict both low frequency and high frequency.

Figure 4 (b) and Figure 5 (b) indicate that the structural forced vibration response in the time domain exhibits complex dynamic characteristics. It is found that the frequency components of the vibration response are mainly concentrated in several frequencies, whereas those of the other frequencies are close to zero. The results show that the main frequency components of the structure are sparse. The response spectrum contains information about the structural dynamic characteristics. This paper primarily focuses on the natural frequency and the forced vibration frequency. The spectrum in Figure 4 (b) and Figure 5 (b) contains two major frequency components with significant amplitudes: a 10Hz component, corresponding to the natural frequency of the structure, and 30Hz and 90Hz components, which represent the forced vibration frequencies of the structure, aligning with the excitation frequency. Based on the response spectrum in Figure 4 (b) and Figure 5 (b), the natural frequency and forced response frequency components predicted by the FNO closely match those calculated by the FEM, with small amplitude errors. This demonstrates that the FNO can accurately capture the structural dynamic inherent characteristics and forced response properties under both low-frequency and high-frequency vibrations. The FNO's accurate frequency component predictions demonstrate its efficiency in mapping inputs into Fourier space and extracting frequency-domain features governing structural dynamic responses.

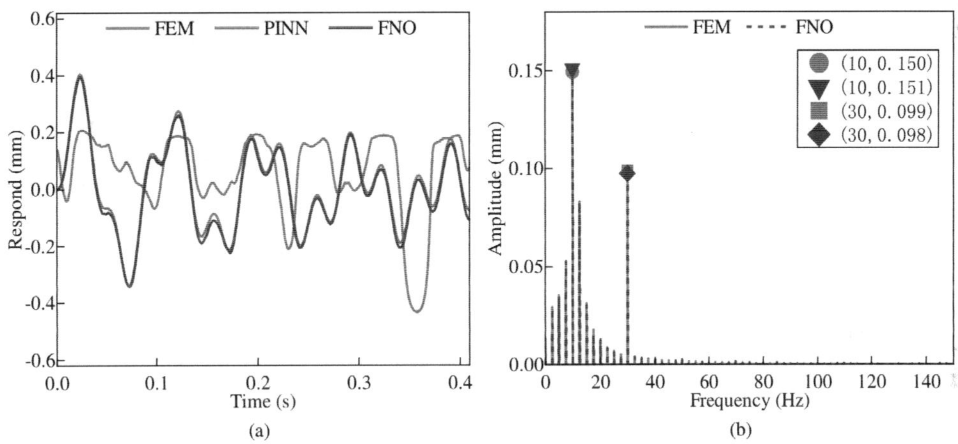

Figure 4. (a) Time-domain response prediction results at 30Hz using PINN and FNO, (b) Response spectrum predicted by FNO

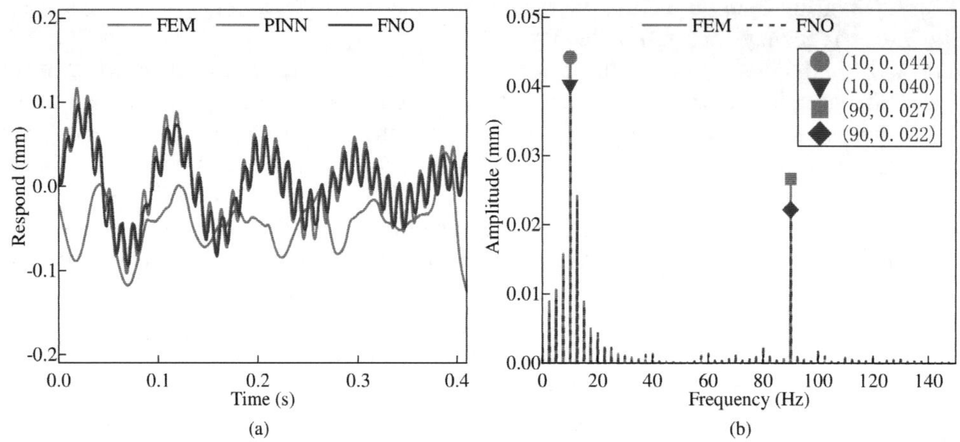

Figure 5. (a) Time-domain response prediction results at 90Hz using PINN and FNO, (b) Response spectrum predicted by FNO

Table 1. Relative L_2 error of FNO predicted response

Excitation frequency	FNO	PINN
30Hz	9.091E-2	1.266
90Hz	3.135E-1	1.540

As shown in Table 2, FNO's prediction time is six orders of magnitude faster than FEM computation and four orders of magnitude faster than PINN computation. Compared to traditional computational methods, the time required by FNO to predict the response is shorter than that of FEM and PINN. This demonstrates FNO's ability to quickly solve structural vibration responses and proves that the FNO method is better than traditional computing techniques in predicting the vibration response of structures.

Table 2. Time consumption of three methods for solving the PDEs in structural dynamic models

Method	FEM	PINN	FNO
Calculation time (s)	4.580E3	4.687E1	3.270E-3

4. Conclusions

This paper proposes an FNO-based data-driven method for predicting structural dynamic responses, leveraging its ability to efficiently resolve frequency-domain PDEs governing such systems. Numerical experiments on the cantilever beam structure demonstrate that the proposed method reduces the prediction error for both low-frequency and high-frequency responses of the structural dynamic model by one to two orders of magnitude compared to the PINN method. The response-generated spectrum confirms the approach's ability to accurately identify inherent dynamic characteristics, notably capturing critical frequency components such as natural and excitation frequencies. The time required for the proposed method to predict the response is significantly lower than that for FEM and PINN. In summary, these results confirm the efficacy and advantages of the proposed approach, delivering accurate structural dynamic response predictions with notable computational efficiency.

FNO has demonstrated excellent accuracy and computational efficiency in solving the PDEs of structural dynamics models. Future research may extend its application to tackle prediction challenges in nonlinear coupled systems and structures exhibiting nonlinear phenomena, exemplified by cracked components.

5. References

[1] Isavand J, Kasaei A and Peplow A, 2024 Enhancing response estimation and system identificati on in structural health monitoring through data-driven approaches *J. Mechanism and Machin e Theory.* 31(1), 57-73.

[2] Tamer A, Schwingshackl C, and Jones A 2023 High-Fidelity Nonlinear Forced Response Analys is of Aero-Engine Turbine Blades With Root Friction Contact. *In International Design Engin eering Technical Conferences and Computers and Information in Engineering Conference. A merican Society of Mechanical Engineers.* Vol. 87387, p. V010T10A010

[3] Jalili D, Jang S and Jadidi M 2024 Physics-informed neural networks for heat transfer predictio n in two-phase flows *J. International Journal of Heat and Mass Transfer.* 221, 125089.

[4] Zhu Z, Wang Z and Feng Y 2025Solution and application of two-dimensional seismic wavefield evolution based on physics-informed neural networks *J. Engineering Applications of Artifici al Intelligence.* 150, 110652.

[5] Lawal Z K, Yassin H, Lai D T C, and Che Idris A 2022 Physics-informed neural network (PINN) evolution and beyond: A systematic literature review and bibliometric analysis *J. Big Data a nd Cognitive Computing.* 6(4), 140.

[6] Xu Z Q J, Zhang Y, Luo T, Xiao Y, and Ma Z 2019 Frequency principle: Fourier analysis sheds l ight on deep neural networks *arXiv preprint* arXiv:1901.06523.

[7] Azizzadenesheli K, Kovachki N, Li Z, Liu-Schiaffini M, Kossaifi J and Anandkumar A 2024 *J. Neural operators for accelerating scientific simulations and design. Nature Reviews Physics.* 1-9.

[8] Li B, Wang H, Feng S, Yang X and Lin Y 2023 *J. Solving seismic wave equations on variable v elocity models with Fourier neural operator. IEEE Transactions on Geoscience and Remote Sensing.* 61, 1-18.

[9] Li Z, Kovachki N, Azizzadenesheli K, Liu B, Bhattacharya K, Stuart A, and Anandkumar A 202 0 Neural operator: Graph kernel network for partial differential equations *arXiv preprint* arXi v:2003.03485.

[10] Li Z, Kovachki N, Azizzadenesheli K, Liu B, Bhattacharya K, Stuart A, and Anandkumar A 202 0 Fourier neural operator for parametric partial differential equations *arXiv preprint* arXiv:20 10.08895.

[11] Moradi S, Duran B, Eftekhar Azam S, and Mofid M 2023 Novel physics-informed artificial neu ral network architectures for system and input identification of structural dynamics PDEs *J. Buildings.* 13(3), 650.

[12] Chen X M, Duan J, Qi H, and Li Y G 2014 Rayleigh damping in Abaqus/explicit dynamic analy sis *J. AMM.* 627: 288–294.

Acknowledgments

This work is sponsored by the National Natural Science Foundation of China (Nos. 52475129 & 52375125).

Numerical investigation of the noise characteristics and localization of aeronautical pitot tubes during cruise conditions of large civil aircraft

Runpei Jiang[1, 2], Peiqing Liu[1,2], Jin Zhang[3,1*]
[1]Key Laboratory of Aero-Acoustics (Beihang University), Ministry of Industry and Information Technology, Beijing 102206, China
[2]School of Aeronautic Science and Engineering, Beihang University, Beijing 102206, China
[3]Sino-French Engineer School, Beihang University, Beijing 100191, China

*Corresponding author's e-mail: jin_zhang@buaa.edu.cn

Abstract. The pitot tube and other components on the fuselage of the large civil aircraft generate significant aerodynamic noise during transonic cruise. In this paper, the Delayed Detached Eddy Simulation method is employed to simulate the flow field of a pitot tube model mounted on the wall composed by a pitot probe and a pitot tube strut at cruising Mach numbers. The far-field noise spectrum and directivity are obtained using the Ffowcs Williams-Hawkings's integral. The results indicate that when the incoming Mach number is 0.85, the shock wave-boundary layer interaction is the primary noise source, with a frequency of approximately 600 Hz, acting as a dipole sound source. The total sound pressure level at a distance of 5 meters in the far field can reach 100 dB. Additionally, the pressure fluctuations of the probe head generate noise with a frequency exceeding 1000 Hz, which increases the total sound pressure level both upstream and downstream by 3 to 5 dB. When the Mach number of the incoming flow is 0.80, the pressure fluctuations at the probe becomes the main noise source, resulting in a total sound pressure level of 80 dB at both the upstream and lateral positions 5 meters away.

1. Introduction

As the aviation industry implements strict standards regarding cabin noise levels during the cruise phase of large civil aircraft, it is essential to conduct a more comprehensive study of the noise generated by pitot tubes and other protruding objects on the fuselage [1,2]. The aerodynamic design of a pitot tube installed on an airliner can typically be simplified to a strut featuring a symmetrical airfoil in cross-section. This strut includes a probe with a rotating body shape aligned with the direction of flow, positioned at the top. The strut exhibits symmetrical characteristics, with a forward-swept airfoil cross-section. The probe, located at the top of the strut, has its axis parallel to the direction of the incoming flow, and its head is streamlined.

Considering the strut component in isolation, a finite-span symmetrical airfoil positioned against a solid wall can serve as a model to represent the pitot tube and other related structures. At low speeds, distinct three-dimensional (3D) flow phenomena occur around the model, including horseshoe vortices at the wall-airfoil junction [3,4] and trailing edge vortices at the wingtip [5,6]. The primary noise phenomena consist of broadband noise generated by trailing edge turbulence and pressure fluctuations [7,8], as well as tonal noise produced by Tollmien-Schlichting (T-S) unstable wave diffraction at the trailing edge of the airfoil [9,10]. Furthermore, when the airfoil exhibits a sweep angle, the spanwise flow at the leading edge can also generate turbulent noise [11,12]. In the transonic cruise regime of contemporary large civil aircraft, Jiang et al. [13] observed that the interaction between the shock wave on the airfoil surface and the boundary layer at the wall-airfoil junction results in oscillatory movement of the shock wave position. This interaction gives rise to the significant phenomenon of alternating vortex shedding and dipole noise, which constitutes the primary source of noise within a specific range of incoming Mach numbers (Ma_∞). When the airfoil exhibits a substantial sweep angle of 35°, the movement of the shock wave is predominantly concentrated at the wing root, particularly at the junction, where dipole noise is most pronounced. In contrast, when the airfoil has a smaller sweep angle, the shock wave on the airfoil surface is localized at the edge and does not engage with the wall boundary layer.

The aerodynamic noise associated with the strut of the pitot tube have been clearly studied. However, research focusing on the aerodynamic noise characteristics of pitot probes in transonic flow remains limited. In this paper, the Delayed Detached Eddy Simulation (DDES) method is employed to numerically simulate the transonic flow field around a complete pitot tube (including the probe) and a pitot tube strut (a swept-forward symmetric finite-span airfoil) mounted on a wall. The far-field noise is calculated using the Ffowcs Williams-Hawkings (FW-H) acoustic analogy method. This study emphasizes the relationship between the flow characteristics near the probe and the resulting far-field noise, providing valuable insights for sound source identification related to fuselage noise and for the design of noise-reducing pitot tube shapes.

2. Methodology Setup

As illustrated in Figure 1, the two computational models utilized in this paper are the complete pitot tube model A (Figure 1a, also referred to as the model with probe) and the pitot strut model B (Figure 1b, also known as the model without probe). The cross-section of the strut is based on the NACA 0012 airfoil, with a chord length $c = 0.1$m, the height of the strut (or the airfoil span) $H = 0.15$m, and a sweep angle $\varphi = 35°$. The probe length $L = 0.25$m, the head of the probe is closed, the minimum outer diameter of the probe $d = 0.006$m and the maximum outer diameter $D = 0.02$m. In numerical calculations, the center of the probe head is designated as the coordinate origin. The direction of the flow is defined as the positive direction of the x-axis, while the direction of the wingspan corresponds to the positive direction of the z-axis. The direction perpendicular to the xz-plane is designated as the y-direction. In the following text, θ is used to indicate direction. The flow direction is defined as 0°, while the positive direction of the y-axis is 90°.

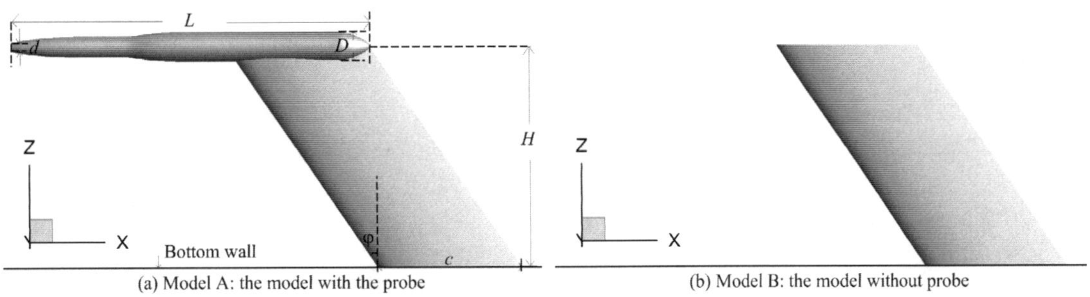

(a) Model A: the model with the probe (b) Model B: the model without probe

Figure 1. Models used in calculation.

The computational domain extends $5c$ in front of the probe, $50c$ at the rear, $10c$ on the sides, and $5c$ at the top. The grid displayed on the surface of the model, the bottom wall, and the grid located in the y = 0 plane are depicted in Figure 2. The grid is structured, with the height of the first layer meeting the criterion of $y^+ < 1$. Both the bottom surface and the model surface are treated as no-slip wall boundary conditions, and the model does not account for the pressure hole. With the exception of the downstream pressure outlet, all other boundary conditions are set to pressure far field. The mesh size of the complete pitot tube model is approximately 3.5×10^7, while the mesh size of the pillar model is about 3.3×10^7.

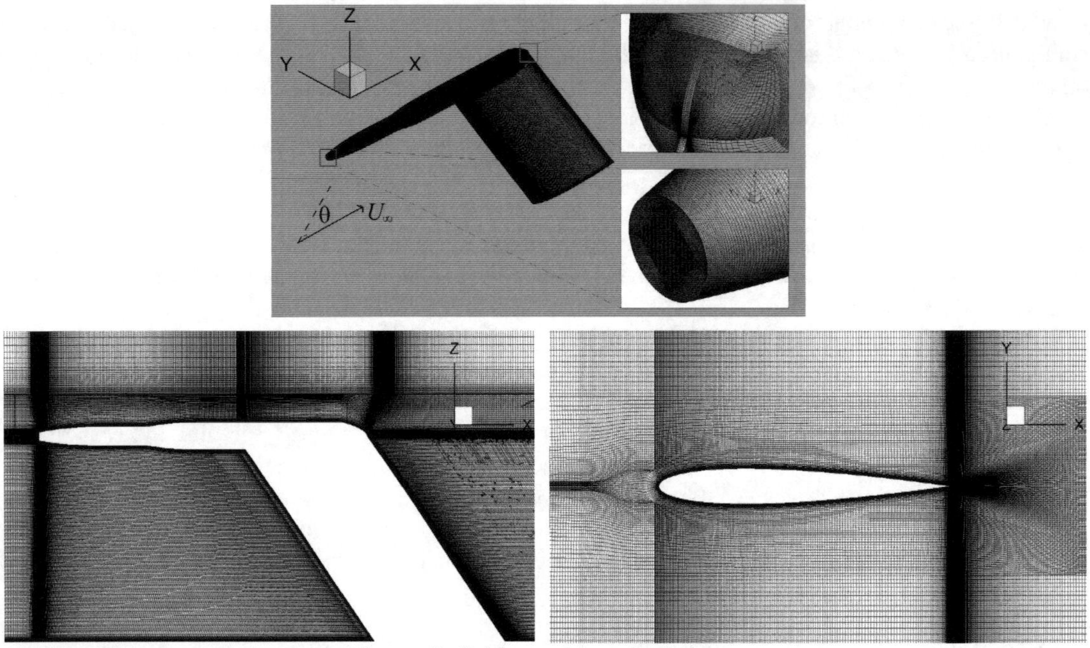

Figure 2. Grids used in calculation.

In examining the influence of Mach number, it has been observed that at $Ma_\infty = 0.80$, there is no significant interaction between the shock wave and the boundary layer at the bottom-strut connection. Conversely, at $Ma_\infty = 0.85$, this interaction is most pronounced [13]. Consequently, these two inflow velocity conditions have been selected for investigation in this study. The probe is oriented to face the direction of the incoming flow, thereby maintaining a strict angle of attack of 0° for the strut. The temperature is set to 223.15 K to replicate the environmental conditions at cruising altitude, during which the ambient sound velocity (a_0) is approximately 300 m/s. The Reynolds numbers based on the strut chord length Re_c are calculated to be 6.9×10^5 and 7.3×10^5 for the two operational conditions. For the unsteady simulation, a time step of 5×10^{-6} s is employed. Initially, a simulation with a physical duration of 0.025 s is conducted to allow the flow to reach a stable periodic state, followed by a simulation of 0.05 s to analyze the sound field and gather statistical data. The sampling frequency is set at 10 kHz, with a frequency resolution of 20 Hz. A sound source control surface is established on the model's surface, and a far-field measurement point is positioned in the z = 0 plane, 5 m (or 50c) from the center of the probe.

3. Results and discussion

3.1. Overall flow field

This section compares the flow field simulation results of Model A and Model B. Firstly, the instantaneous flow fields around the two models at $Ma_\infty = 0.85$ and 0.80 are compared in Figure 3, where the iso-surface of $Q = 1\times10^7$ is colored by the instantaneous pressure coefficient C_P. The semi-transparent purple iso-surface, $(U/a_0) \cdot(\nabla p/\|\nabla p\|) = 16$, represents the region where pressure changes dramatically within the flow field. This value serves as an indicator of the shock wave's position. Only the model surface streamlines at the junctions are depicted here. It can be observed that under the same Ma_∞, the probe has no significant influence on the flow pattern at the bottom-strut junction; however, it does increase the vorticity near the bottom surface directly below the probe. When $Ma_\infty = 0.85$, vortex shedding is clearly evident at the junction of both models, and a shock wave also appears around the rear edge of the probe. In contrast, when $Ma_\infty = 0.80$, no flow separation or vortex shedding occurs in the junction region.

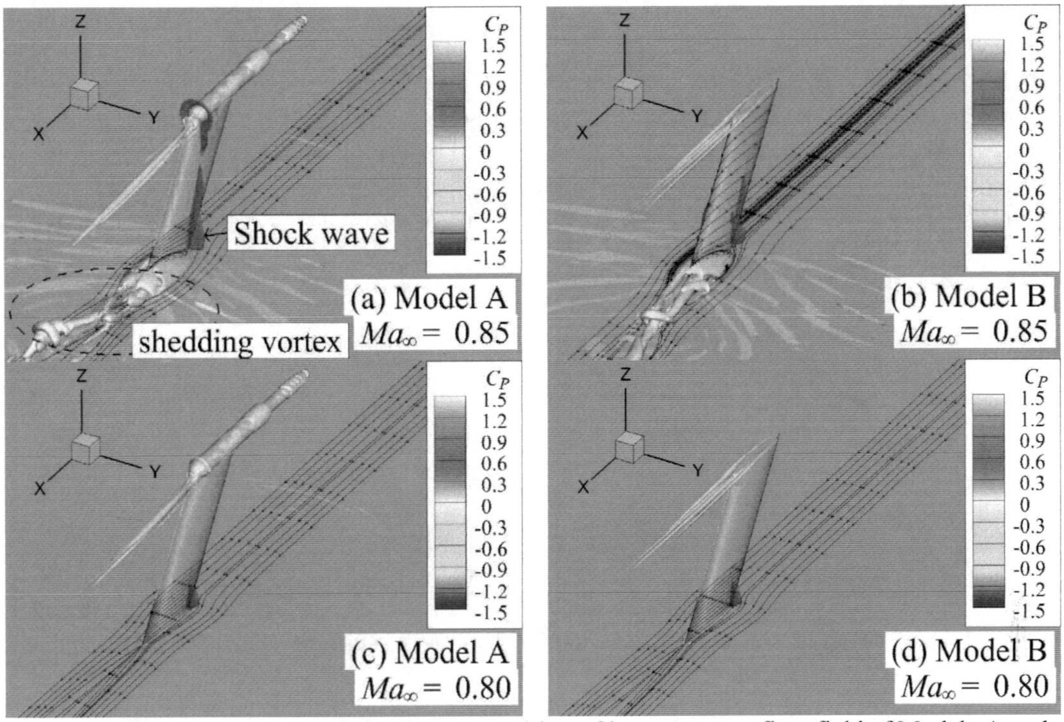

Figure 3. Vortex structure and shock wave position of instantaneous flow field of Models A and B ($Ma_\infty = 0.85$ and $Ma_\infty = 0.80$)

Figures 4 and 5 compare the time-averaged pressure coefficient $\overline{C_P}$ and the root mean square of the fluctuating pressure coefficient $C_{P_{rms}}$ on the surfaces of each case. From Figure 4, it is evident that the time-averaged pressure on the strut, both with and without the probe, is similar at $Ma_\infty = 0.85$ and 0.80. At $Ma_\infty = 0.80$, the shock wave at the wing root of the strut is perpendicular to the base surface. However, at $Ma_\infty = 0.85$, the spanwise effect of the forward-swept wing increases the intensity of the shock wave at the wing root, leading to flow separation. This results in the shock wave at the junction tilting toward the leading edge of the airfoil near the bottom surface. The pressure distribution on the probe is influenced by its shape, resulting in the formation of a shock wave at the tail of the probe. Furthermore, the presence or absence of the probe does not affect the average pressure coefficient of the strut.

In Figure 5, the logarithm of the root mean square fluctuating pressure coefficient is presented for enhanced clarity. It is evident that at $Ma_\infty = 0.85$, the higher values of fluctuating pressure on the strut surface are concentrated near the shock position and the trailing edge of the junction. This observation indicates the movement of the shock position and the occurrence of surface vortex shedding in this region. The probe model exhibits significant fluctuating pressure on its surface, particularly around the probe head. The shock wave position at the trailing edge of the probe also experiences higher pressure fluctuations; however, these fluctuations are considerably weaker than those observed at the probe head and the junction. The movement of the shock wave position in this area is not pronounced, and the interaction with the shock wave at the wing root is minimal. When $Ma_\infty = 0.80$, there is almost no pressure fluctuation on the surface of model B, while the pressure fluctuations in model A are confined to the probe and the probe-strut junction. The influence of the probe on the fluctuating pressure on the strut is limited solely to the probe-strut junction.

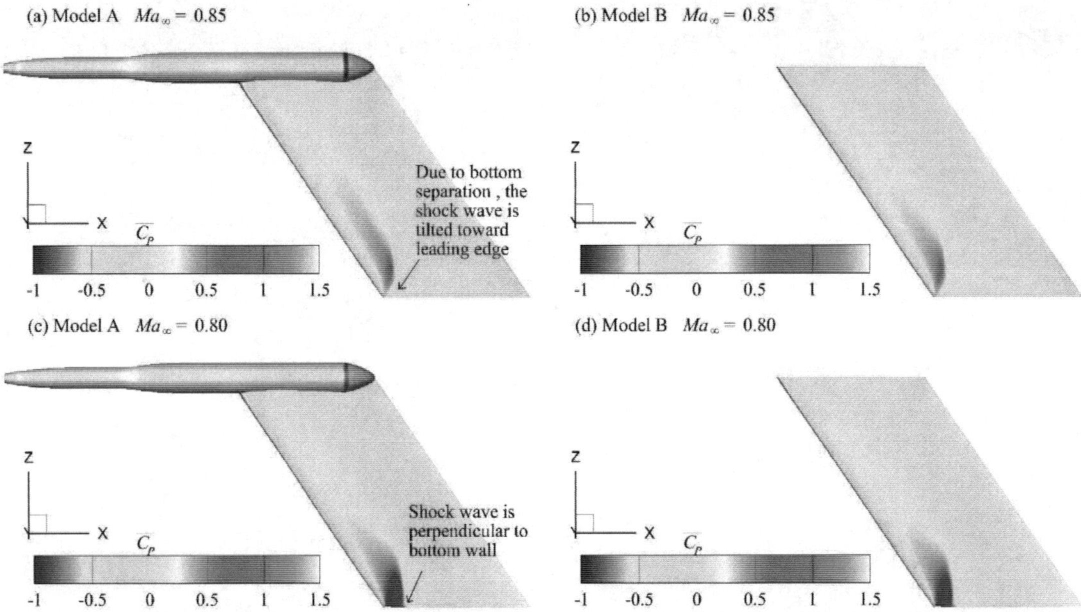

Figure 4. Time-averaged pressure coefficient on surfaces of Models A and B (Ma_∞ = 0.85 and Ma_∞ = 0.80).

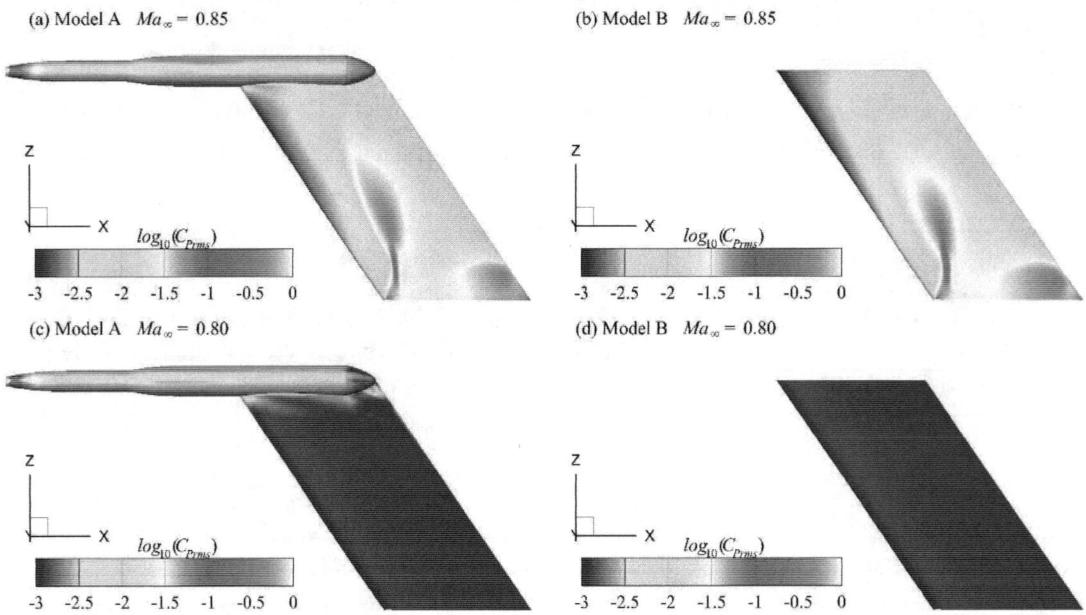

Figure 5. Logarithm of the root mean square pressure coefficient on the surfaces of Models A and B (Ma_∞ = 0.85 and Ma_∞ = 0.80).

3.2. Flow field details near probe head and tail

Figure 6 illustrates the flow field characteristics of the probe head at the $Q = 5\times10^8$ iso-surface. Under both conditions of Mach number, vortices are observed to propagate downstream along the wall adjacent to the probe. Specifically, at Ma_∞ = 0.85, irregular vortices manifest on both the upper and lower sides of the probe. It is posited that this phenomenon arises from the oscillation of the shock wave position at the pillar, which induces significant disturbances in the flow field surrounding the probe, thereby resulting in a marked disparity in the frequency of vortices on the upper and lower sides. Conversely, at Ma_∞ = 0.80, the vortices surrounding the probe exhibit a more regular pattern, and the frequency

 doi:10.1088/1742-6596/3041/1/012013

difference between the vortices on the upper and lower sides is minimal. This is attributed to the absence of pronounced shock wave position oscillation at the pillar (refer to Figures 4 and 5).

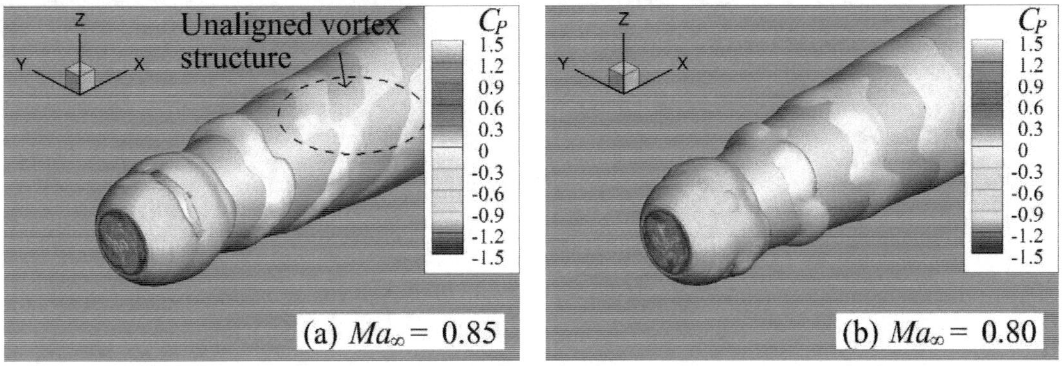

Figure 6. Detail of instantaneous flow field vortex structure at probe head.

Figure 7 presents the flow lines and Mach number distribution around the probe in the $y = 0$ plane. Under both operational conditions, separation vortices are generated at the leading edge of the probe and propagate backward along the probe wall. Notably, at $Ma_\infty = 0.85$, the vortices on either side of the probe display asymmetry, further corroborating that the vortex shedding around the probe occurs at different frequencies at various positions on both sides.

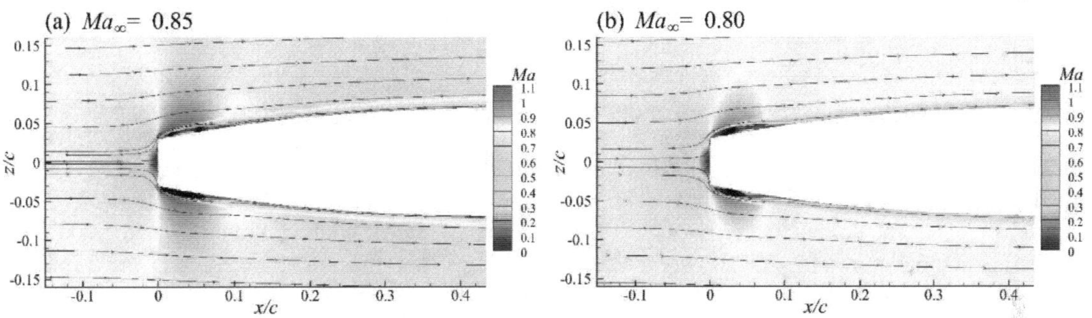

Figure 7. Mach number and instantaneous streamline distribution near probe head in $y = 0$ plane.

Figures 8 and 9 illustrate the flow field details at the tail of the probe. Figure 8 presents the iso-surface of $Q = 5 \times 10^8$. It is evident that when $Ma_\infty = 0.85$, irregular vortex shedding occurs near the tail end, whereas no vortex shedding is observed behind the shock wave. In contrast, when $Ma_\infty = 0.80$, the vortex shedding at the tail is significantly diminished.

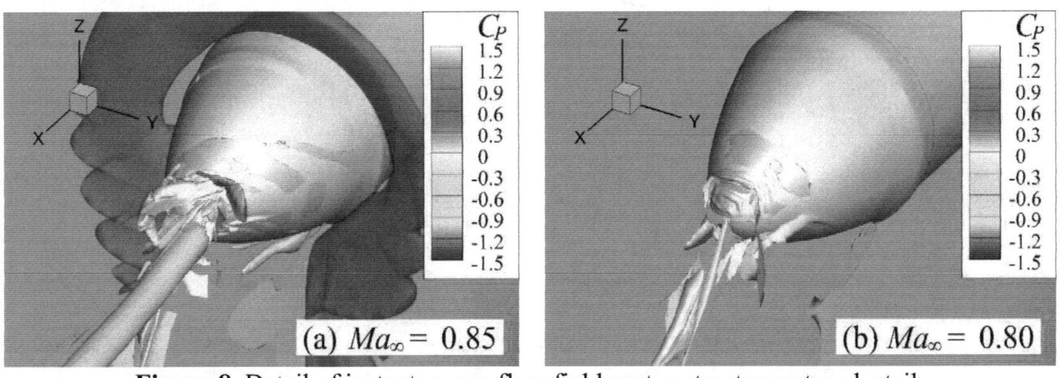

Figure 8. Detail of instantaneous flow field vortex structures at probe tail.

The flow lines and Mach number distribution in the $z = 0$ plane are depicted in Figure 9. It is apparent that at $Ma_\infty = 0.85$, a complex and asymmetric vortex shedding phenomenon is present at the tail of the probe. However, at $Ma_\infty = 0.80$, there is no noticeable vortex shedding at the tail, and only a limited, attached separation vortex exists near the tail.

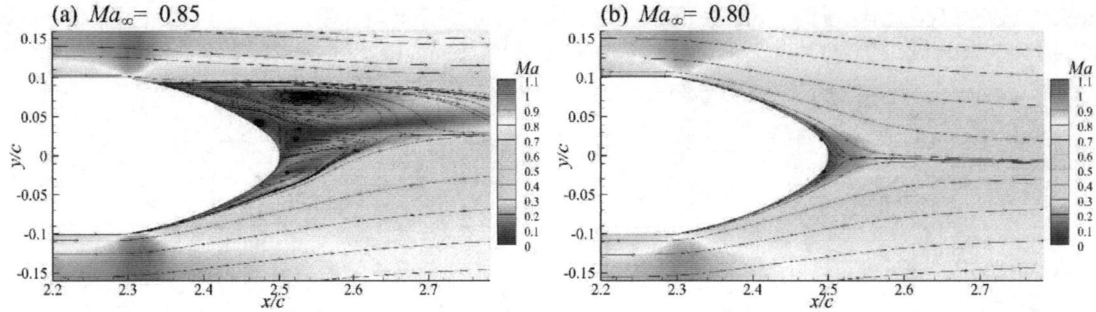

Figure 9. Mach number and instantaneous streamline distribution near probe tail in $z = 0$ plane.

To summarize, there are two primary sources of pressure fluctuations on the surface of the pitot tube in transonic flow. The first source is the probe itself, where pressure fluctuations predominantly occur at the head section. At a Mach number of $Ma_\infty = 0.85$, a notable disparity is observed in the frequency of vortices on the upper and lower surfaces of the probe. The second source relates to the oscillation of the shock wave position at the bottom-strut junction, as well as the vortex shedding occurring at the trailing edge of the strut; this phenomenon is present only at $Ma_\infty = 0.85$ and not at $Ma_\infty = 0.80$. Furthermore, the interaction between the probe and the pressure fluctuations at the bottom-strut junction is not pronounced. The presence or absence of the probe has minimal impact on the average surface pressure and fluctuation pressure at the bottom-strut junction, with significant effects observable only at the probe-strut junction.

3.3. Far-field noise directivity and spectrum

Figure 10 compares the far-field noise overall sound pressure level (OASPL) directivity of the two models at $Ma_\infty = 0.85$ and 0.80. When $Ma_\infty = 0.85$, the OASPL of Model A is 5 dB and 3 dB greater downstream ($\theta = 0°$) and upstream ($\theta = 180°$), respectively, compared to Model B. Additionally, Model A exhibits an OASPL that is approximately 1 dB higher in the side directions ($\theta = 90°$ and 270°) than that of Model B. The noise directivity of both models displays dipole characteristics. At $Ma_\infty = 0.80$, the OASPL of Model A is about 20 dB higher in all directions than that of Model B, and the directivity curve of Model B shows distinct serrations. An analysis of surface pressure fluctuations indicates that the strut does not generate significant noise, while the noise from the probe primarily originates from the probe head. Furthermore, at $\theta = 0°$ and 180°, the OASPL of Model A at $Ma_\infty = 0.80$ is approximately 2 dB and 6 dB lower than that at $Ma_\infty = 0.85$, with the difference exceeding 20 dB at $\theta = 90°$ and 270°.

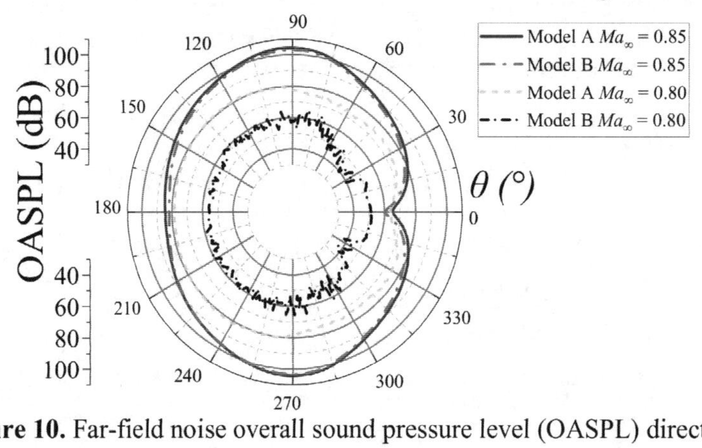

Figure 10. Far-field noise overall sound pressure level (OASPL) directivity.

Figure 11 compares the noise power spectral density (PSD) at a measurement point located 5 meters away in the direction of $\theta = 90°$. When $Ma_\infty = 0.80$, the noise from the probe-free model is negligible across all frequency bands. Although the model with the probe also exhibits low noise levels, it is significantly higher than that of the probe-free model, particularly in the frequency range above 1000

Hz, with a peak occurring near 3100 Hz. The corresponding Strouhal numbers $St = fl/U_\infty$, where f is the frequency, U_∞ is the incoming flow velocity, and l is the reference length. By using the strut airfoil chord length c, thickness $T = 0.12c = 0.012$m, d and D as reference lengths, they are respectively $St_c = 1.192$, $St_T = 0.143$, $St_d = 0.072$, and $St_D = 0.238$. When Ma_∞ increases to 0.85, the spectra of the two models become relatively similar. The peak value for the probe model occurs at 600 Hz, which is approximately 5 dB/Hz higher than that of the non-probe model at 570 Hz, corresponding to Strouhal numbers of $St_c = 0.235$, $St_T = 0.028$, $St_d = 0.014$, and $St_D = 0.047$. In addition, the spectral density of both models increases when f = 1100 Hz, and the spectral density of Model B being higher than that of Model A at this frequency.

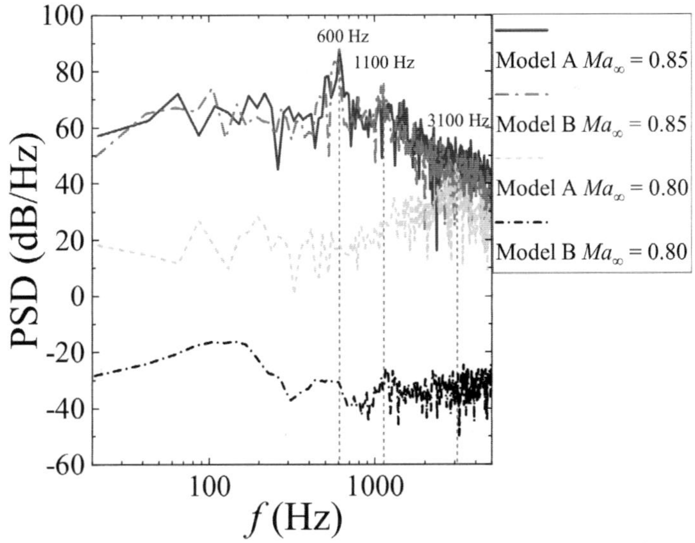

Figure 11. Power spectral density (PSD) of far-field noise at $\theta = 90°$.

By comparing the noise characteristics and flow field, it is evident that when $Ma_\infty = 0.85$, a peak noise frequency of approximately 600 Hz is produced by vortex shedding at the bottom-strut junction and shock boundary layer interference. The noise generated by the probe is distributed across a frequency range above 1000 Hz, with the most pronounced frequency occurring around 3100 Hz.

4. Conclusion

In this paper, DDES is employed to simulate the flow field around a pitot tube model under the cruise conditions of large civil aircraft. Far-field noise is obtained by FW-H equation. The characteristics and source locations of the noise are analyzed, leading to the following conclusions:

(1) At $Ma_\infty = 0.85$, the interaction between the shock wave and the boundary layer in the bottom-strut junction becomes pronounced, leading to vortex shedding and the generation of sonorous dipole noise around 600 Hz in the far field. This phenomenon is less noticeable when the incoming Mach number is 0.80.

(2) When Ma_∞ is in the range of 0.80 to 0.85, the probe head generates significant pressure fluctuations. This phenomenon corresponds to frequency bands above 1000 Hz, especially around 3100 Hz, in the far-field noise. The resulting noise propagates in all directions, increasing noise levels of about 3dB both upstream and downstream of the incoming flow.

(3) The flow coupling between the probe and the strut is not obvious and there is no significant noise phenomenon associated with it. While the probe tail may generate vortex shedding, it also does not produce any notable noise phenomena.

5. References

[1] Hu N, Buchholz H, Herr M, Spehr C and Haxter S 2013 Contributions of different aeroacoustic sources to aircraft cabin noise *19th AIAA/CEAS Aeroacoustics Conference* (Berlin: German Aerospace Center) AIAA 2013-2030

Journal of Physics: Conference Series **3041** (2025) 012013 doi:10.1088/1742-6596/3041/1/012013

[2] Hassan M, Tai J C, Denney R K and Benjamin Havrilesko B 2014 En route jet aircraft noise analysis *AIAA/3AF Aircraft Noise and Emissions Reduction Symposium* (Atlanta: American Institute of Aeronautics and Astronautics) AIAA 2014-3163
[3] Mehta R 1984 Effect of wing nose shape on the flow in a wing body junction *Aeronaut. J.* **88** 456
[4] Simpson R L 2001 Junction flows *Annu. Rev. Fluid Mech.* **33** 415–443
[5] Giuni M and Green R B 2013 Vortex formation on squared and rounded tip *Aerosp. Sci. Technol.* **29** 191–199.
[6] Karakus C, Akilli H and Sahin B 2008 Formation, structure, and development of the near-field wing tip vortices *P. I. MECH. ENG. G-J. AER.* **222** 13–22
[7] Moreau D J, Brooks L A and Doolan C J. 2011 Broadband trailing edge noise from a sharp-edged strut *J. Acoust. Soc. Am.* **129** 2820–2829
[8] Arbey H and Bataille J 1983 Noise generated by airfoil profiles placed in a uniform laminar flow *J. Fluid Mech.* **134** 33–47
[9] Moreau D J, Prime Z, Porteous R and Valeau V 2014 Flow-induced noise of a wall-mounted finite airfoil at low-to-moderate Reynolds number *J. Sound Vib.* **333** 6924–6941
[10] Moreau D J and Doolan C J 2016 Tonal noise production from a wall-mounted finite airfoil *J. Sound Vib.* **363** 199–224
[11] Giez J, Vion L, Roger M and Moreau S 2016 Effect of the edge and tip vortex on airfoil selfnoise and Turbulence Impingement Noise *22nd AIAA/CEAS Aeroacoustics Conference* (Lyon: American Institute of Aeronautics and Astronautics) AIAA 2016-2996
[12] Yakhina G R, Roger M, Finez A, Bouley S, Baron V, Moreau S and Giez J 2020 Localization of swept free-tip airfoil noise sources by microphone array processing. *AIAA J.* **58**(8) 3414-3425
[13] Jiang R P, Liu P Q, Zhang J and Guo H. 2024 Numerical study on far-field noise characteristic generated by wall-mounted swept finite-span airfoil within transonic flow *Aerospace-Basel* **11**(8) 645

Numerical study on the suppression and separation boundary layer of low-pressure turbine wavy suction surface

Qi Wang[1,2], Timothy H. Lee[1,2], Pan Liu[1,3], Bangwei Liu[1], Yutao Zhuang[1] and Jiahuan Cui[1,4,*]

[1]ZJU-UIUC Institute, Zhejiang University, Haining, 314400, China
[2]College of Energy Engineering, Zhejiang University, Hangzhou, 310027, China
[3]Shenyang Aircraft Design and Research Institute, Shenyang, 110066, China
[4]School of Aeronautics and Astronautics, Zhejiang University, Hangzhou, 310007, China

*Corresponding author: ZJU-UIUC Institute, Zhejiang University, Haining, 314400, Zhejiang Province, China. E-mail address: jiahuancui@intl.zju.edu.cn (J. Cui)

Abstract. This study examines the effect of a wavy suction surface on boundary layer separation and its performance under mainstream turbulence disturbances.The feasibility and reliability of the flat plate simulation were validated by comparing the simulation results with experimental data from Zhang et al.The study optimized the dimensions of the wavy suction surface, drawing on the work of Liu et al. on blade cascades. The results indicate that a wavy suction surface with an amplitude of 0.9 mm, a wavelength of 50 mm, and a maximum amplitude at x/l = 0.68 most effectively suppresses boundary layer separation. Under 5% mainstream turbulence, the flat plate boundary layer undergoes premature transition, which partially mitigates separation, moderate turbulence helps to reduce separation losses. The wavy suction surface generates a counter-rotating vortex structure, which enhances the energy exchange between the mainstream flow and the boundary layer, thereby further suppressing separation and reducing losses. Even with a turbulence intensity of 5%, the wavy suction surface effectively reduces boundary layer separation and associated losses.These findings provide a theoretical foundation for the application of wavy suction surfaces in low-pressure turbine blades, highlighting their potential to control separation and reduce vortex-induced losses.

1. Introduction

The low-pressure turbine (LPT) constitutes approximately 30% of an aircraft's total weight and 15% of its total cost[1], making it a critical component of aircraft engines. Research has shown that reducing the weight of various turbofan engine components by 10% can significantly reduce the operating costs of civil aircraft. Furthermore, the efficiency of the low-pressure turbine has a direct impact on engine fuel consumption. Consequently, reducing the weight of the LPT while simultaneously improving its efficiency has become a key focus in engine design. Over the past decade, high-load low-pressure turbines have been developed to reduce weight while maintaining the same level of efficiency. However, as the load on the turbine increases, boundary layer separation becomes more pronounced, especially at higher altitudes where Reynolds numbers are lower. Large separation bubbles on the suction surface can lead to substantial losses, and if the flow does not reattach to the suction surface before reaching the trailing edge, it can deviate from the blades, negatively affecting the load and efficiency of downstream turbine blades. Therefore, effective control of the boundary layer separation on the suction surface is essential.

Due to the significant impact of suction surface separation bubbles on aerodynamic performance, extensive research has been conducted on methods to control these separation bubbles. These studies are generally categorized into active and passive control techniques. Compared to active control, passive control methods do not require external energy input. Zhang et al.[2] investigated various methods for suppressing separation bubbles on the suction surfaces of ultra-high-load high-pressure turbines. They found that using rear-loaded blades and implementing passive control measures, such as strategically placed protrusions on the suction surface, could reduce the total pressure loss of ultra-high-load blades to levels comparable to traditional designs. Early research on leading-edge modifications was inspired by humpback whales, where sinusoidal protrusions on the leading edge of whale fins were found to enhance maneuverability[3][4][5]. Subsequent studies sought to verify the aerodynamic benefits of such undulations on airfoils. These studies revealed that reverse-rotating vortices formed between the undulations play a crucial role in improving aerodynamic performance[6]. Research has shown that applying such undulations to the leading edge of an airfoil can enhance its performance, and based on these findings, the reverse-rotating vortices generated by a wavy suction surface can effectively suppress boundary layer separation on turbine blades. Liu et al.[7] demonstrated that undulating suction surfaces can significantly reduce boundary layer separation in high-load, low-pressure turbines.

Due to the high cost and complexity of testing with annular cascades, most studies on low-pressure turbine flow mechanisms use planar cascades. Although planar cascades simplify the geometry compared to annular cascades, boundary layer measurements are more challenging due to the absence of a curved surface. To overcome this, researchers have proposed using flat plates to model boundary layer development on planar cascades[8][9], as shown in Figure 2. The goal of this study is to investigate the inhibitory effect of a wavy suction surface on boundary layer separation through numerical simulations.

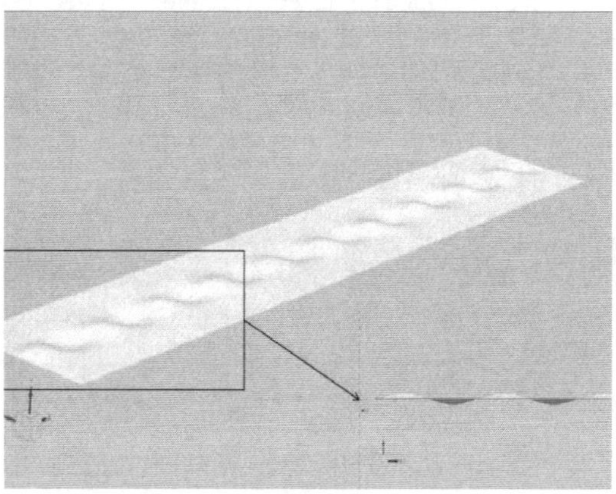

Figure 1. Schematic diagram of wavy suction surface

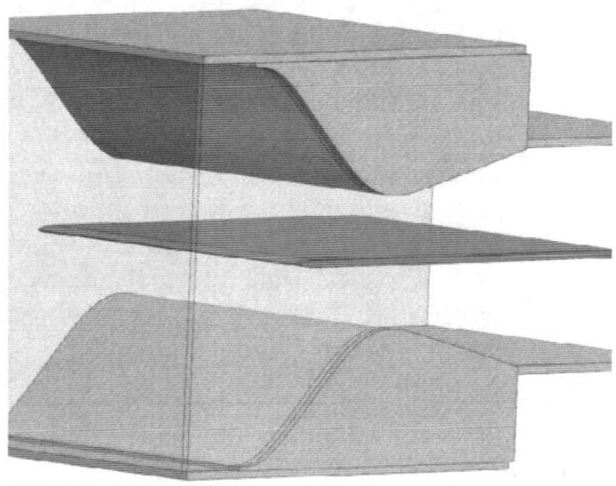

Figure 2. Flat plate modeling test bench

2. Numerical calculation framework

2.1 Computation model

The research object of this article is the flat plate model of the T106C blade cascade model of the ultra-high load low-pressure turbine[10] obtained after flat plate modeling treatment. The specifications and parameters of the tablet are shown in Table 2.1, and the schematic diagram is shown in Figure 3. The specific parameters of the wavy suction surface added to the tablet in this article will be detailed in Section 4.

Journal of Physics: Conference Series **3041** (2025) 012014 doi:10.1088/1742-6596/3041/1/012014

Table 2.1. Parameters of the flat model

Parameters	Value
Reynolds number (Re)	1.3×10^5-3.9×10^5
Board length (C)	570 mm
Effective board length (Cx)	367 mm
Width (L)	500 mm
Thickness (h)	10 mm
Turbulence intensity (Tu)	0.5%,5%

2.2 Boundary conditions

The computational example's inlet boundary is situated roughly 0.5 Cx upstream of the plate's leading edge. The entrance boundary's influence from the endwall is confined to the vicinity of the proximal wall region. To attain a Reynolds number of $1.3\text{x}10^5$, based on the effective plate length and exit velocity, an appropriate mainstream velocity should be applied at the entrance. The outlet boundary is positioned approximately 0.3 Cx downstream of the blade's trailing edge. Non-slip wall conditions were implemented on the flat surface and end walls, while periodic boundary conditions were applied in the pitch direction.

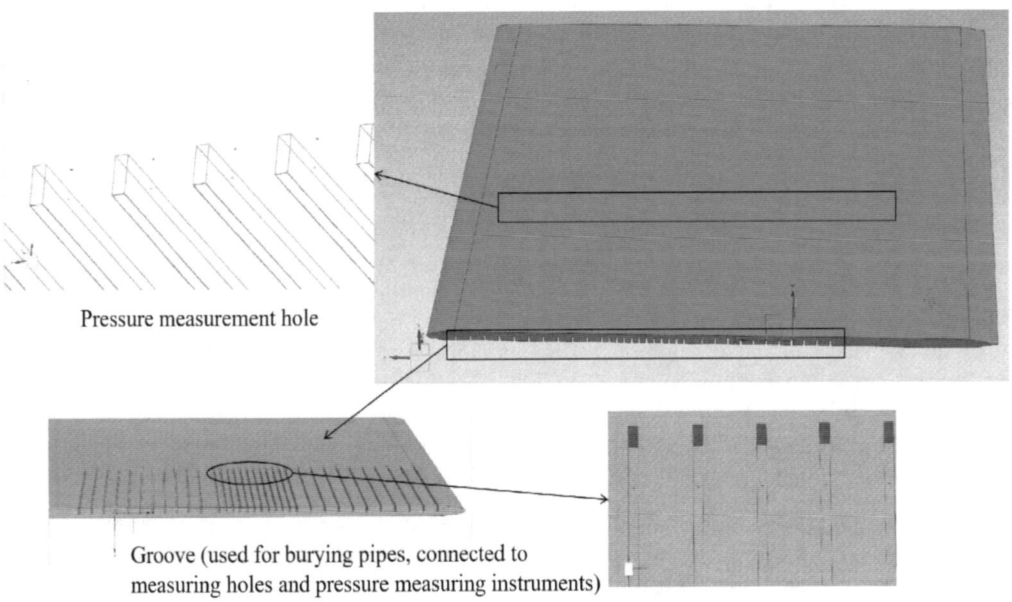

Figure 3. Schematic diagram of flat plate

2.3 Mesh and solver model

The computational domain is discretized using a high-quality structured grid, comprising approximately 520,000 elements, as shown in Figure 4.

ANSYS-CFX software based on the finite volume method for numerical calculations was used in this article. In order to predict flow separation and boundary layer transition phenomena at low Reynolds numbers, a transition SST model combining the Gamma Theta transition model[11] and SST turbulence model[12] will be used for simulation calculations in steady-state mode.

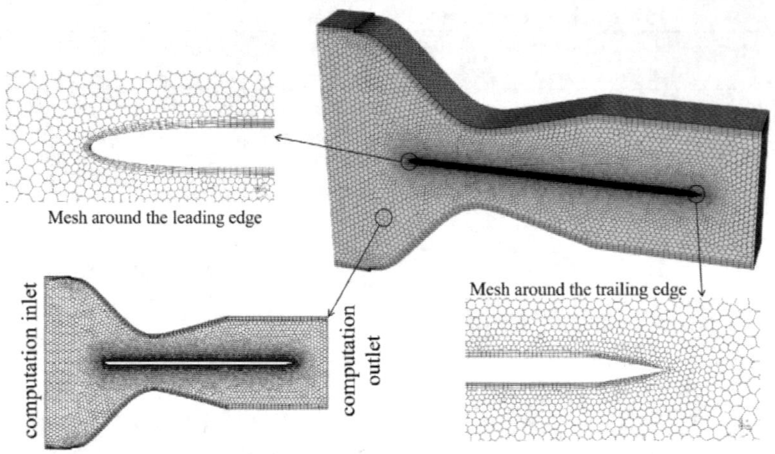

Figure 4. Distribution of three-dimensional computational grid

2.4 Grid independence verification

The grid that has not been refined is referred to as the original grid. For the original grid, approximately 500,000 grid cells were used for both the flat plate and wavy flat plate cases. To assess grid independence, the original grid was refined, and a coarser grid with fewer cells was also tested. The refined grids consisted of approximately 1 million and 4 million cells, while the coarser grid had around 320,000 cells. Grid independence will be verified by comparing the pressure coefficient (Cp) distributions. The calculation formula for Cp is as follows:

$$C_p = \frac{P_{oin} - P}{P_{oin} - P_{TE}} \tag{3.1}$$

Among them, P_{oin} is the total inlet pressure measured 0.1 times the effective plate length from the leading edge, P is the static pressure on the flat plate, and P_{TE} is the static pressure measured 1.3 times the effective plate length from the leading edge, near the trailing edge.

Taking a Reynolds number of 3.9×10^5 as an example, Figure 5 compares the Cp distribution along the flat plate surface under the same operating conditions for four different grid resolutions.It is clear that the Cp distribution remains virtually unchanged after modifying the grid. Analysis shows that the differences are less than 1%, indicating that the results are no longer sensitive to grid resolution and thus grid independence has been achieved.Furthermore, the reliability of the simulation data has been validated through comparison with experimental results.

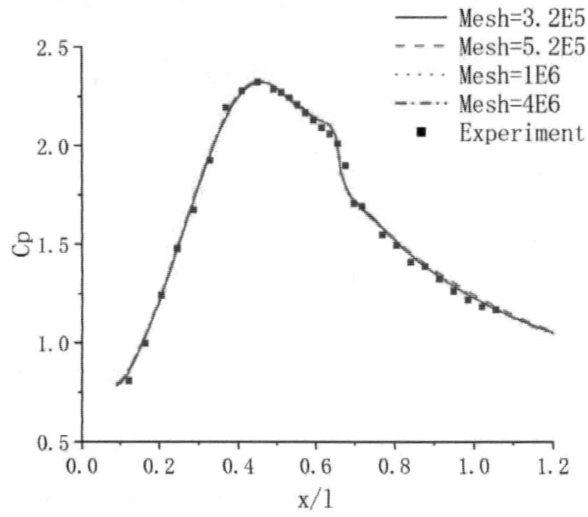

Figure 5. Grid independence verification pressure coefficient distribution curve

3. Results and discussion of flat plate test

The inlet velocities of 5 m/s, 8 m/s, and 12 m/s are applied, resulting in a Reynolds number range of 1.3×10^5 to 3.9×10^5. Within this range, significant flow separation is observed on the flat plate, accompanied by distinct separation bubbles, which provides a conducive environment for further study.

The separation bubble on the suction side of low-pressure turbine blades is a primary source of losses within the blade passage. The presence or absence of these separation bubbles can be identified by examining the Cp distribution trend on the blade surface. Figure 6 compares the pressure coefficient distributions on the flat plate at different Reynolds numbers. Zhang et al.[9] performed experimental tests using a flat plate simulation setup. At the same Reynolds number, the pressure coefficient distribution on the plate closely matched the experimental results, validating the feasibility and reliability of the flat plate simulation.

From the figure, it is evident that, for all three Reynolds numbers, a "prominent platform" which is the thing between 0.6 and 0.8 for x/l appears in the pressure coefficient distribution. This indicates boundary layer separation on the suction side during the flow process. In the separated region, the pressure is lower than in the region away from the separation, creating a negative pressure zone. This negative pressure generates suction, applying an outward force on the plate surface and significantly reducing static pressure within that region. As a result, the rate of decrease in the pressure coefficient becomes slower in this area. If transition occurs in the boundary layer, the negative pressure in the separated region decreases, the static pressure on the surface increases, and the rate of decrease in the pressure coefficient accelerates, forming the "prominent platform."

As the Reynolds number increases, the position of the separation bubble shifts toward the flow direction, and its size gradually decreases. This suggests that at higher Reynolds numbers, the flow becomes more stable, and the degree of flow separation diminishes.

Figure 7 compares the pressure coefficient distributions on the flat plate at the same Reynolds number but with varying turbulence intensities. At a turbulence intensity of 5%, the "prominent platform" caused by flow separation almost disappears. This suggests that mainstream turbulence, as a passive control method for ultra-high-load low-pressure turbines, can effectively suppress boundary layer separation by promoting an earlier transition in the boundary layer. As a result, this helps reduce flow separation and minimize losses.

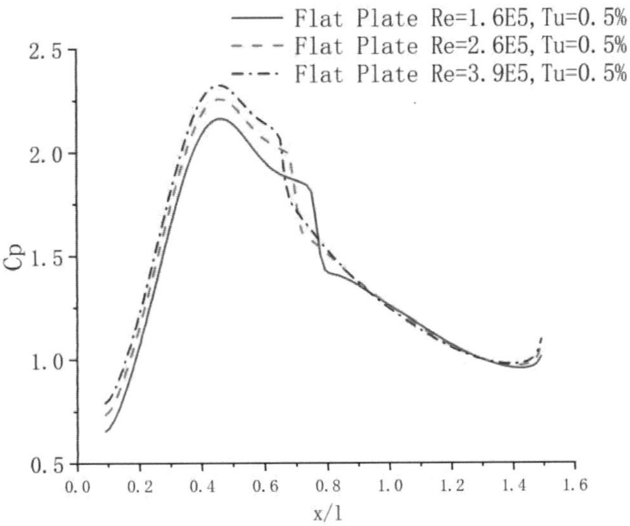

Figure 6. Differences in pressure coefficient distribution on flat plates at different Reynolds numbers

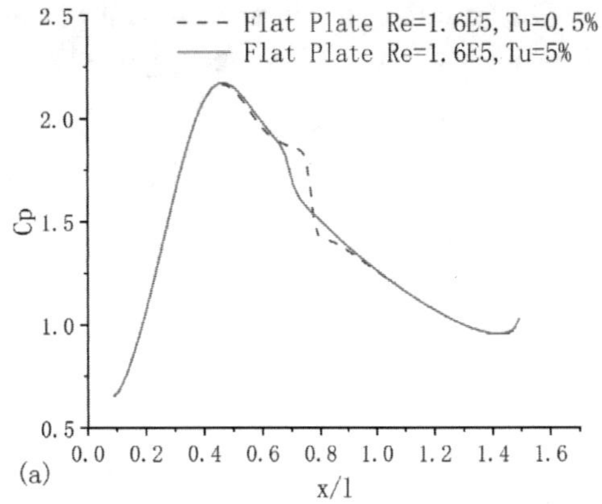

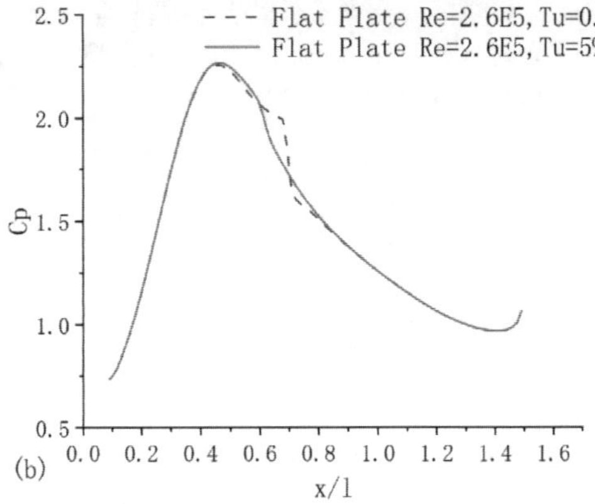

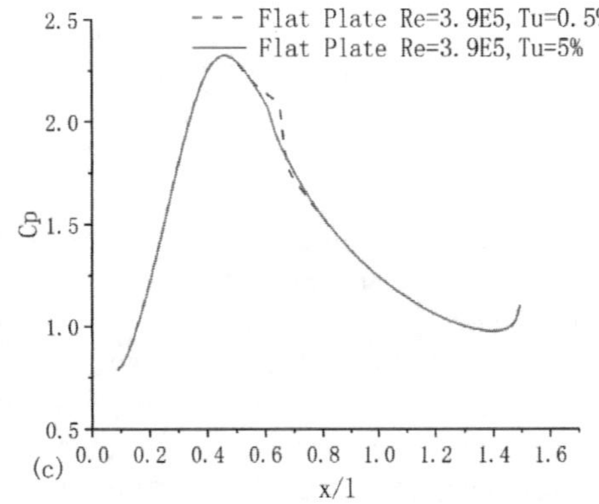

Figure 7. Differences in pressure coefficient distribution of flat plates with different turbulence intensities under the same Reynolds number

4. Results and discussion of flat plate test with wave shaped suction surface

Liu et al.[7] have carried out extensive numerical simulations on wavy suction surfaces using a blade cascade model and developed a well-parameterized model for the wavy suction surface. The flat plate model used in this study is derived from the T106C blade flat plate simulation, adopting the wavy shape from their work. The model is normalized by the effective blade length, with the maximum amplitude of the wavy suction surface occurring at x/l = 0.68. The amplitude, A, is set to 2.4 mm, and the wavelength is 50 mm, as shown in Figure 8.

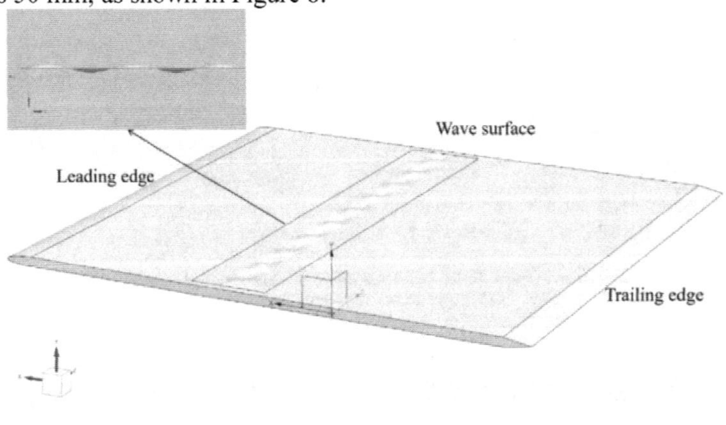

Figure 8. Schematic diagram of wavy plate

Simulations were conducted using these parameters and compared with the case where no wavy suction surface was applied, as shown in Figure 9. It is clear that adding the wavy suction surface introduces a significant and unreasonable negative pressure region. This occurs because the established wavy amplitude is too large, causing higher velocities at the wave crest, which results in a low-pressure zone, while the trough of the wave leads to flow separation. This alters the flow path and creates local low-pressure regions.

To address this, the amplitude parameter was optimized by reducing the amplitude in 0.5 mm increments. Figure 10 compares the pressure distribution of the wavy suction surface for amplitudes ranging from 0.9 mm to 2.4 mm. It can be seen that, when the amplitude is reduced to 0.9 mm, the negative pressure region caused by an overly large amplitude disappears, and the negative pressure region resulting from flow separation gradually vanishes as well.

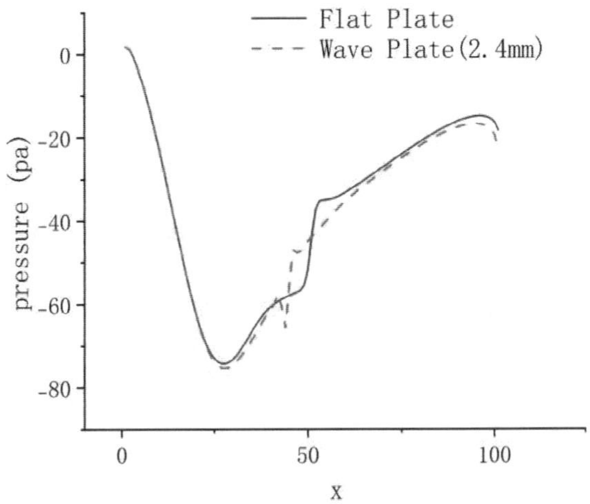

Figure 9. The difference in pressure distribution between a flat plate and a wave shaped suction surface with an amplitude of 2.4 mm added

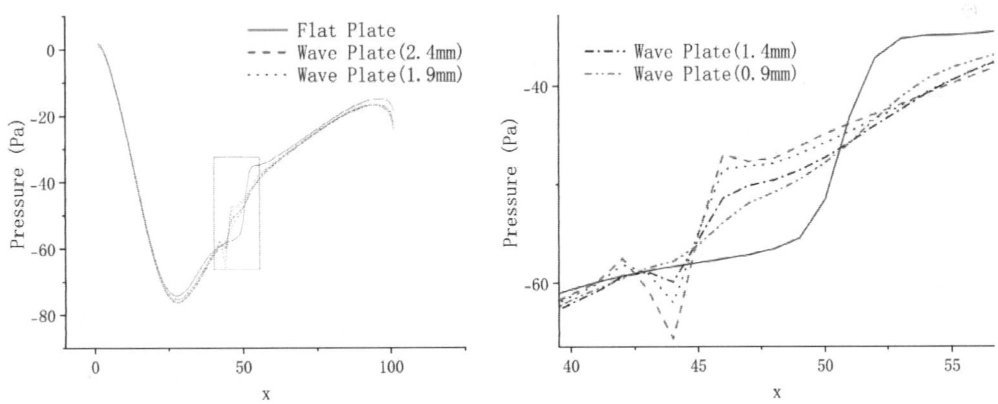

Figure 10. Left: Differences in pressure distribution on wave plates with wave suction surfaces of different amplitudes. Right: Enlarged view of the highlighted section in red

The subsequent work was conducted using a wavy suction surface with an amplitude of 0.9 mm and a wavelength of 50 mm. To mitigate the influence of wall-end effects in the wavy suction surface control scheme, spanwise averaging was applied. Figure 11 compares the pressure coefficient distributions on the wavy flat plate at different Reynolds numbers. It is clear that the "prominent platform" almost disappears under the influence of the wavy suction surface.

The wavy suction surface generates an induced counter-rotating vortex structure, which introduces instability into the boundary layer flow. This significantly advances the transition point and effectively suppresses flow separation in the boundary layer, improving overall flow performance.

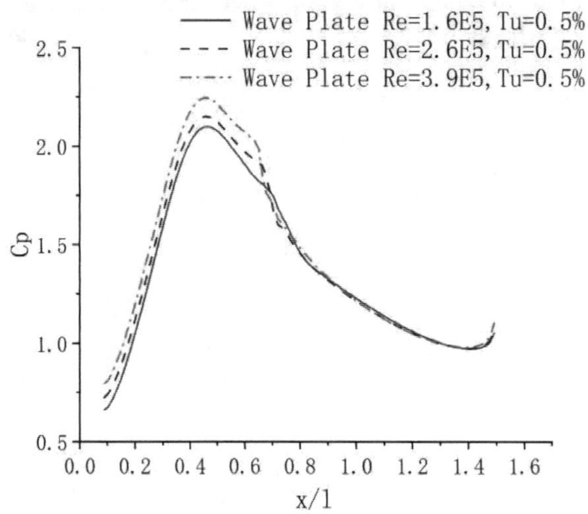

Figure 11. Differences in pressure coefficient distribution on wave plates at different Reynolds numbers

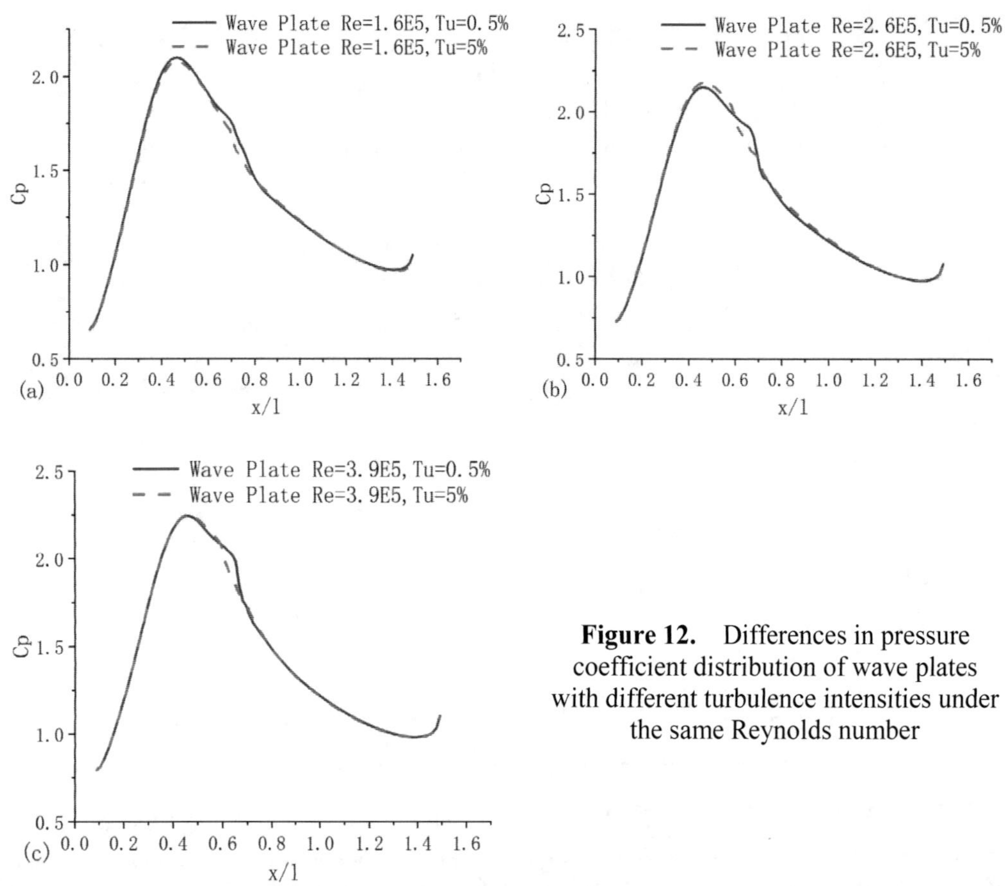

Figure 12. Differences in pressure coefficient distribution of wave plates with different turbulence intensities under the same Reynolds number

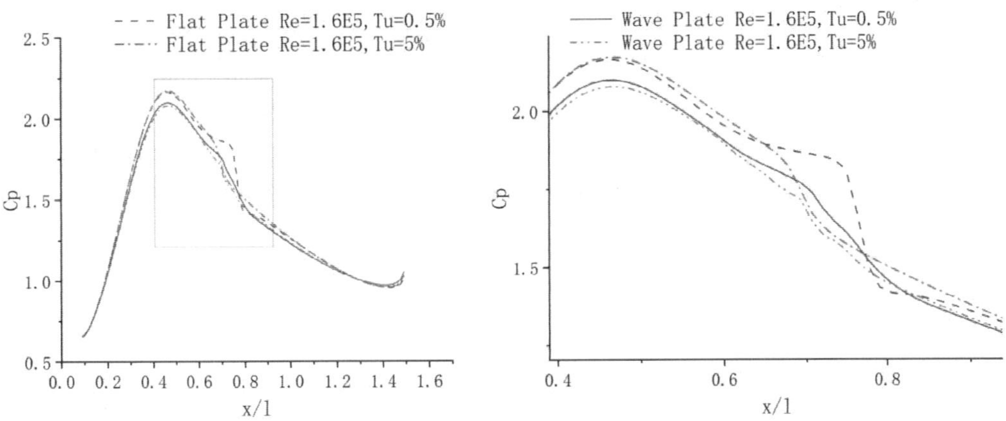

Figure 13. Left: Differences in pressure coefficient distribution of wave plates with different turbulence intensities under the same Reynolds number Right: Enlarged view of the section highlighted in red

Figure 12 compares the pressure coefficient distributions on the wavy flat plate under different turbulence intensities. Taking a Reynolds number of 1.6×10^5 as an example, Figure 13 presents a comparison of the pressure coefficient distributions on both the flat plate and the wavy flat plate at varying turbulence intensities. It is clear that, even in the highly disturbed environment created by mainstream turbulence, the wavy suction surface continues to effectively suppress boundary layer separation.

5. Conclusion

This paper first validates the feasibility and reliability of the flat plate simulation experiments by comparing the simulation results with the experiments conducted by Zhang et al. in Section 3. The analysis focuses on the impact of mainstream turbulence on boundary layer separation before adding the wavy suction surface. Next, by incorporating the wavy suction surface and referencing the parameters used by Liu et al. in blade cascades, the optimal dimensions for the wavy suction surface on the flat plate are determined. Finally, the effect of the wavy suction surface on boundary layer separation is compared and analyzed, particularly under the strong disturbances of mainstream turbulence. The conclusions are as follows:

1. By normalizing with the effective blade length, the wavy suction surface with an amplitude of 0.9 mm, a wavelength of 50 mm, and a maximum amplitude at $x/l = 0.68$ provides the most effective suppression of boundary layer separation.
2. In the absence of a wavy suction surface, under 5% mainstream turbulence intensity, the flat plate boundary layer experiences premature transition, leading to partial suppression of boundary layer separation. Moderate incoming turbulence can help suppress separation and reduce separation losses.
3. In the absence of increased mainstream turbulence intensity, the wavy suction surface generates a counter-rotating vortex structure, which enhances energy exchange between the mainstream flow and the boundary layer, effectively suppressing boundary layer separation and reducing separation losses.
4. Even under the strong turbulence environment with 5% turbulence intensity, the wavy suction surface continues to effectively suppress boundary layer separation and reduce separation losses.

References

[1] D. Wisler, The technical and economic relevance of understanding blade row interactions effects in turbomachinery, Von Karman Inst. *Fluid Dynamis* Lect. Ser. (1998).

[2] Vera M, Zhang X F, Hodson H, Harvey N. Separation and transition control on an aft-loaded ultra-high-lift LP turbine blade at low Reynolds numbers: High-speed validation[J]. *Journal of Turbomachinery*, 2007, **129**(2): 340 - 347.

[3] F. Fish, J. Battle, Hydrodynamic Design of the Humpback Flipper, vol. **225**, 1995,

https://doi.org/10.1002/jmor.1052250105.
[4] J. Favier, A. Pinelli, U. Piomelli, Control of the separated flow around an airfoil using a wavy leading edge inspired by humpback whale flippers, Compt. Rendus Mec. **340** (1) (2012) 107–114, https://doi.org/10.1016/j.crme.2011.11.004.
[5] M. Zhang M, F. Wang G, Z. Xu J, Aerodynamic Control of Low-Reynolds-Number Airfoil with Leading-Edge Protuberances, vol. **51**, 2013, https://doi.org/10.2514/1.J052319.
[6] K.L. Hansen, R.M. Kelso, B.B. Dally, Performance variations of leading-edge tubercles for distinct airfoil profiles, AIAA J. **49** (1) (2011) 185–194, https://doi.org/10.2514/1.J050631.
[7] Liu, P., Sun, W., Zhou, Y., Yan, Y., & Cui, J. (2023). Numerical study of flow in low-pressure turbines with undulated suction surface. *Case Studies in Thermal Engineering*, **51**, Article 103595. https://doi.org/10.1016/j.csite.2023.103595
[8] Coull J D, Hodson H P. Unsteady boundary-layer transition in low-pressure turbines[J]. Journal of Fluid Mechanics, 2011, **681**: 370 - 410.
[9] Zhang X F, Vera M, Hodson H, Harvey N. Separation and transition control on an aft-loaded ultra-high-lift LP turbine blade at low reynolds numbers: Low-speed investigation[J]. *Journal of Turbomachinery*, 2006, **128**(3): 517–527.
[10] X.F. Zhang, Separation and Transition Control on Ultra-high-lift Low Pressure Turbine Blades in Unsteady Flow[D].Cambridge University. 2005.
[11] R.B. Langtry, F.R. Menter, Correlation-based transition modeling for unstructured parallelized computational fluid dynamics codes, AIAA J. **47** (12) (2009) 2894–2906, https://doi.org/10.2514/1.42362.
[12] F.R. Menter, Two-equation eddy-viscosity turbulence models for engineering applications, AIAA J. **32** (8) (1994) 1598–1605, https://doi.org/10.2514/3.12149.

Acknowledgments

This study was supported in part by State Key Laboratory for Aerodynamics, the Zhejiang University/University of Illinois at Urbana-Champaign Institute and National Natural Science Foundation of China (Grant No. 52106060 and 92152202). It was led by Supervisor Jiahuan Cui.

Analysis of the Droplet collection efficiency of the Ringsail parachute

Donghao Li*, Juan Fang, Ruirui Liu, Tingting Geng

Jiuquan Satellite Launch Center, Jiuquan 732750, China

*1179005985@qq.com

Abstract. A numerical model is established for calculating the water droplet collection efficiency on the ringsail parachute during spacecraft recovery, revealing distribution patterns of droplet accumulation on parachute surfaces. The methodology consists of two sequential phases: First, the airflow field around the ringsail parachute was simulated using computational fluid dynamics (CFD) methods. Subsequently, the droplet-phase governing equations were solved to obtain the surface collection efficiency, thereby mapping the water droplet impingement distribution – a critical foundation for evaluating hydrodynamic effects on parachute integrity. The results indicate that water droplets are primarily concentrated in the gap between the parachute and its edge, with an increase in the diameter of the water droplets and a significant expansion of the impact area on the upper surface.

1. Introduction

Ringsail parachutes are widely employed for spacecraft recovery, especially during cloud penetration. Clouds consist of atmospheric suspensions of water droplets and ice crystals. As the parachute passes through clouds, impacting droplets increase its weight, potentially causing structural collapse, performance degradation, and safety risks. At high-altitude low temperatures, droplet freezing upon contact may further compromise the parachute's functionality.

In recent years, there have been numerous studies on the characteristics of water droplet impacts [1-2]. The research primarily focuses on icing calculations for aircraft. There are two methods: Lagrange method and the Eulerian method [3-4]. The Lagrange method is used to establish a motion equation for each water droplet based on the streaming field. By calculating the motion trajectory, we can determine the impact characteristics and surface collection efficiency. The Eulerian flow of air containing water is treated as a two-phase flow of liquid water. A governing equation for this two-phase flow is established and subsequently solved using a finite volume method, allowing us to obtain the impact characteristics of water droplets along their trajectory and on the surface of the component. This article presents a numerical model of ringsail parachutes employing the Eulerian method, with three-dimensional water droplet collection efficiency calculations conducted to characterize droplet impingement patterns across varying diameters. The methodology establishes a computational framework for analyzing hydrodynamic interactions between multiphase flows and parachute structures during spacecraft recovery.

Journal of Physics: Conference Series **3041** (2025) 012015 doi:10.1088/1742-6596/3041/1/012015

2. Calculation of the flow field

2.1. The governing equation

$\Omega_t \subset \mathbb{R}^{n_{sd}}$ represents the fluid domain, Γ_t represents the boundary, $t \in (0,T)$ represents the time domain, the incompressible Navier-Stokes (N-S) equations are expressed as follows[5]:

$$\rho\left(\frac{\partial \vec{u}}{\partial t} + \vec{u}\cdot\nabla\vec{u} - \vec{f}\right) - \nabla\cdot\vec{\sigma} = \vec{0} \tag{1}$$

$$\nabla\cdot\vec{u} = 0 \tag{2}$$

$$\vec{\sigma}(p,\vec{u}) = -p\vec{I} + 2\mu\vec{\varepsilon}(\vec{u}) \tag{3}$$

$$\vec{\varepsilon}(\vec{u}) = \frac{1}{2}\left((\nabla\vec{u}) + (\nabla\vec{u})^T\right) \tag{4}$$

In the above formula, $\rho, \vec{u}, \vec{f}, p, \vec{I}, \mu$ represent independently density, speed, external volume force, pressure, pressure, tension and viscosity, $\vec{\sigma}(p,\vec{u}), \vec{\varepsilon}(\vec{u})$ represent stress tension and stress rate tension.

The inlet is the velocity inlet boundary condition, the velocity is 10m/s, the outlet is the pressure outlet boundary condition, and the parachute wall is the no-slip wall boundary condition.

3. Calculation of droplet impact characteristics

3.1. The governing equation

By introducing the water drop volume fraction α, which is defined as the volume proportion of the water drop phase in the space, the governing equation of the water drop phase can be established, whose continuity equation and momentum equation are listed:

$$\frac{\partial(\rho_d\alpha)}{\partial t} + \nabla\cdot(\rho_d\alpha\vec{u}_d) = 0 \tag{5}$$

$$\frac{\partial(\rho_d\alpha\vec{u}_d)}{\partial t} + \nabla\cdot(\rho_d\alpha\vec{u}_d\times\vec{u}_d) = \rho_d\alpha K(\vec{u}-\vec{u}_d) + \rho_d\alpha\vec{g} \tag{6}$$

ρ_d represents the drop density, $\vec{u}_d$ represents the drop velocity, where the inertia coefficient K is defined as:

$$K = \frac{18\mu}{\rho_d d_d^2}\cdot\frac{C_D \mathrm{Re}}{24} \tag{7}$$

In the above formula, μ represents the aerodynamic viscosity coefficient, d_d represents the droplet diameter, C_D represents the droplet drag coefficient, Re represents the relative Reynolds number.

$$C_D = \begin{cases} 24\left(1+0.15\mathrm{Re}^{0.687}\right)/\mathrm{Re} & \mathrm{Re}\le 1000 \\ 0.44 & \mathrm{Re} > 1000 \end{cases} \tag{8}$$

$$\mathrm{Re} = \rho\left|\vec{u}-\vec{u}_d\right| d_d / \mu \tag{9}$$

3.2. Calculation method

In the calculation of water droplets, due to the small volume fraction of water droplets, the influence of water droplets on the air can be ignored, and only the influence of air on water droplets can be considered. Therefore, the air flow field can be calculated first, and then the control equation of water droplets can be solved on the basis of the obtained air flow field, and then the water droplets collection efficiency on the surface of the object can be calculated through the following formula:

$$\beta = \frac{\alpha u_{dn}}{\alpha_\infty u_\infty} \tag{10}$$

β represents the water droplets collection efficiency on the surface of the object, α and u_{dn} represent the volume fraction of the water drop at the surface of the object and the surface velocity of the vertical object, α_∞ and u_∞ represent the volume fraction and velocity of the water drop at the far end. The mass of water droplets collected per unit time on an object surface can be expressed as:

$$\dot{m} = \alpha u_{dn} \rho_d A \tag{11}$$

4. Results and analysis

4.1. Flow field calculation results

The numerical simulation of the ringsail parachute indicates that the drag coefficient in a steady state is 0.813, while the average drag coefficient observed in the experiment is 0.8, resulting in a relative error of 1.625%.

As shown in Figures 1 and 2, the pressure distribution and streamline patterns demonstrate that the airflow velocity decreases as it approaches the canopy and is deflected outward in a radial direction along the canopy. As the airflow moves downstream through the lower edge of the parachute, it becomes separated due to the influence of the inverse pressure gradient, resulting in the formation of a significant recirculation area at the rear of the canopy, which serves as the primary vortex of the entire flow field. The high pressure within the parachute, combined with the contraction of the flow tube, causes the fluid to be ejected at high speed through the top opening of the parachute, creating a jet. Under the influence of the main vortex, the jet speed gradually decreases and disperses to the sides. Additionally, smaller jets are present at the ring seam and sail seam. The substantial velocity difference between the jet and the surrounding airflow generates a strong shear effect, leading to the formation of numerous small vortices that adhere to the surface of the parachute.

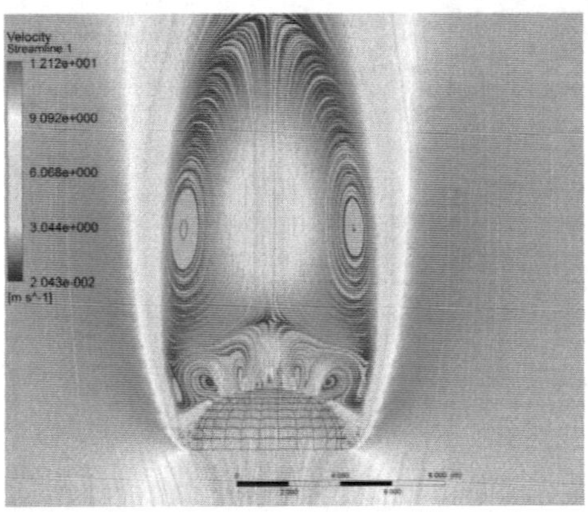

Figure 1. The streamline around the parachute.

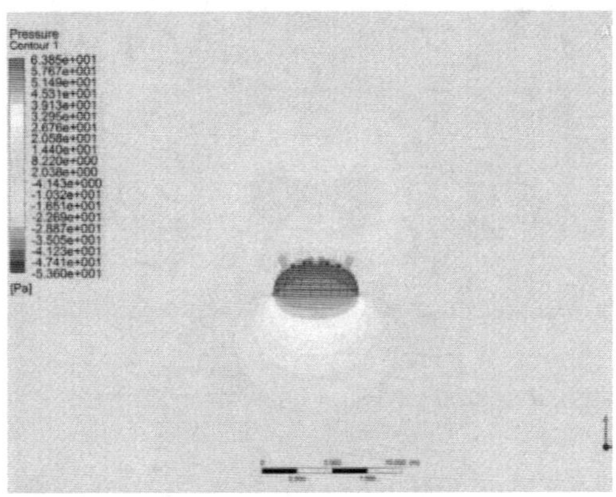

Figure 2. The pressure distribution around the parachute.

Journal of Physics: Conference Series **3041** (2025) 012015 doi:10.1088/1742-6596/3041/1/012015

4.2. Calculation results of water drop collection efficiency

Diameter of 20 microns and liquid water content of 1 $g \cdot m^{-3}$ are used for calculation. The resulting water droplet volume fraction is shown in the Figure 3.

It can be observed that the volume fraction of water droplets is minimal at the main vortex of the parachute. The parachute creates a small shielding area on its rear side due to its shielding effect. However, the presence of vortices leads to backflow on the upper surface, which does not entirely prevent the water droplets from passing through; consequently, some water droplets remain. In contrast, the volume fraction of water droplets on the lower surface is greater due to the aggregation effect of the droplets.

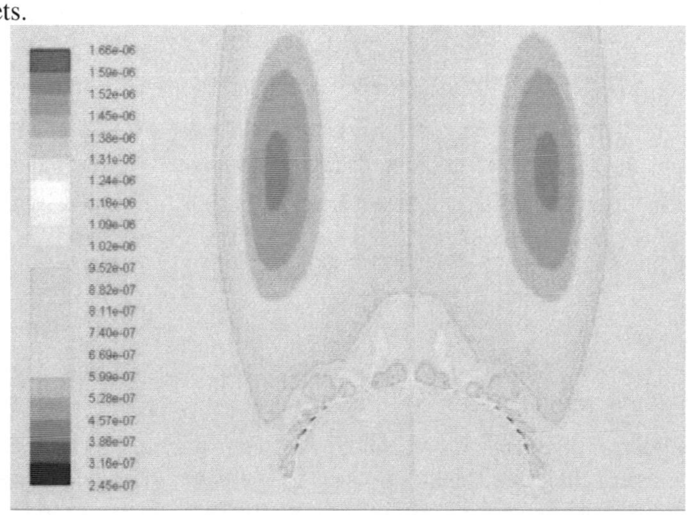

Figure 3. The volume fraction of water around the parachute.

The water droplet collection efficiency on the surface of the parachute is illustrated in Figures 4 and 5. It is evident that the front water droplets primarily impact the ring seam and sail seam, while the rear water droplets are mainly concentrated at the edge of the first sail, with a small number of droplets present on the rest of the parachute. This phenomenon occurs because, as water droplets approach the surface of the parachute, they are deflected by the airflow. Due to inertia, the droplets strike the edge, whereas, at the rear, the action of vortices causes the droplets to return and impact the upper surface of the parachute.

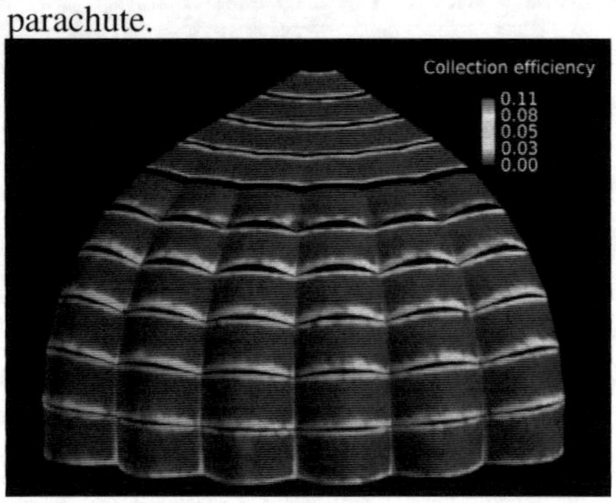

Figure 4. The water droplet collection efficiency on the parachute lower surface.

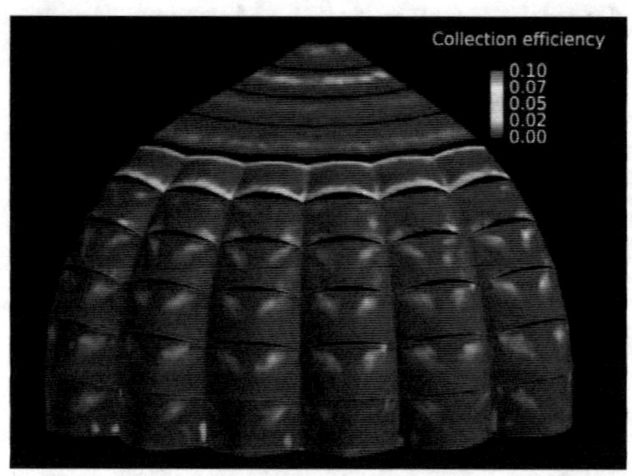

Figure 5. The water droplet collection efficiency on the parachute upper surface.

The collection efficiency for surface water droplet impact from droplets measuring 40 microns, 60 microns, and 80 microns have been calculated (see Figures 6 to 11). It is evident that as the diameter of the water droplets increases, the impact characteristics on the surface beneath the canopy show minimal variation, primarily concentrated around the sail seam, while the impact at the ring seam decreases. Conversely, on the back of the canopy, the impact range of the water droplets on the ring gradually expands, with a significant number of droplets also impacting the second and third sails. Notably, for the 80-micron droplets, impacts are observed across nearly the entire surface of the parachute band. An analysis of droplet dynamics indicates that droplets are influenced by both air resistance and gravity. When the droplet diameter is very small, air resistance predominates, rendering the effect of gravity negligible. However, as the diameter increases, the influence of gravity becomes more significant. Due to the high internal pressure within the parachute, the air velocity is considerably low when the water droplets approach the parachute ring. Consequently, the air force acting on the droplets is minimal, making it difficult for them to impact the lower surface of the parachute ring. As the droplet diameter increases, the frequency of impacts on the lower surface of the parachute ring decreases, with the primary impact location shifting to near the sail seam. The flow diagram illustrates numerous vortices on the upper surface of the parachute, which create backflow effects, causing droplets to fall onto the upper surface under the influence of gravity. For smaller water droplets, the force of air is significantly greater than that of gravity, resulting in their tendency to be carried away by the airflow rather than falling onto the upper surface of the parachute.

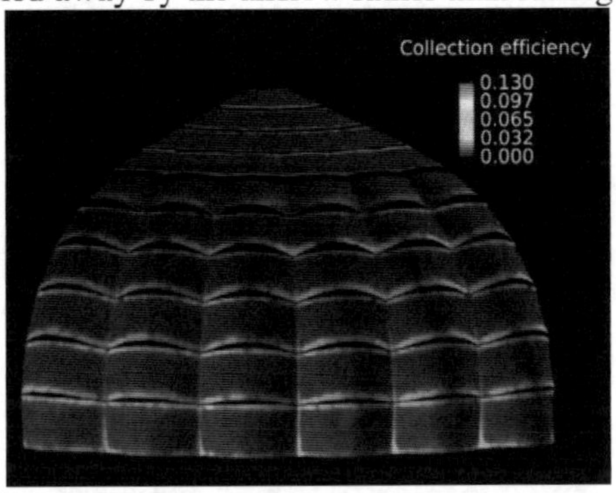

Figure 6. The collection efficiency of 40-micron water droplets on the parachute lower surface.

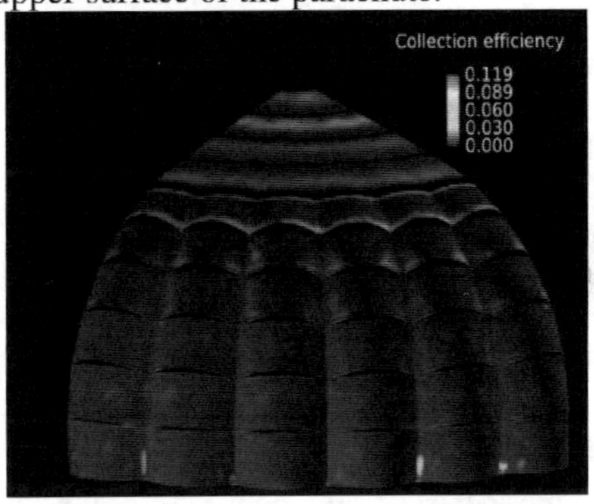

Figure 7. The collection efficiency of 40-micron water droplets on the parachute upper surface.

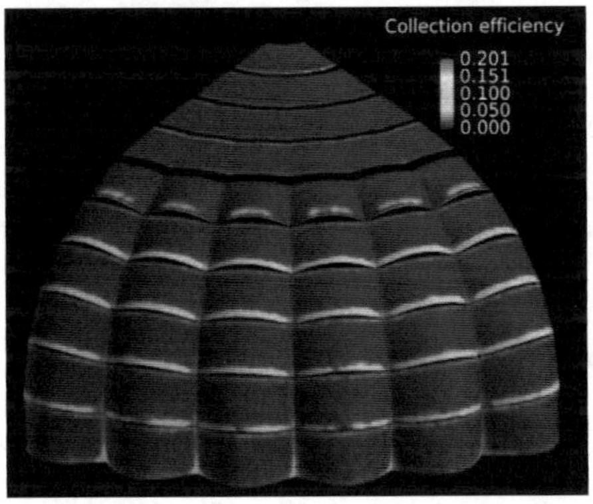

Figure 8. The collection efficiency of 60-micron water droplets on the parachute lower surface.

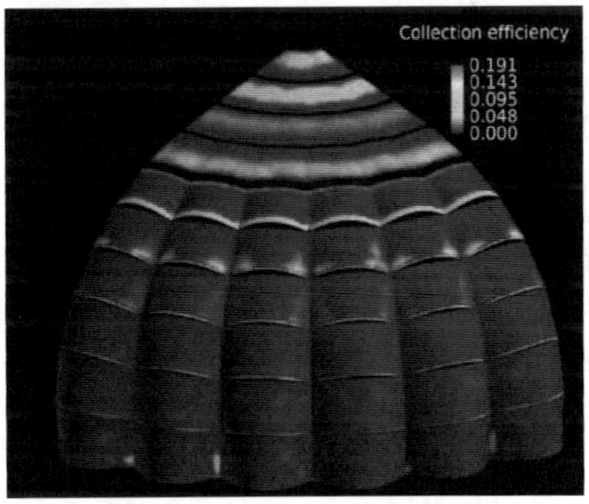

Figure 9. The collection efficiency of 60-micron water droplets on the parachute upper surface.

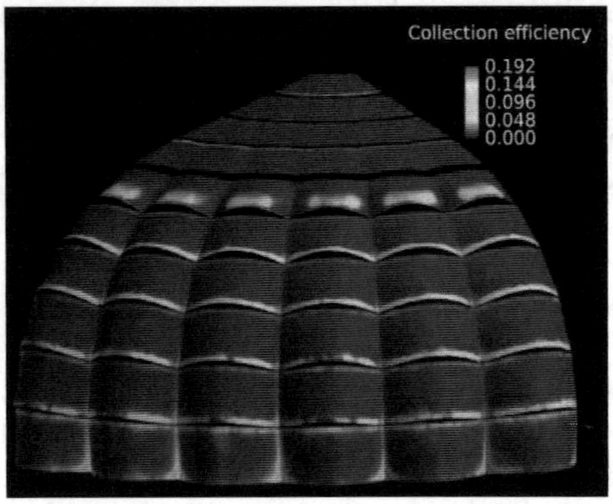

Figure 10. The collection efficiency of 80-micron water droplets on the parachute lower surface.

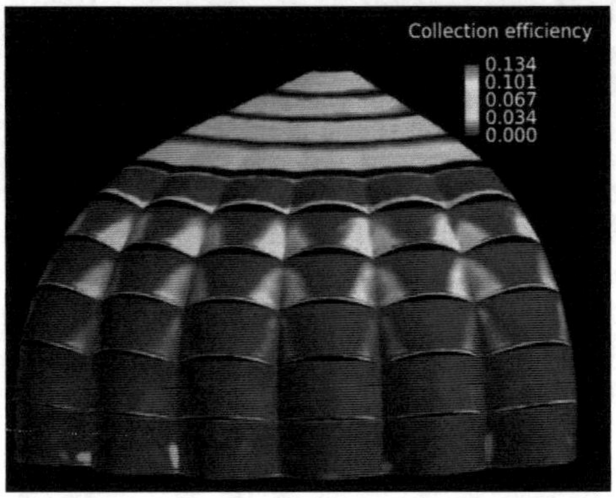

Figure 11. The collection efficiency of 80-micron water droplets on the parachute upper surface.

5. Conclusion

This article develops a numerical model for calculating the water droplet collection efficiency on ringsail parachutes, revealing distinct distribution patterns of droplet accumulation across the canopy surface. Key findings include:

1. Droplet impingement is predominantly localized at sail seams and ring seams on the lower canopy surface. The collection efficiency of these droplets decreases as their diameter increases.
2. Small water droplets on the upper surface of the canopy tend to gather at the edge of the first sail. With the diameter of the water droplets increases, both the impact area and the collection efficiency on the surface of the belt significantly rise. This phenomenon is primarily attributed to the increased diameter and mass of the droplets, as well as the influence of gravity.

Water droplets collected by the canopy may increase its mass and alter its structure, which could affect the recovery process. This paper establishes a foundation for future research on the effects of parachutes.

6. References

[1] BIAN Qingyong, ZHU Chengxiang, ZHAO Ning, et al. Numerical investigation on droplet impact and freezing [J] . Transactions of Nanjing University of Aeronautics and Astronautics, 2023,40(2): 179-192.

[2] ZHAO Wenzhao, NING Yijun, WU Yuan, et al. Mesh impact analysis of Eulerian method for droplet impingement characteristics under aircraft icing conditions [J] . Transactions of Nanjing University of Aeronautics and Astronautics, 2023,40(2):148-158

[3] SHEN X B, ZHAO W Z, LIN G P, et al. Accuracy analysis of Eulerian method for droplet impingement characteristics under aircraft icing conditions[J]. Journal of Beijing University of Aeronautics and Astronautics, 2023, 49(8):1912-1921

[4] REN Jinghao, YI Xian, WANG Qiang, et al. Lagrangian simulation method of droplet collection efficiency for complex configu ration[J]. Journal of Aerospace Power, 2020, 35(12):2553-2561

[5] Stein, K. R., Benney, R. J., Kalro, V., Tezduyar, T. E., Leonard, J., Accorsi, M..Parachute Fluid-Structure Interactions: 3D Computation [J]. Computer Methods in Applied Mechanics and Engineering. Vol.190, 2000, pp., 373-386

A numerical study on the parachute inflation process at a high angle of attack using the arbitrary Lagrangian-Eulerian method

Jian Jiang[1], Chang Cai[2], Tengyuan Wang[2], Yan Wang[1,*]

[1]College of Aerospace Engineering, Nanjing University of Aeronautics and Astronautics, Nanjing, 210016, China
[2]CAS Laboratory of Wind Energy Utilization, Institute of Engineering Thermophysics, Chinese Academy of Sciences, Beijing, 100190, China

[*]aerowangy@nuaa.edu.cn

Abstract. High-altitude wind energy is widely distributed, and parachutes have great potential to capture them. This paper focuses on the aerodynamic inflation process of a high-altitude parachute at a high angle of attack, which is essential for wind energy harvest. First, the arbitrary Lagrangian-Eulerian method (ALE) method is verified by simulating the main inflation phase of the C-9 parachute. Good agreements with the experimental data are achieved quantitatively. Then, numerical simulations of the inflation process of a parachute with an attack angle of 35°and a wind speed of 20ms^{-1} at an altitude of 1500m high are carried out. The evolution of the morphology of the parachute, the stress distribution, and the flow field characteristics are demonstrated and discussed.

1. Introduction

With the rapid development of industry, non-renewable energy has profoundly impacted the environment, and a shift in the energy structure is imminent. High-altitude wind energy is a renewable energy that has so far been largely unexploited by humankind. High wind speeds, wide distribution, sustainability, and perennial continuity characterize this kind of energy [1]. As an inflatable aerodynamic decelerator, the parachute has a wide range of applications due to its advantages of good aerodynamic deceleration performance [2]. On the other hand, the arrangement of the parachute system also has a high potential for harvesting wind energy at high altitudes. It has many advantages, such as low cost and ease of recovery. However, during the energy harvest process, the inflation process of the parachute for harvesting energy is very different from that used in the deceleration process. Numerical studies on this are very limited due to the complexity of the fluid-structural interaction (FSI) and the slow development of wind energy harvest technique with parachutes.

In fact, parachute inflation involves large structural deformations over a short period, which is a problem of structural dynamics with rapid large deformation involving both geometric and material nonlinearity [3]. Most studies focused on numerical simulations of subsonic parachutes based on the ALE method [4], which combines the advantages of the Lagrangian and Eulerian methods. It can describe the movement of structural boundaries without severe mesh distortion, which is beneficial for dealing with significant deformation problems. As decelerators, parachutes inflate with a certain angle of attack before opening during an airdrop landing. In 2002, Irvin Aerospace's Taylor considered a cruciform drag parachute using the ALE method of the LS-DYNA software to study its dynamic load and the tension of the suspension lines after inflation [5]. Taylor also used this software in 2003 to simulate a quarter model of a cruciform drag parachute and studied the changes in the drag coefficient of the parachute fabric during inflation when the boundary position, flow field velocity, and payload of the numerical simulation were different [6]. After introducing the LS-DYNA algorithm for the air permeability of thin shell elements, Benjamin used this software in 2007 and added the air permeability algorithm to simulate the TP8 parachute [7]. The study found that adding air permeability reduced the numerical oscillation of the parachute system significantly. In the same year, Lingard et al. employed LS-DYNA software to simulate the main inflation phase of the disk-gap-band parachute within the Mach number range of 0.4 to 2.8 [8]. They analyzed its inflation performance and flow characteristics due to the need for the Mars exploration mission. More and more scholars are adopting the ALE method to simulate the various parachutes' main inflation phase. Cheng et al. simulated the C-9 parachute numerically and compared the parachute shape with the airdrop test [9]. Gao et al. simulated the main inflation phase of the ring-sail parachute during a low-speed airdrop and analyzed the flow field characteristics of its inflation phase [10].

However, during the inflation process, the air inlet of the parachute quickly aligns with the incoming flow direction. Few studies consider the inflation process of a single parachute when maintaining a certain angle of attack, which motivates the present study.

In this paper, numerical simulations of the parachute inflation with an attack angle of 35° and a wind speed of 20ms^{-1} at an altitude of 1500m high are carried out using the ALE method. The morphology and evolution of the canopy in the inflation process are clearly demonstrated, and the flow field characteristics are also discussed.

2. Mathematical model

2.1. Flow field and structural governing equations

2.1.1. Fluid dynamics. Under the ALE method description, the relationship between the material and spatial coordinate systems for each physical variable is as follows [11]:

$$\frac{\partial f(\boldsymbol{X}_i,t)}{\partial t} = \frac{\partial f(\boldsymbol{x}_i,t)}{\partial t} + (\boldsymbol{v}_i - \boldsymbol{w}_i)\frac{\partial f(\boldsymbol{x}_i,t)}{\partial t} \tag{1}$$

Where $\boldsymbol{X}_i$ and $\boldsymbol{x}_i$ are the Lagrangian and Eulerian coordinates, respectively, $\boldsymbol{v}_i$ and $\boldsymbol{w}_i$ are fluid velocity and mesh velocities.

Consider that the fluid domain Ω_f with boundary Γ_f and the structure domain Ω_s with boundary Γ_s. At low speeds, the air can be treated as an incompressible fluid. The Navier-Stokes equations based on the ALE description are as follows [11]:

$$\frac{\partial \rho_f}{\partial t} = -\rho_f \frac{\partial \boldsymbol{v}_i}{\partial \boldsymbol{x}_i} - (\boldsymbol{v}_i - \boldsymbol{w}_i)\frac{\partial \rho_f}{\partial \boldsymbol{x}_i} \tag{2}$$

$$\rho_f \frac{\partial \boldsymbol{v}_i}{\partial t} = \boldsymbol{\sigma}_{ij,j} + \rho_f \boldsymbol{f}_i - \rho_f (\boldsymbol{v}_j - \boldsymbol{w}_j)\frac{\partial \boldsymbol{v}_i}{\partial \boldsymbol{x}_j} \tag{3}$$

$$\rho_f \frac{\partial e}{\partial t} = \boldsymbol{\sigma}_{ij}\boldsymbol{v}_{i,j} + \rho_f \boldsymbol{f}_i \boldsymbol{v}_i - \rho_f (\boldsymbol{v}_j - \boldsymbol{w}_j)\frac{\partial e}{\partial \boldsymbol{x}_j} \tag{4}$$

Where:

ρ_f = fluid density; $\boldsymbol{f}_i$ = body force; e = specific internal energy.

$\boldsymbol{\sigma}_{ij}$ is the stress tensor defined as follows:

$$\boldsymbol{\sigma}_{ij} = -p\boldsymbol{\delta}_{ij} + \mu(\boldsymbol{v}_{i,j} + \boldsymbol{v}_{j,i}) \tag{5}$$

Where:

μ = fluid dynamic viscosity; p = fluid pressure; $\boldsymbol{\delta}_{ij}$ = Kronecker's delta function.

The boundary conditions are as follows:

$$\boldsymbol{v}_i = \boldsymbol{v}_i^D \quad on \ \Gamma_f^D \tag{6}$$

$$\boldsymbol{\sigma}_{ij}\vec{n} = 0 \quad on \ \Gamma_f^N \tag{7}$$

Where Γ_f^D and Γ_f^N are Dirichlet and Neumann boundary conditions, respectively.

2.1.2. Structure dynamics. A parachute canopy is a flexible, large deformable body with typical non-linear dynamic characteristics. The governing equation of the structure domain is represented by:

$$\rho_s \frac{\mathbf{d}\boldsymbol{v}_s}{\mathbf{d}t} = \boldsymbol{\sigma}_{ij,j} + \rho_s \boldsymbol{f}_i \quad in\ \Omega_s \tag{8}$$

Where:
$\boldsymbol{v}_s$ = velocity of structure; ρ_s = material density.

2.2. Fluid–structure interaction

The above governing equations are solved using an explicit central difference method with second-order accuracy. For each node, the displacement and speed are updated according to the following equations:

$$\boldsymbol{x}^{n+1} = \boldsymbol{x}^n + \Delta t \cdot \boldsymbol{u}^{n+1/2} \tag{9}$$

$$\boldsymbol{u}^{n+1/2} = \boldsymbol{u}^{n-1/2} + \Delta t \cdot \boldsymbol{M}^{-1} \cdot (\boldsymbol{F}_{in} + \boldsymbol{F}_{ex}) \tag{10}$$

Where:
$\boldsymbol{x}$ = displacement vector; $\boldsymbol{u}$ = speed vector; $\boldsymbol{M}$ = mass matrix; $\boldsymbol{F}_{in}$ and $\boldsymbol{F}_{ex}$ are internal and external force vectors. Fluid–structure interaction is achieved through the penalty function:

$$\boldsymbol{F} = k \cdot \boldsymbol{d} \tag{11}$$

In equation (11), k is the stiffness factor based on the characteristics of the master and slave node mass model, the speed and displacement are adjusted based on $\boldsymbol{F}$, which is a part of $\boldsymbol{F}_{ex}$.

The fluid-structure interaction first solves the fluid governing equations to calculate the velocity and pressure in the fluid domain and then calculates the force exerted by the fluid on the structure to solve the structural deformation equation. Figure 1 shows the process of simulating the main inflation phase of a parachute using the ALE method.

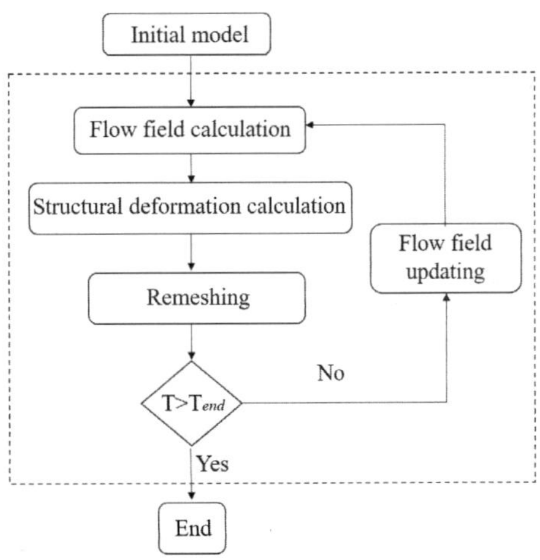

Figure 1. Fluid-structure interaction process.

3. Numerical validations

This paper used the C-9 parachute, a typical flat circular parachute, as an example to verify the validity of the ALE method. Due to the complex folding modeling of flexible fabrics, only weft folding is considered in this paper, and radial folding is not considered for the time being. In other words, we only consider the inflation stage after the parachute system is straightened, and the initial model is shown in Figure 2. The model structure used in this paper mainly consists of a parachute canopy, suspension lines, and radial reinforcement belts. The model size and material parameters are sourced from the article (Fan) [12] and will not be given here. The incoming air velocity is 80 ms^{-1}, and the air density is 1.18 kg m^{-3}.

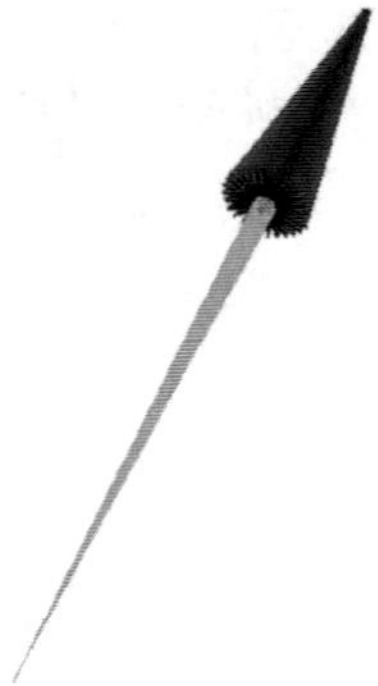

Figure 2. The initial configuration of the C-9 parachute model.

Figure 3 and Figure 4 compare the numerical simulation of the canopy's inflation process with the test result [13]. The shape of the inflated canopy is essentially the same as the test results.

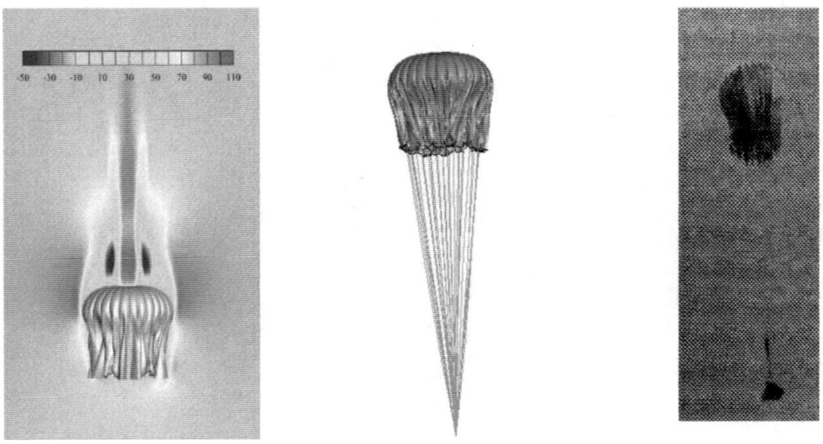

Figure 3. The shape of the canopy during inflation.

During the inflation process, vortices are symmetrically distributed on the top of the canopy. As the canopy inflates, the vortices gradually become asymmetrically distributed.

Figure 4. Canopy inflated state.

Figure 5 shows the evolution of the drag force. After the initial transient, the drag force is stabilized, and its normalized value is 0.77. This drag coefficient is consistent with the reference data [14], ranging from 0.65 to 0.9. Therefore, the ALE method is able to better analyze the canopy inflation situation with satisfactory accuracy.

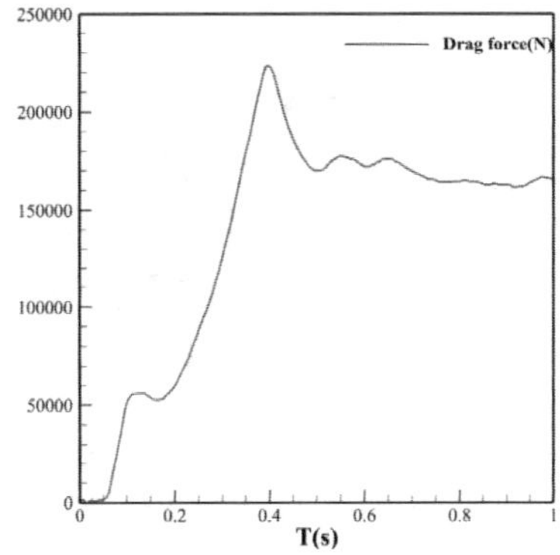

Figure 5. Drag curve.

4. Inflation of a parachute at high angle attack and high altitude

The simulation of a high-altitude parachute still uses a C-9 parachute placed at an altitude of 1500m, with an angle of attack of 35° and an inflow speed of 20ms^{-1}. The parachute's structure is always subject to gravity. Figure 6 shows a slice of the flow field mesh, where D_0 represents the nominal diameter of the canopy.

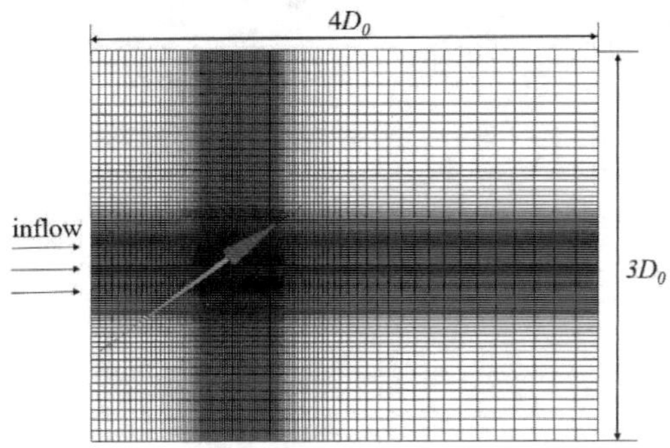

Figure 6. Flow field mesh.

Figure 7 shows the evolution of the morphology during inflation. The inflow air first brings the canopy into contact with the cable, restricting it and preventing it from moving to the right. The air then fills the lower part of the canopy first and finally drives the upper part of the canopy to inflate and expand. After a period of coupling, it reaches a full state, at which time the shape of the parachute will not change significantly.

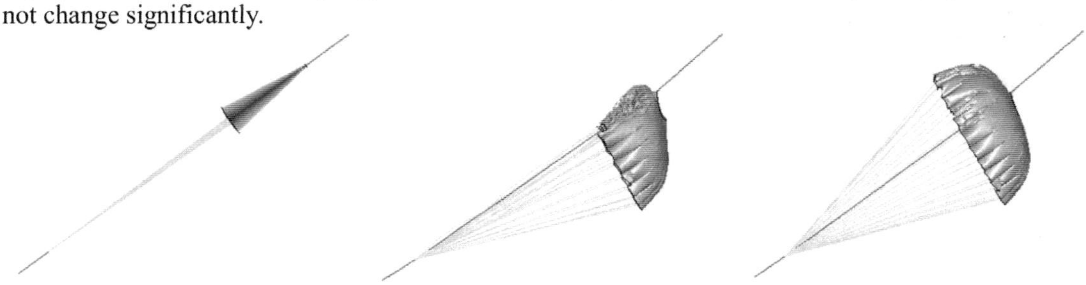

(a) (b) (c)

Figure 7. Changes in shape of canopy during inflation process.

Figure 8 shows the surface stress distribution of the fully inflated canopy. The stress is greatest at the top of the canopy where it contacts the cable, gradually decreasing along the radial direction of the canopy, and also greater at the connection between the canopy and the suspension line.

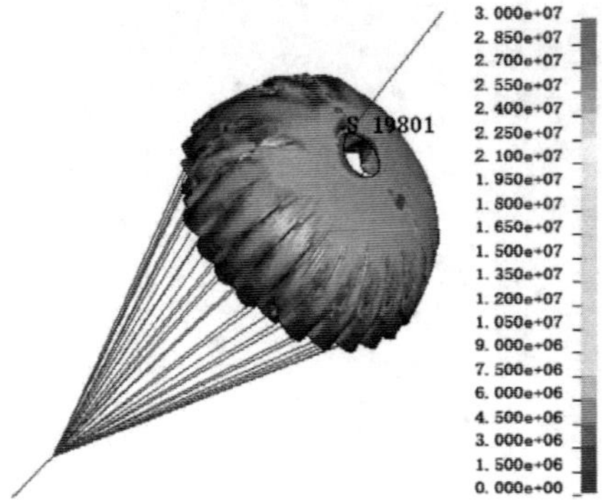

Figure 8. Surface stress distribution of the canopy in inflated state.

Figure 9 and Figure 10 show the velocity contour of the flow field at the vertical center section of the canopy during inflation. It can be observed that when the canopy is not fully inflated, the air velocity at the outer edge of the canopy is higher. As the canopy fills, the air flows out from the top at a higher velocity. During the process of the canopy's inflation, the vortex at the top of the canopy is always in an asymmetrical distribution state. As the canopy gradually fills, the vortex also spreads out.

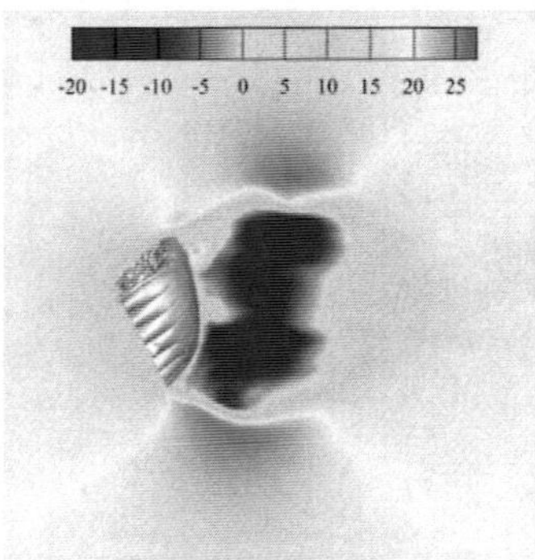

Figure 9. Velocity distribution of the intermediate state

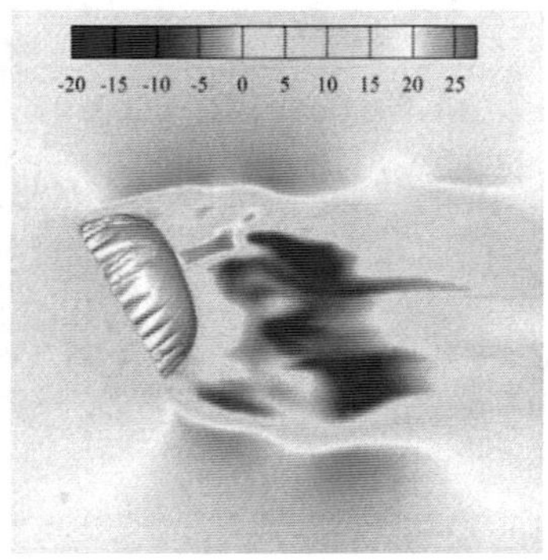

Figure 10. Velocity distribution of the full inflation state

Figure 11 shows the cable axial force. Under the wind, the canopy comes into contact with the cable, at which time the cable is subjected to the force of the parachute line and to a squeezing force of the canopy on the cable. The maximum axial force on the cable is 45675N. As the canopy gradually inflates, the squeezing force of the canopy on the cable gradually decreases, as does the cable axial force.

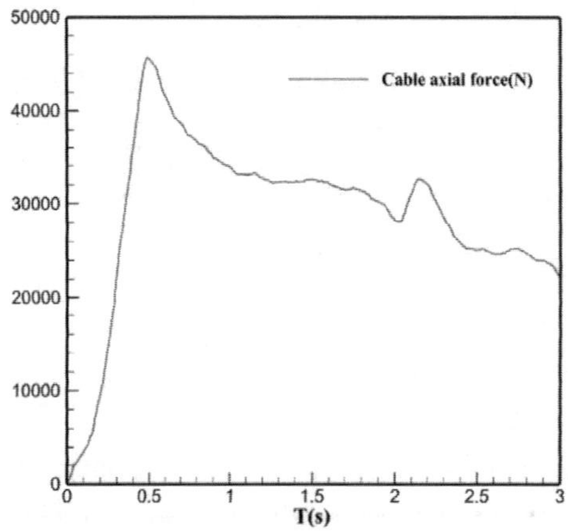

Figure 11. The cable axial force curve

5. Conclusion

This paper uses the ALE method to consider the parachute inflation at a high angle of attack for wind energy harvesting at high altitudes. Based on this method, it simulates the opening process of a flat circular canopy under the combined action of gravity and a wind speed of $20 ms^{-1}$ at an angle of attack of 35°. The inflation process, structural response, and flow field characteristics are obtained and analyzed. It shows that the bottom of the parachute is first filled, and then the top is filled. The stress at the top hole of the canopy is the largest where it contacts the cable, followed by the connection between the suspension lines and the canopy. The flow field at the top of the parachute always shows an asymmetrically distributed vortex and continues to grow. The method used in this paper can predict the dynamic behavior of high-altitude working parachutes under complex flows, which may help design the structure of high-altitude wind energy generation systems.

6. References

[1] Zhang J, Zhang M, Li Y, Qin J, Wei K and Song L 2020 Analysis of wind characteristics and wind energy potential in complex mountainous region in southwest China *J. Cleaner Prod.* **274** 123036

[2] Zhu H, Tao J, Sun Q, Sun H, Duan F, Chen Z, Zeng X and Soulat D 2023 Effect of shear modulus on the inflation deformation of parachutes based on fluid-structure interaction simulation *Sustainability* **15** 5396

[3] Stein K, Benney R, Kalro V, Tezduyar T, Leonard J and Accorsi M 2000 Parachute fluid–structure interactions: 3-D Computation *Comput. Methods in Appl. Mech. and Eng.* **190** pp 373-86

[4] Zhang G, Wang S, Sun Z and Xiao Q 2022 Research developments in numerical methods of fluid-structure interactions in naval architecture and ocean engineering *Chin. J. ship Res.* **17** pp 52-73

[5] Taylor A 2002 Developments in the application of LS-DYNA to fluid structure interaction (FSI) problems in recovery system design and analysis *7th Int. LS-DYNA Users Conf.* (Detroit: LSTC) pp 10-7

[6] Taylor T 2003 An investigation of the apparent mass of parachutes under post inflation dynamic loading through the use of fluid structure interaction simulations *17th AIAA Aerodyn. Decelerator Systems Tech. Conf. and Semin.* (Monterey: AIAA) pp 2104

[7] Aquelet N, Tutt B 2007 Euler-Lagrange coupling for porous parachute canopy analysis *The Int. J. Multiphys.* **1** pp 53-68

[8] Lingard J, Darley M and Underwood J 2007 Simulation of the Mars science laboratory parachute performance and dynamics *19th AIAA Aerodyn. Decelerator Systems Tech. Conf. and Semin.* (Williamsburg: AIAA) pp 2507

[9] Cheng H, Yu L, Rong W and Jia H 2012 A numerical study of parachute inflation based on a mixed method *Aviation* **16** pp 115-23

[10] Gao X, Zhang Q, Tang Q and Tao Y 2013 Fluid-structure interaction simulation of parachute in low speed airdrop *Proc. World Cong. Eng.* (London: Imperial Coll London) **3** pp 1923-28

[11] Souli M, Ouahsine A and Lewin L 2000 ALE formulation for fluid–structure interaction problems *Comput. Methods in Appl. Mech. and Eng.* **190** pp 659-75

[12] Fan Y and Xia J 2014 Simulation of 3D parachute fluid–structure interaction based on nonlinear finite element method and preconditioning finite volume method *Chin. J. Aeronaut.* **27** pp 1373-83

[13] Lee C 1984 Experimental investigation of full-scale and model parachute opening *8th Aerodyn. Decelerator and Balloon Tech. Conf.* (Hyannis: AIAA) pp 215-23

[14] Tian J, Wang Y and Wang B 2021 *Aircraft parachute technology* (Beijing: Beijing University of Aeronautics and Astronautics) pp 49-50

Acknowledgements

This work is supported by the National Key R&D Program of China (No. 2023YFB4203401) and the High Performance Computing Platform of Nanjing University of Aeronautics and Astronautics.

Design and performance analysis of a finned tube heat exchanger

Pan Jin [1,3], **Jin Feng** [2,4]

[1]Jiangsu Maritime Institute, Nanjing 211170, China

[2] College of Energy and Power, Nanjing University of Aeronautics and Astronautics, Nanjing 210016 , China

[3]pbrj@163.com [4]fengjin_22@nuaa.edu.cn

Abstract. This paper designs two kinds of finned tube exchangers with wall cooling system. Performance is simulated by Fluent software. The results show that the designed finned tube heat exchanger has a convective heat transfer coefficient of 1332.8 W/(m^2·K) and 1.3% total pressure loss. The high enthalpy airflow's temperature decreases about 32% of the inlet air temperature by cooling down. The temperature of the wall keeps within the permitted area of the metal material.

1.Introduction

With the increasing of the flight Mach number that can be achieved by the hypersonic vehicle, the aerodynamic heating of the surface of the aircraft is becoming more and more intense, resulting in a continuous rise in temperature. If the temperature exceeds the working limit temperature of the component material, it will not only shorten the working life, but even the material will be burned and the normal flight of the aircraft can not be guaranteed[1]-[5].

Therefore, in hypersonic flight, due to the strong effect of aerodynamic heating, it is necessary to cool the high-enthalpy incoming air[6]. In order to introduce regenerative active cooling into the intake cooling system, it is necessary to design an efficient heat exchanger that takes up little space. In this paper, a compact oil-air cooling tube fin heat exchanger with wall cooling structure under high heat flow condition is designed, and the performance of the heat exchanger is analyzed by numerical simulation.

2.Engineering design of gas-liquid heat exchanger

2.1. Design method combining engineering calculation and numerical simulation

This paper adopts a design method combining engineering calculation and numerical simulation. Firstly, using engineering design methods, calculate the heat transfer coefficient, heat transfer area, and other parameters of the heat exchanger according to the required indicators. Then, using the results obtained from engineering calculations, guide the structural design of the heat exchanger, model it using commercial software, partition the grid, and perform numerical simulations. If the target requirements cannot be met, further modify the heat exchanger model and conduct numerical simulations again until the optimal solution is achieved.

Journal of Physics: Conference Series **3041** (2025) 012017 doi:10.1088/1742-6596/3041/1/012017

2.2. Numerical calculation model

After conducting engineering calculations on some basic parameters of the heat exchanger, CFD software was used to simulate the flow and heat transfer of the heat exchanger, obtaining details of the internal air and kerosene flow, pressure distribution, temperature distribution, and so on.

For the tube fin heat exchanger studied in this article, the standard k - ε model can meet the accuracy and reliability of calculations.

3. Numerical simulation of flow and heat transfer in tube fin heat exchanger

Based on the aforementioned heat exchanger design method, a compact heat exchanger - tube fin heat exchanger was designed. After reasonable design, the tube fin heat exchanger can combine the wall cooling cycle with the internal oil circuit of the heat exchanger into a circulation system.

This paper considers two wall cooling heat transfer schemes and uses Fluent software to numerically simulate their flow and heat transfer characteristics.

3.1. The first wall cooling scheme for tube fin heat exchanger

3.1.1. Structural design of heat exchanger. The external structure of the designed tube fin heat exchanger is shown in Figure 1, consisting of a shell, heat exchange tubes, straight fins, oil circuit baffles, etc. The flow unit of the heat exchanger and wall cooling system is shown in Figure 2. At the junction of two adjacent flow units, in the liquid (kerosene is selected in this article) interlayer on both sides, one side is equipped with a baffle and the other side is not equipped with a baffle. Kerosene flows in from the outlet of the previous flow unit and enters the upper wall channel of this unit; After being blocked by the baffle, a portion of the fluid flows into the heat exchange tube to exchange heat with the air, while another portion enters the wall cooling channels on both sides to cool the side walls; Three fluids converge in the lower wall channel, and due to the absence of baffles in the cooling channel on the lower wall, kerosene flows into the next heat exchange unit to continue heat exchange. This completes the circulation of kerosene in a flow unit, where in each cycle, the four walls of the heat exchanger are cooled by kerosene.

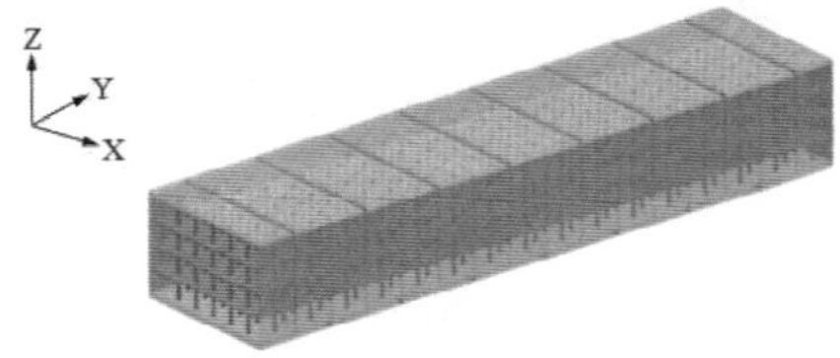

Figure 1. Outline of tube fin heat exchanger

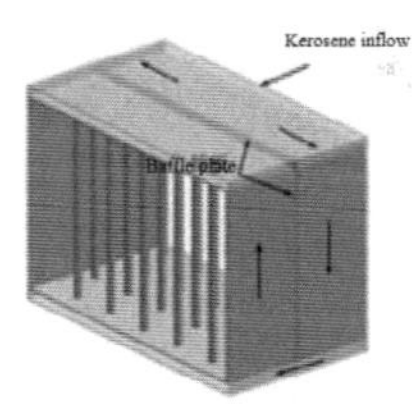

Figure 2. Heat exchanger unit

3.1.2. Verification of grid independence and selection of boundary conditions

3.1.2.1 Grid independence verification. Due to the large number of grids in the complete computational model, this paper adopts the method of conducting grid experiments on heat exchanger unit cells, that is, extracting a unit cell from the heat exchanger model, modifying the number of nodes in the grid, and then conducting experimental simulations on unit cell models with different numbers of grids to apply the results to the complete computational model. This not only ensures the accuracy of the calculation, but also saves computing resources.

The flow and heat transfer process inside the heat exchanger mainly occurs in the thermal boundary layer near the tube wall and fins. Therefore, the mesh division of the boundary layer determines the accuracy of the numerical simulation of the heat exchanger and also determines the size of the mesh. The grid experiment simulated five groups of different grids, with parameter settings as shown in Table 1.

Table 1. Experimental Grid Parameter Settings

serial number	set boundary layers or not	First layer boundary layer scale,	boundary layer growth ratio,	number of grids
1	N		0	1. 01×106
2	Y	0. 07	2	1. 20×106
3	Y	0. 07	1. 5	2. 11×106
4	Y	0. 05	1. 5	3. 32×106
5	Y	0. 05	1. 2	5. 24×106

Taking into account both computational accuracy and grid size, this article adopts the third set of settings for grid partitioning of future models.

3.1.2.2 Selection of Turbulence Models and Boundary Conditions. The fluid in the shell side of the heat exchanger is hot air, and the fluid in the tube side is kerosene. The standard k - ε turbulence model is selected, and the physical parameters of kerosene and air are set as constants in the calculation process. Define both the inlet of air and kerosene as mass flow inlet, air as incompressible fluid, outlet as pressure outlet, and the outer wall of the heat exchanger as adiabatic wall.

3.1.3 Overall numerical simulation of tube fin gas-liquid heat exchanger

3.1.3.1 Analysis of flow characteristics of tube fin heat exchanger. After calculation, the total pressure loss of high-temperature gas in the heat exchanger is 1.3%. On average, 17 planes are selected along the direction of gas flow in the heat exchanger, and the total pressure decreases linearly along the direction of gas flow on the shell side.

3.1.3.2 Analysis of Heat Transfer Characteristics of Tube fin Heat Exchanger. The gas outlet temperature drops to 606K, and the average heat transfer coefficient of the heat exchanger is K=1332.8 w/(m2 · K). On average, 17 planes are selected according to the direction of gas flow inside the heat exchanger, and the temperature gradually decreases with the direction of gas flow, but the decrease becomes smaller and smaller. This is because temperature difference is the driving force for heat transfer. According to Fourier's law, the smaller the temperature difference, the smaller the dT/dR, and the smaller the heat transfer. This also conforms to the basic law of heat exchange.

The temperature distribution of gas in the X-axis plane is shown in Figure 3, and the temperature distribution of gas in the Z-axis plane is shown in Figure 4. It can be seen that the heat transfer process not only occurs mainly near the heat exchange tubes, but also strongly occurs near the cooled walls and fins. This indicates that the presence of fins effectively improves the heat transfer efficiency; The wall cooling mechanism not only cools the wall, but also provides additional heat transfer area, enhancing the heat transfer effect.

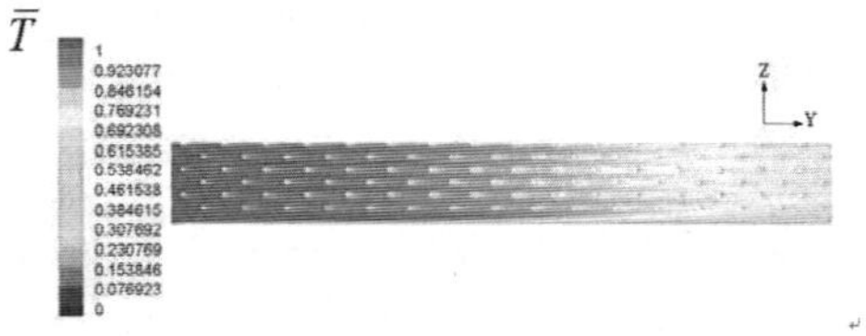

Figure 3. Non dimensional temperature distribution of gas in the X-axis plane

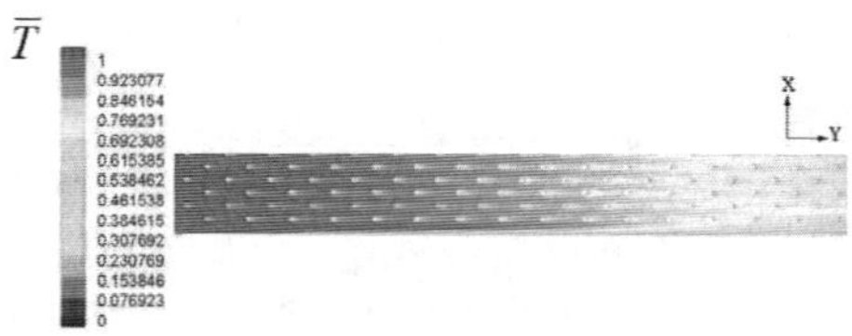

Figure 4. Non dimensional temperature distribution of gas in the Z-plane

3.1.3.3 Temperature distribution of heat exchange tube wall. The magnified local temperature distribution of gas, liquid, and pipe wall at the inlet is shown in Figure 5. Due to the counterflow flow mode of the heat exchanger, the highest temperature of the liquid and the tube wall occurs at the gas inlet. After calculation, the highest temperature of the liquid is 411K, and the temperature of the tube wall is within the allowable range of the metal material.

Journal of Physics: Conference Series **3041** (2025) 012017 doi:10.1088/1742-6596/3041/1/012017

3.1.3.4 Temperature distribution of fins. The temperature distribution of the fins is shown in Figure 6. The highest temperature of the fins occurs at the inlet, near the central axis, with a maximum temperature of about 1000K, which does not exceed the working limit of the material; near the wall, the temperature of the fins is affected by wall cooling, resulting in a lower temperature.

The average temperature of the fins is approximately 600K to 700K, with a total area of 0.54m^2. Therefore, after adding fins, it provides a secondary heat exchange surface for the heat exchanger with objective temperature and area. Although the secondary heat transfer surface is not a true "cold source", its temperature is closer to that of a fluid with a high convective heat transfer coefficient (here kerosene), so adding fins indirectly increases the heat transfer area.

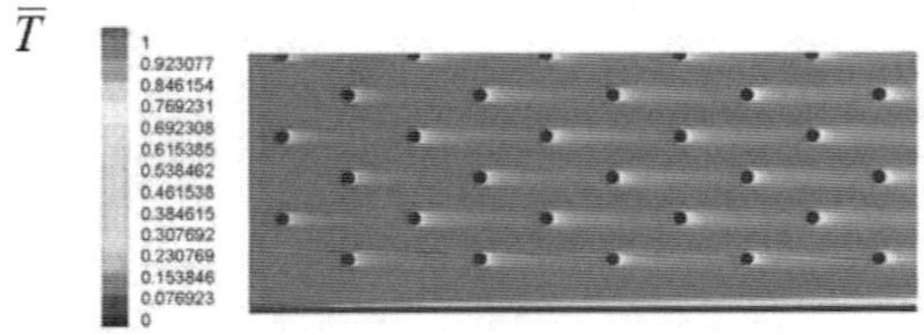

Figure 5. Local enlarged view of temperature distribution of gas, liquid and pipe wall at inlet

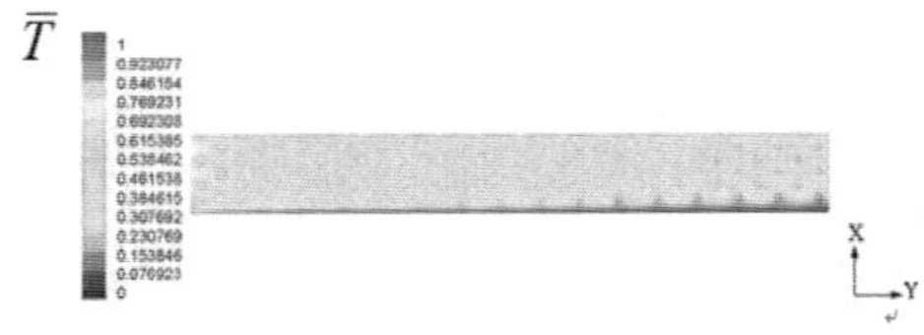

Figure 6. Temperature distribution of fins

Due to the long flow path in the tube side of the tube fin heat exchanger, the total pressure loss reached 60%. To reduce the total pressure loss, a second wall cooling scheme, the tube fin heat exchanger, was designed.

3.2. Design of tube fin heat exchanger for the second wall cooling scheme

The external structure of the second wall cooling scheme designed for the tube fin heat exchanger is shown in Figure 7, consisting of a shell, heat exchange tubes, straight fins, oil circuit baffles, etc. The flow unit of the heat exchanger and wall cooling system is shown in Figure 8.

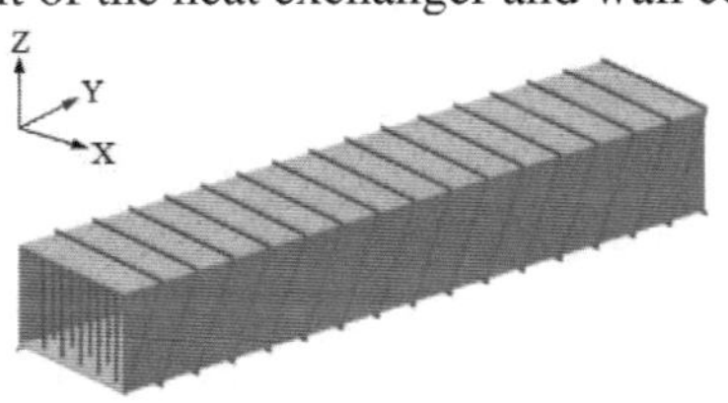

Figure 7. Structure of the first wall cooling scheme

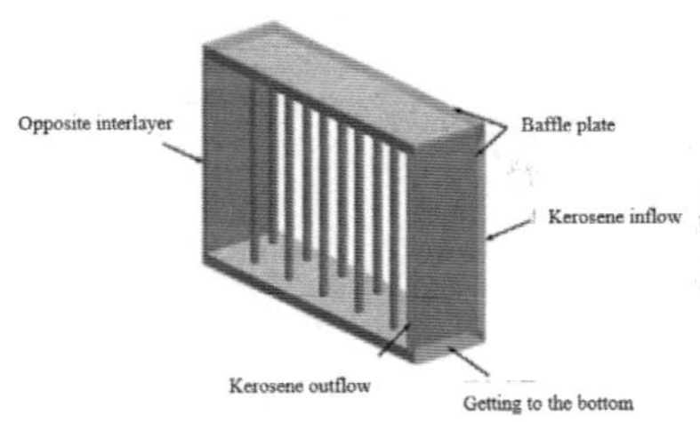

Figure 8. Flow unit diagram

From the figure 9 and10, it can be seen that the flow pattern of liquid in the heat exchanger is as follows: the upper and lower walls of each flow unit, as well as one of the side walls, are equipped with vertical baffles to block the connection with the front and rear unit cells, while the other side wall is equipped with an inclined baffle, which is connected to the adjacent baffles of the upper and lower walls at the beginning and end. The liquid flows into the outlet of the previous unit and is subjected to the tissue flow of the built-in inclined baffle, flowing into the cooling interlayer on the lower wall. Most of the fluid enters the heat exchange tube under pressure and exchanges heat with the air inside the heat exchanger; A small portion of the fluid is pressed into the cooling interlayer on the other side wall, and then collects with the kerosene in the upper wall channel of the heat exchange tube and flows into the side wall together. Under the action of the inclined baffle, it flows into the next heat exchange unit. This completes the circulation of the liquid within a flow unit, and in each cycle, all four walls of the heat exchanger are cooled by the liquid.

3.3. Comparison of wall cooling effects between two schemes

The temperature distribution of the upper, lower, and side walls of the two schemes is shown in Figures 9 and 10. It can be seen that after cooling, the temperature of the wall surface is basically maintained within 550K.

Figure 9. Temperature distribution of upper and lower walls

Figure 10. Temperature Distribution Diagram of side wall

4. Conclusion

This paper designs a compact gas-liquid cooling heat exchanger with a wall cooling structure. The performance of the heat exchanger was analyzed through numerical simulation research, as follows:

(1) The design of the heat exchanger was carried out by combining engineering calculations with numerical simulations.

(2) The temperature drop of the tube fin heat exchanger designed in this paper is 32% of the original temperature, the average heat transfer coefficient is K=1332.8W/(m2 · K), and the total pressure loss of the shell side fluid air is 1.3%. The fins in the tube fin heat exchanger provide a secondary heat transfer surface with considerable area and temperature, indirectly increasing the heat transfer area and improving the heat transfer efficiency of the heat exchanger.

(3) Two wall cooling schemes have been designed, integrating the wall cooling flow path with the heat exchanger tube flow path. After cooling, the wall temperature is within the allowable range of the metal material, and the cooling effect is good.

References

[1] Hu Q, Yao M, Wang C, et al. Adaptive backstepping control for air-breathing hypersonic vehicles with input nonlinearities. *Aerospace Science & Technology*, 2018, 73: 289-299.

[2] Fujii K, Watanabe S, Kurotaki T, et al. Aerodynamic heating measurements on nose and elevon of hypersonic flight experiment vehicle. *Journal of Spacecraft & Rockets*, 2015, 38(1): 8-14.

[3] Tao Z, Cheng Z, Zhu J, et al. Effect of turbulence models on predicting convective heat transfer to hydrocarbon fuel at supercritical pressure. *Chinese Journal of Aeronautics*, 2016, 29(5): 1247-1261.

[4] Hou Y Z, Zheng J L, Dong W. Transient test of aerodynamic heating for hypersonic vehicle. *Journal of Aerospace Power*, 2010, 25(2): 343-347.

[5] Liu S， Xu X， Yang Q， et al. Numerical simulation of supercritical catalytic steam reforming of aviation kerosene coupling with coking and heat transfer in mini-channel. *International Journal of Thermal Sciences*， 2019,137（3）： 199-214.

[6] Li Haowei， Lin Guowei， Feng Weijun， et al. Review on flow and heat transfer characteristics of hydrocarbon fuels and progress in re-generative cooling technology in hydrocarbon-fueled scramjet. *Aerospace Technology*， 2022（6）： 57-66.

Flow Heat Transfer and Thermal Stress Characterization of Elliptical Kelvin Structures Based on CT Reconstruction

Kangjie Liu, Zhaoda Zhang, Yu Liu*, Mingrui Sun, Jun Zhang and Guanghan Yan

Key Laboratory of Ocean Energy Utilization and Energy Conservation of Ministry of Education, Dalian University of Technology, Dalian, China

*liuyu@dlut.edu.cn

Abstract. Currently, most lattice structures are fabricated using additive manufacturing technology. However, the test specimens produced by this method often exhibit changes in surface roughness and structural morphology due to the bonding of metal powders, which can compromise the accuracy of numerical simulations. In this study, selective laser melting (SLM) technology is employed to manufacture ellipsoidal Kelvin test specimens. The actual elliptical Kelvin model was reconstructed using X-ray computed tomography (CT), and numerical simulations were conducted to compare the flow, heat transfer, and thermal stress characteristics between the reconstructed model and the design model. The results show that at a Reynolds number of 5000, the Nusselt number and friction factor of the actual model are 31.4% and 90.6% higher, respectively, compared to the design model. And the overall heat transfer performance of the actual model is found to be 22.8% lower. Furthermore, due to significant morphological changes in the metal 3D-printed test specimens, the thermal stress distribution in the actual model differs from that of the design model.

1. Introduction

Open-cell metal foams offer advantages such as a large specific surface area and high thermal conductivity, making them highly promising for the development of compact heat exchangers[1]. However, the existing metal foam fabrication process is challenging to control and optimize. Additionally, the topology of metal foam is difficult to design, and the shape of its skeleton is hard to fix. As a result, researchers have turned to ideal lattice structures as alternatives to metal foams[2], such as Kelvin, Octet-truss, BCC, and Weaire-Phelan, among others. These idealized lattice structures exhibit excellent heat exchange capabilities, the ordered structures make them more convenient for study[3-5], Among them, the Kelvin structure, which has a tetrakaidecahedron unit cell geometry, is simple and easy to arrange, making it an ideal candidate to replace metal foam for heat transfer research[6].
Subsequently, researchers made improvements to the Kelvin structure, one of which involved extending the structure in the flow direction. Kong et al.[2] found that the Kelvin anisotropic foams are more prone to inducing turbulence. When the Reynolds number exceeds 750, the Kelvin anisotropic foam structure outperforms the real foam. Sun et al.[7] investigated the Ellipsoidal Kelvin Structure (EKS) through both simulation and experimental studies. They found that the pressure drop of the EKS was 38.6% lower than that of the traditional Kelvin structure, while the Nusselt number was 10% lower. However, the overall heat transfer performance, characterized by the j/f value, increased by 50.9%. The EKS thus demonstrates excellent heat transfer and pressure drop characteristics, offering significant advantages for practical applications.

Most lattice structures, including EKS, are fabricated using additive manufacturing technology. In particular, porous structures used in heat exchangers require advanced selective laser melting (SLM) technology for metal printing. This process employs an energy source to selectively melt metal powders, forming a three-dimensional structure[8]. The resulting test pieces typically exhibit dense organization, high precision, and excellent mechanical properties[9], making them advantageous for producing small metallic test specimens. However, this method have certain drawbacks. For instance, some structures may end up heavier than the original design model due to the partial melting of loose powder beneath the part, particularly when scanning along the vector sintering contour trajectory. As shown in the SEM images of Al-Ketan et al[10], a significant amount of powder adheres to the bottom surface of the structure, increasing the relative density of the test piece while also causing a noticeable increase in surface roughness, which deviates from the designed shape. In a study by Liu et al.[11], lattice samples fabricated by the SLM technique exhibited defects such as strut waviness, variations in strut thickness, and strut oversizing or undersizing. Furthermore, Deng et al.[12] used a rough surface modeling method to simulate the actual 3D-printed surface topography. Their findings revealed that the heat transfer capacity of rectangular channels on 3D-printed rough surfaces increased by 8.99% compared to smooth surfaces, while the pressure loss increased by 25.02%. These observations suggest that the hydraulic-thermal properties of lattice structures, as well as the thermal stress conditions derived from CAD design models, may contain significant errors. Therefore, it is essential to study and validate the properties of actual 3D-printed test pieces.

In summary, to investigate the impact of metal 3D printing defects and surface roughness on the performance of novel porous media heat exchangers, the elliptical Kelvin structure—known for its superior hydrothermal performance—was selected as the research focus. The real elliptical Kelvin model was reconstructed using X-ray computed tomography. Numerical simulations were conducted to evaluate the thermofluidic behavior of both the real and designed models under turbulent flow conditions. Additionally, quantitative analyses were performed to assess the performance of the models. This study provides valuable insights for the design and practical application of porous media heat exchangers.

2. Manufacturing and CT reconfiguration

2.1. Design and manufacture of test pieces

An elliptical Kelvin structure with a/b = 2 was chosen for this study. Each elliptical cell has dimensions a × b × c = 3.84 mm × 1.92 mm × 1.92 mm, a porosity of 0.9 and a rod diameter of 0.31 mm. Additionally, a flat plate with a thickness of 0.5 mm was incorporated into the design of the EKS to simulate the spacer typically found in heat exchangers, as illustrated in figure 1.

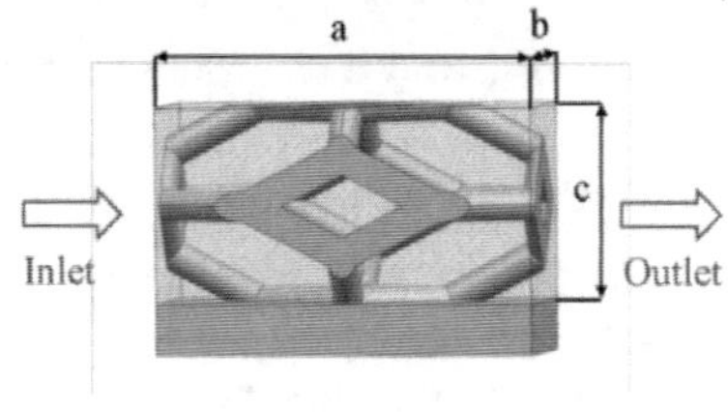

Figure 1. Ellipsoidal Kelvin structure.

The model was fabricated using additive manufacturing techniques and the parameters of the printing process are shown in table 1 The material used in this study is GH4169. A batch of test samples was printed simultaneously to evaluate the material's physical properties. The results of these tests are summarized in table 2 To address the issues of mesh volume and computational efficiency in the subsequent simulations, we selected a set of representative structures for study. Therefore, the model used in the following analysis has dimensions of 7.68 mm × 3.84 mm × 2.42 mm, corresponding to a 2×2 array as shown in figure 1.

Table 1. Parameters used in additive manufacturing.

Matter	Value/lectotype
Scanning pitch (mm)	0.08
Powder layer thickness (mm)	0.02
Scanning speed (mm/s)	900
Laser power (W)	135

Table 2. Physical parameters of GH4169.

Physicality	Value
Density (kg/m³)	8250
Thermal conductivity (W/(m·°C))	12.15
Specific heat capacity (J/(kg·°C))	460.61
Coefficient of linear expansion (10^(-6)/°C)	6.78
Young's modulus (Gpa)	183
Shear modulus (Gpa)	93
Poisson's ratio	0.3

2.2 CT scanning and reconstruction

The printed test piece was scanned using CT equipment with a voltage of 160 kV, a voxel size of 0.011 mm and a line size of 0.023 mm. After scanning, the reconstruction of the real structure is done using ImageJ and Simpleware software. figure 2 presents the 3D model of the EKS produced through reconstructed additive manufacturing technology. The morphology of the actual printed structure is clearly visible, revealing a rough surface with protrusions and depressions. To facilitate comparison, a CAD model with the same porosity as the actual printed structure is included. Both the design and the actual models are characterized parametrically, as summarized in table 3 The parameters used for this characterization include model size, porosity, rod diameter, equivalent diameter, and total heat transfer area.

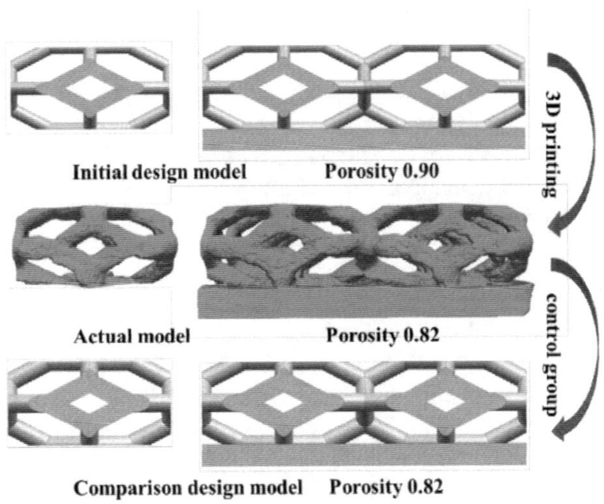

Figure 2. Diagram of initial design model, actual model and comparison design model.

Table 3. Designed and actual structural parameters.

Model	Model size (mm^3)	Porosity (ε)	Rod diameter (mm)	Characteristic length dh (mm)	Total heat transfer area (mm^2)
Initial design model	7.68×3.84×2.42	0.90	0.31	2.11	107.35
Actual model	7.77×3.86×2.51	0.82	0.37-0.53	1.58	144.42
Comparison design model	7.68×3.84×2.42	0.82	0.43	1.39	162.78

3. Simulation

3.1 Numerical simulation methods

Fluent is used to simulate the heat transfer in fluid flow. The model employed in the simulation is shown in figure 3 In this setup, the fluid flows along the x-axis. A fluid domain of 7.68 mm is allocated at both the inlet and outlet to prevent backflow during calculations and to allow the flow to fully develop. The inlet is defined as a velocity inlet, with the velocity set in the range of 10–60 m/s, while the outlet is modeled with a pressure boundary condition, maintaining a pressure of 0 Pa. Symmetric boundary conditions are applied at both ends along the y-axis. The bottom surface of the flat metal plate is heated, with a heat flux of 13,000 W/m² applied. The air properties are set to the default values from the Fluent database, and the material of the metal skeleton is specified as GH4169.

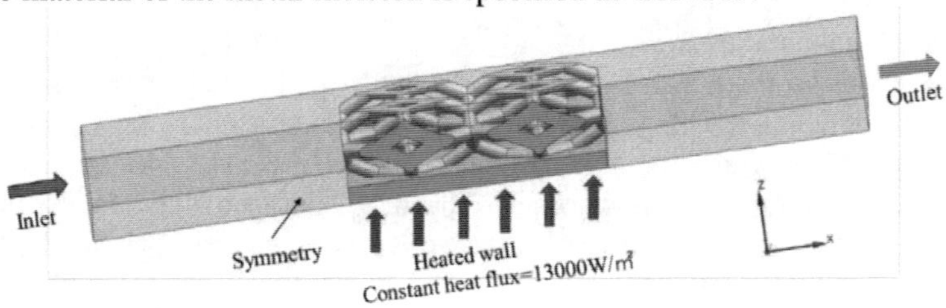

Figure 3. Simulation model diagram with an example of the designed structure.

The SST k-ω model is employed to address the limitations of the traditional k-ε model under certain flow conditions, particularly in the near-wall free-flow region, thus enabling a broader range of accurate applications. Pang et al.[13] found that the SST k-ω model provides more accurate predictions of high Reynolds number flows around a cylinder. Additionally, this model outperforms other models in capturing the effects of surface roughness. The k-ω model is also more precise than others in this regard. A dimensionless wall distance, **y+**, of less than 1 was selected for the simulations. Steady-state calculations were performed to statistically account for temperature and pressure in the steady-flow case. The relevant equations are as follows:

$$\frac{\partial}{\partial x}(\rho u_i)=0 \tag{1}$$

$$\frac{\partial}{\partial x_i}\left(\rho u_i u_j\right)=-\frac{\partial p}{\partial x_i}+\frac{\partial}{\partial x_j}\left(\mu\frac{\partial u_i}{\partial x_j}\right) \tag{2}$$

$$\frac{\partial}{\partial x_i}\left(\rho u_i c_p T\right)=\frac{\partial}{\partial x_i}\left(\lambda_f\frac{\partial T}{\partial x_i}\right) \tag{3}$$

$$k\frac{\partial^2}{\partial x_i^2}(T)=0 \tag{4}$$

$$\frac{\partial}{\partial x_j}\left(\rho u_j\frac{\partial k}{\partial x_j}-\left(\mu+\sigma_{k3}\mu_t\right)\frac{\partial k}{\partial x_j}\right)=P_k-\beta'\rho\omega k \tag{5}$$

Journal of Physics: Conference Series **3041** (2025) 012018 doi:10.1088/1742-6596/3041/1/012018

$$\frac{\partial}{\partial x_j}\left(\rho u_j\omega-(\mu+\sigma_{\omega 3}\mu_t)\frac{\partial \omega}{\partial x_j}\right)=2(1-F_1)\rho\sigma_{\omega 2}\frac{1}{\omega}\frac{\partial k}{\partial x_j}\frac{\partial \omega}{\partial x_j}+\alpha_3\frac{\rho}{\mu_t}P_k-\beta_3\rho\omega^2 \quad (6)$$

where the k is the thermal conductive of solid. The symbol ω denotes the turbulence frequency, while P_k represents the production rate of tur-bulence kinetic energy. The definition of the other variable and constants values are provided in Ref[14].

The model used in the heat-fluid-solid simulation focuses on the metal skeleton. The temperature field obtained from the flow heat transfer simulation is imported into Workbench for further analysis, where the structure is allowed to expand freely without any fixed constraints.

From Newton's second law, the governing equations for the solid can be derived as follows:

$$\rho_s\ddot{d}_s=\nabla\cdot\sigma_s+f_s \quad (7)$$

Where ρ_s is the solid density, σ_s is the Cauchy stress tensor, f_s is the volumetric force vector, and $\ddot{d}_s$ is the solid domain acceleration vector.

Thermal deformation term due to temperature difference:

$$f_T=\alpha_T\cdot\nabla \mathrm{T} \quad (8)$$

where α_T is the temperature-dependent coefficient of thermal expansion.

$$\begin{cases}\tau_f\cdot n_f=\tau_s\cdot n_s\\ \mathrm{d}_f=d_s\\ \mathrm{q}_f=q_s\\ \mathrm{T}_f=T_s\end{cases} \quad (9)$$

The above equation is the conservation of variables between the fluid and the solid, including stress, displacement, heat flow, temperature, where f denotes the fluid and s denotes the solid[15].

3.2 Grid-independent verification

The CT reconstruction model is meshed using Simpleware software. The mesh for the design model is created using ICEM and Meshing software. During the meshing process, local refinement is applied to certain areas of the model to ensure mesh quality. A visualization of the refined mesh is shown in figure 4. The grid-independence validation determines that the Actual model mesh size in the CFD grid is 4.21million and the Design model mesh size is 1.12million. The Actual model mesh size in the FE grid is 1.20million and the Design model mesh size is 0.42million.

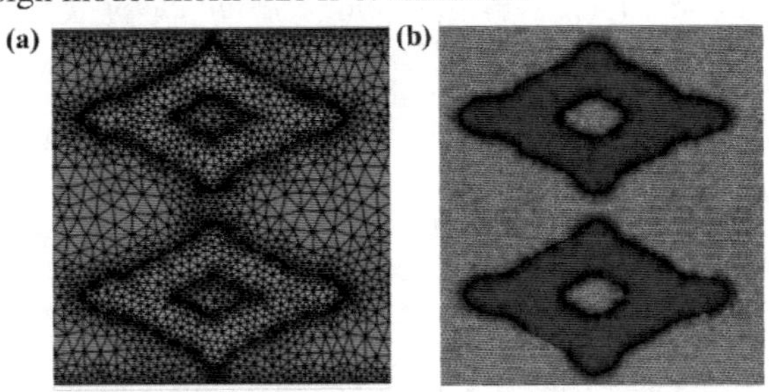

Figure 4. (a) Design model mesh detail (b) Actual model mesh detail.

4. Results and discussion

4.1 Flow Heat Transfer Characteristics

EKS with similar porosity are also presented by Sun et al.[16]. The data were compared as shown in figure 5. Both the pressure drop and heat transfer coefficient exhibit a consistent upward trend. In this study, the model accounts for an increase in base plate friction, which leads to a higher pressure drop compared to the values reported in the reference literature. Additionally, due to the spaced base plate heating configuration, the heat transfer coefficient is lower. Overall, the comparison results fall within reasonable predictions and align with the observed trend.

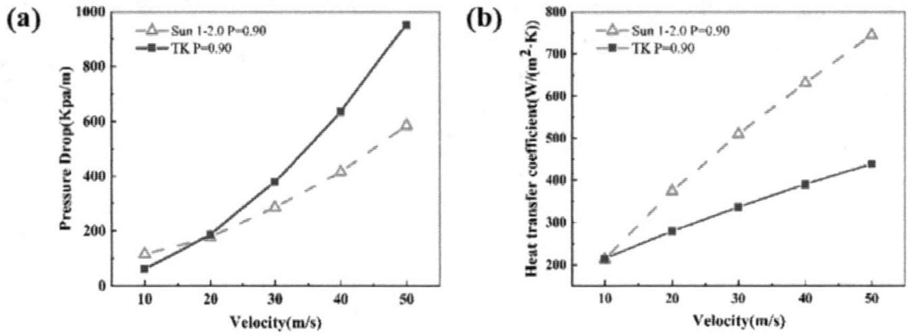

Figure 5. Comparison with simulation results of sun et al.[16] (a) pressure drop (b) heat transfer coefficient.

To evaluate the flow heat transfer performance of the ellipsoid Kelvin structure, the Nusselt number (Nu), the friction factor (f), and the heat transfer factor (j) are used. The corresponding calculation formulas are as follows:

$$f = \frac{\Delta P}{\Delta L} \cdot \frac{2d_h}{\rho u^2} \tag{10}$$

$$j = \frac{Nu}{RePr^{1/3}}, \tag{11}$$

Where u is the velocity of the fluid, Re is the Reynolds number and Pr is the Prandtl number.

The characteristics of the designed and actual models are shown in figure 7. The pressure drop plots in figure 6(a), the pressure drops of all three groups of models increase with the increase of velocity, and at a velocity of 60 m/s, the pressure drop of the CT reconstructed 0.82 porosity model increases by 164.2% and 61.2%, respectively, compared with that of the designed 0.9 porosity and 0.82 porosity models. In figure 6(b), it can be seen that the Nusselt number of the CT reconstructed model is higher than that of the other two groups of models, and the Nusselt number of the CT reconstructed model is increased by 49.3% and 31.4% compared to the designed models with 0.9 porosity and 0.82 porosity, respectively, at a Reynolds number of 5000. Figure 6(c) shows the friction factor of the three models at different Reynolds numbers. At a Reynolds number of 5000, the friction factor of the CT reconstructed model is 91.8% higher than that of the 0.9 porosity design model and 90.6% higher than that of the 0.82 porosity design model. Figure 6(d) illustrates the overall heat transfer performance at different Reynolds numbers. As shown in the Figure, the heat transfer performance of the actual model is lower than that of the designed model. Specifically, at a Reynolds number of 5000, the overall heat transfer performance of the CT-reconstructed model is 31.6% and 22.8% lower than that of the designed model with porosities of 0.9 and 0.82, respectively. This suggests that the heat transfer performance of the actual printed ellipsoidal Kelvin model is inferior to that of the original design, despite exhibiting similar friction losses. Consequently, this indicates that a heat exchanger fabricated with the ellipsoidal Kelvin unit as the core will not achieve the desired performance outcomes.

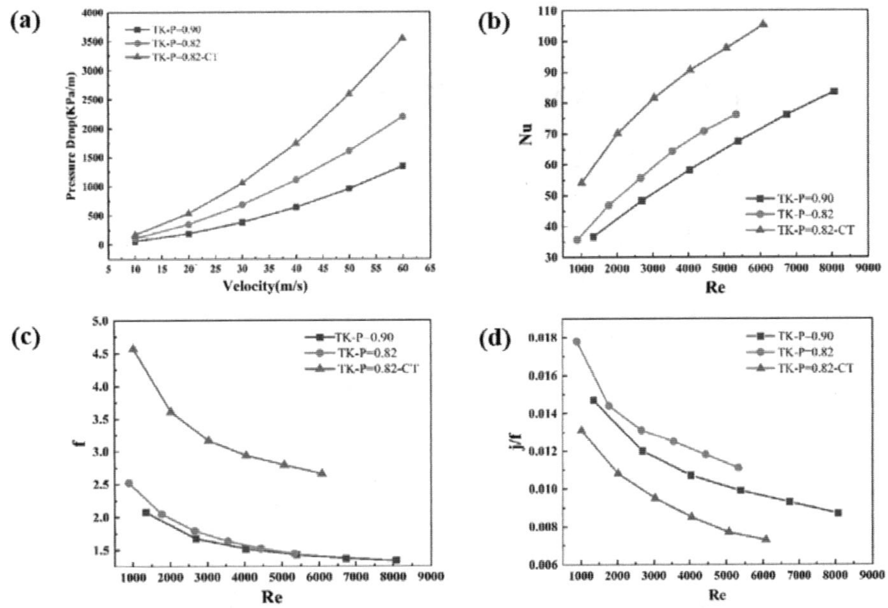

Figure 6. Comparison of initial design model, actual model, and comparison design model. (a) Pressure drop, (b) Nussle number, (c) Friction factor, and (d) Overall heat transfer coefficient.

4.2 Analysis of flow heat transfer mechanism

The hydraulic and thermal properties of the actual model differ significantly from those of the design model. This section presents a comparative study of the two models, both having the same porosity of 0.82, to investigate the differences in flow and heat transfer characteristics. To facilitate the analysis, a cross-sectional plane along the flow direction, located 2 mm in the positive direction of the z-axis, has been selected for the study.

Figure 7 illustrates the pressure distribution along the selected cross-section of both models under a flow condition with an inlet velocity of 40 m/s. From the interface, it is evident that the actual model exhibits a deviation from the regular symmetry, showing an irregular structure compared to the design model. The cross-section of the original design model features an elliptical Kelvin shape at the intersection of the circular pillars. This configuration results in a combination of several elliptical cross-sections, which closely resemble a rounded or elliptical skeleton. In contrast, the actual printed model exhibits significant distortion, deviating from the intended design.

The pressure distributions shown in figure 7 reveal localized areas of high pressure along the windward side of the skeleton in the flow direction for both models. This suggests that the flow is obstructed as it passes through the skeleton, causing a rapid decrease in flow velocity. To assess the pressure drop more effectively, a straight line was drawn at the center of the cross-section, intersecting the midpoints of the two skeletons. The pressure drop of the fluid after passing through the two skeletons was then calculated. The analysis indicates that the pressure drops in the actual model are 30.3% and 59.5% higher than those in the design model at the corresponding intersection points. This demonstrates that the actual model creates greater resistance to fluid flow, resulting in more significant pressure losses compared to the design model.

Journal of Physics: Conference Series **3041** (2025) 012018 doi:10.1088/1742-6596/3041/1/012018

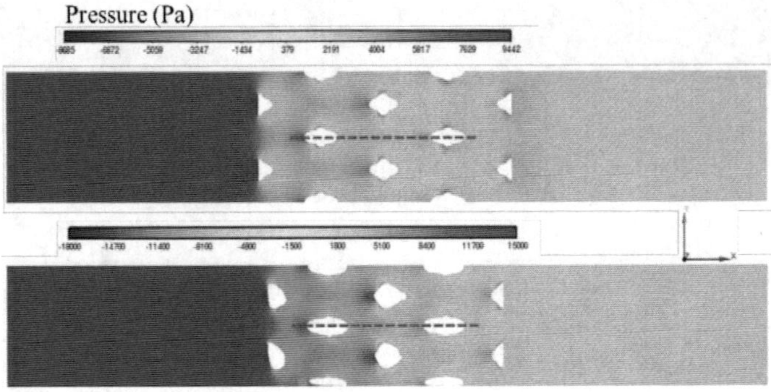

Figure 7. Cloud diagram of pressure drop distribution between design model and actual model.

Figure 8 illustrates the flow velocity distribution for both models. As the fluid enters through the inlet, its velocity increases significantly while passing through the skeleton. In the actual model, the average fluid velocity is 48.7 m/s, compared to 46.7 m/s in the design model. In the cross-sectional view, the fluid velocity is also higher in the actual model, reaching 62.4 m/s, compared to 53.9 m/s in the design model. The higher flow velocity and rougher surface of the skeleton in the actual model result in a larger Nusselt number (Nu) compared to the design model. Near the skeleton, the flow velocity increases more significantly in the design model. To quantify this change more accurately, two straight lines (A and B) are drawn along the cross-section, perpendicular to the flow direction, to compare the flow velocity at the two boundaries. Under the influence of the skeleton, the flow velocity increases by 31.6% in the design model, whereas the increase is 19.8% in the actual model. This indicates a more pronounced velocity increase in the design model, particularly in the middle section. This effect is especially noticeable near the elliptical skeleton's contact areas, where the flow velocity increases as the fluid flows around the skeleton. This also contributes to the observed pressure reduction, as discussed earlier.

In the design model, a noticeable pressure drop occurs near the elliptical cross-section of the skeleton, perpendicular to the flow direction, which does not occur in the actual model. In the design model, the fluid flows along the sides of the skeleton, and due to the elliptical cross-section, the fluid's cross-sectional area increases on both sides of the skeleton. This expansion causes the flow rate to rise, resulting in the formation of a distinct low-pressure zone. In contrast, the elliptical cross-section has a minimal effect in the actual model due to printing defects, leading to a less significant pressure drop on the windward side of the structure. As noted in the previous analysis, there is a significant difference in the flow characteristics between the actual and design models during the fluid flow process.

The actual model, due to severe bonding of the 3D-printed metal powder at the skeleton's inlet section, experiences a reduction in the effective cross-sectional area. As a result, the flow rate at the skeleton's inlet is higher than in the design model. However, the flow rate increase in the middle section of the design model leads to a gradual narrowing of the gap between the two models. This suggests that manufacturing-related factors have caused significant changes in the flow characteristics of the actual model. The high flow velocity and rough surface of the skeleton, due to the reduced cross-sectional area at the inlet of the actual model, result in a higher Nusselt number (Nu) compared to the design model. However, these same factors—high flow velocity and surface roughness—also cause an increase in fluid pressure drop and friction factor, ultimately leading to a lower overall heat transfer capacity than that of the design model. In practical applications of metal 3D printing with an EKS, treatments such as acid etching and sandblasting can be applied to reduce the porosity and surface roughness of the test piece, bringing it closer to the design model.

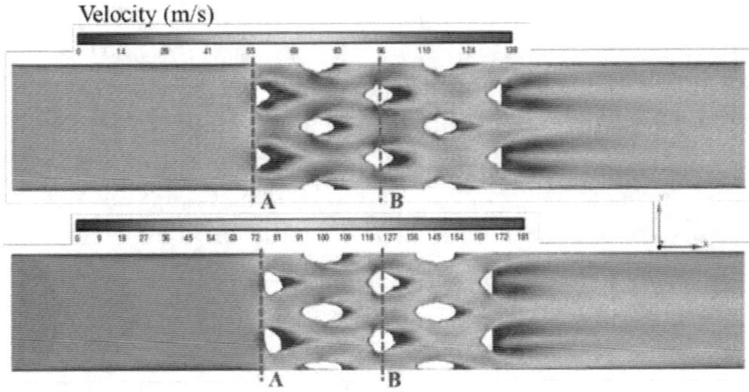

Figure 8. Cloud plot of flow velocity distribution against the design model and the actual model

4.3 Thermal stress characterization

As shown in figure 9, under a flow rate of 10 m/s, the temperature distribution of the skeleton is displayed in the form of a cloud diagram. The temperature range of the design model is between 29.5°C and 76.6°C, with an average temperature of 64.1°C, while the temperature range of the actual model is from 36.2°C to 66.3°C, with an average temperature of 56.4°C. The actual model exhibits better temperature uniformity, which can be attributed to the enhanced convective heat transfer caused by the roughness of the metal 3D-printed structure.

As the fluid flows along the X-axis, the temperature distribution on the skeleton surface shows that the temperature is relatively low near the inlet side of the cooling fluid, gradually increasing in the direction of flow. Similarly, perpendicular to the flow direction, the temperature of the heated surface is higher due to the upward transfer of heat along the bottom surface. As a result, the temperature of the skeleton decreases gradually along the height direction. The cloud diagram clearly shows two regions on the bottom plate with relatively low temperatures. These regions are located near the ellipsoidal Kelvin skeleton. The heat conduction from the flat plate to the ellipsoidal Kelvin skeleton facilitates the upward transmission of the skeleton temperature, while simultaneously creating fluid disturbances that significantly enhance heat transfer. The drop in temperature on the flat plate serves as strong evidence of the important role that the skeleton plays in enhancing heat transfer.

From the perspective of temperature distribution, the maximum temperature difference between the minimum and maximum temperatures in the design model and the actual model is 47.1°C and 30.1°C, respectively. This temperature difference results in material expansion and generates thermal stresses, as shown in the equivalent thermal stress distribution of the design model. The maximum stress occurs at the cross-section of the pillars on the inlet side. Other regions with relatively high thermal stresses, which are similar in shape, are located at the intersections of semi-circular pillars. These regions are characterized by structural discontinuities and large temperature gradients, leading to localized high thermal stresses.

Figure 9. Skeleton temperature distribution against design model and actual model.

Figure 10 presents the thermal stress distribution of the actual model, which is influenced by temperature. It can be observed that the location of the maximum thermal stress has shifted. In the actual model, there is no stress concentration at the intersection of the ellipsoidal Kelvin skeleton struts, as seen in the design model. This is due to the melting of the metal powder in the strut intersections, which did not form distinct boundaries but instead resulted in a smooth transition compared to the design model. The maximum thermal stress in the actual model occurs at the connection between the bottom of the ellipsoidal Kelvin skeleton and the upper surface of the base plate. This is attributed to localized protrusions and depressions caused by printing defects at the connection point between the strut and the flat plate. Due to the poor continuity at this connection during structural expansion, large thermal stresses are generated.

Since the heat transfer performance of the design model is not as efficient as that of the actual model, the temperature difference in the skeleton is larger in the design model. However, the maximum thermal stress is nearly the same in both models. Therefore, even with the same temperature distribution as the design model, the actual 3D-printed structure may still experience increased local thermal stresses due to additive manufacturing defects. This is an important consideration for practical use.

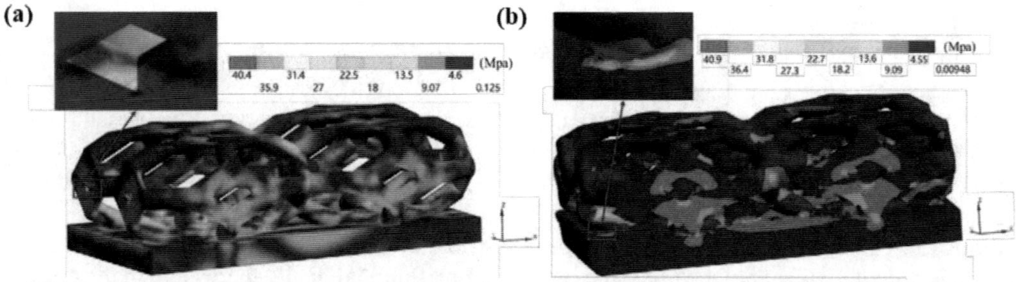

Figure 10. Thermal stress cloud of the skeleton against the design model and the actual model.

5. Conclusion

In this paper, the ellipsoidal Kelvin skeleton was fabricated using Selective Laser Melting (SLM) technology, and the actual model of the Ellipsoidal Kelvin Structure was reconstructed via computed tomography (CT). It was observed that the surface of the skeleton exhibited numerous protrusions and depressions, caused by the bonding of metal powder during the printing process. The design model, which had the same porosity, was re-drawn for comparison. The hydraulic-thermal properties, flow heat transfer mechanism, and thermal stress of both models were investigated using flow heat transfer simulations and heat flow solid simulations, respectively. The conclusions are as follows:

At an inlet fluid flow rate of 60 m/s, the pressure drop of the actual model with 0.82 porosity was found to be 61.2% higher than that of the design model. At a Reynolds number of 5000, the Nusselt number of the actual model was 31.4% higher than that of the design model, and the friction factor was 90.6% higher. However, the overall heat transfer performance of the actual model decreased by 22.8% compared to the design model.

Regarding thermal stress, the maximum thermal stresses in both models were found to be similar. In the design model, the maximum thermal stress occurred at the intersection of the struts in the ellipsoidal Kelvin skeleton, primarily due to the discontinuity between the two cylindrical surfaces. In contrast, the actual model did not exhibit a distinct interface at the intersection due to the melting of the metal powder, which resulted in a more continuous structure and reduced thermal stress in that region. The thermal stress in the actual model was primarily concentrated at the interface between the skeleton and the ground. There was a significant difference in the location of the maximum thermal stress between the two models.

6. References

[1] Hu H, Zhao Y and Li Y 2023 Research progress on flow and heat transfer characteristics of fluids in metal foams *Renewable and Sustainable Energy Reviews* **171** 113010

[2] Kong X, Zhang H, Du Y, Wang X and Xiao G 2024 A study of pore scale flow and conjugate heat transfer characteristics in real and Kelvin anisotropic foams *Int J Heat Mass Tran* **221** 125024
[3] Sun M R, Yan G H, Liang Y Q, Zhao J F and Song Y C 2024 The investigation of anisotropic kelvin cells: Forced convective heat transfer *Energy* **292**
[4] Ekade P and Krishnan S 2019 Fluid flow and heat transfer characteristics of octet truss lattice geometry *Int J Therm Sci* **137** 253-261
[5] Shahrzadi M, Emami M D and Akbarzadeh A H 2022 Heat transfer in BCC lattice materials: Conduction, convection, and radiation *Compos Struct* **284**
[6] Iasiello M, Cunsolo S, Bianco N, Chiu W K S and Naso V 2017 Developing thermal flow in open-cell foams *Int J Therm Sci* **111** 129-137
[7] Sun M, Yan G, Hu C, Zhao J, Duan F and Song Y 2023 Thermal and hydraulic behaviours of Kelvin cells from metallic three-dimensional printing *Appl Therm Eng* **219** 119433
[8] Lv J J, Jia C Z, Yang J C and Zhang Y F 2017 *PROCEEDINGS OF THE 2017 7TH INTERNATIONAL CONFERENCE ON MANUFACTURING SCIENCE AND ENGINEERING (ICMSE 2017)* vol 128, ed Z Y Jiang (7th International Conference on Manufacturing Science and Engineering (ICMSE): pp 256-259
[9] Bai Y C, Fu F, Xiao Z F, Zhang M K, Wang D, Yang Y Q and Song C H 2018 Progress in selective laser melting equipment, related biomedical metallic materials and applications *J Zhejiang Univ-Sc A* **19(2)** 122-136
[10] Al-Ketan O, Rowshan R and Abu Al-Rub R K 2018 Topology-mechanical property relationship of 3D printed strut, skeletal, and sheet based periodic metallic cellular materials *Addit Manuf* **19** 167-183
[11] Liu L, Kamm P, García-Moreno F, Banhart J and Pasini D 2017 Elastic and failure response of imperfect three-dimensional metallic lattices: the role of geometric defects induced by Selective Laser Melting *J Mech Phys Solids* **107** 160-184
[12] Deng Y P, Fu L, Liu Y S, Jiang X, Cui Y and Wu D F 2022 Investigation on Heat Transfer Characteristics of Rectangular Channels with Internal Rough Surface Naturally Formed by Selective Laser Melting Three-Dimensional Printing *J Fluid Eng-T Asme* **144(11)**
[13] Pang A, Skote M and Lim S Y 2016 Modelling high Re flow around a 2D cylindrical bluff body using the k-ω (SST) turbulence model *Prog Comput Fluid Dy* **16(1)** 48-57
[14] Sun M R, Yang L, Hu C Z, Zhao J F, Tang D W and Song Y C 2021 Simulation of forced convective heat transfer in Kelvin cells with optimized skeletons *Int J Heat Mass Tran* **165**
[15] Cai, H. W. (2018). Thermal Stress Analysis and Structural Optimization of Fixed Tube-Sheet Heat Exchanger (Master's thesis). Nanchang University. Retrieved from http://www.wanfangdata.com.cn/
[16] Sun M R, Zhang L X, Hu C Z, Zhao J F, Tang D W and Song Y C 2022 Forced convective heat transfer in optimized kelvin cells to enhance overall performance *Energy* **242**

Research on on-orbit temperature analysis and thermal design optimization of geosynchronous orbit cryogenic infrared camera

Xu Nana, Yu Feng, Wang Yang and Liu Chang

Beijing Institute of Space Mechanics and Electricity, China

Email：xunn_85@126.com

Abstract. With the development of space sensors, infrared cameras are increasingly being used in deep space exploration and earth observation missions. It is necessary to reduce the temperature of the opto-mechanical body to a low temperature and maintain high stability while reducing the detector to a low temperature when the detection target is weak. In this paper, according to the temperature control requirement of a geosynchronous orbit cryogenic optical infrared camera, the thermal control design is carried out, and the results of different thermal control schemes are compared. Finally, through the integrated design of satellite payload, the high stable temperature control of the fully transmissive cryogenic optical system was realized by means of attitude adjustment & beveled cut hood & indirect radiation refrigeration. The temperature data during the camera's three-year lifespan on orbit is analyzed and the correctness of the camera's thermal control design is validated. Also the optimization suggestions for some of the thermal control schemes used in the camera are analyzed, which provides support for further improving the temperature control accuracy of high-orbit low-temperature infrared cameras and reducing thermal control resources.

1. Introduction

Infrared cameras are increasingly used in deep space exploration and earth observation missions because of their all-day observation, better penetration, and the ability to observe targets that are difficult to detect by visible light systems[1-3]. The geosynchronous orbit has the advantages of wide monitoring range and high temporal resolution, which can continuously observe the targets in the shooting area. China has also successively launched some geosynchronous orbit satellites such as GF-4 and FY-4 et al to achieve continuous observation of the earth[4-7]. Some of them are infrared cameras, but most are normal temperature optics & low temperature detectors. When the weak target is detected, the opto-mechanical background thermal radiation will become the main factor affecting the sensitivity of the detector, and the temperature of the opto-mechanical body needs to be reduced to a low temperature.

Combined with the characteristics of the thermal environment of a geosynchronous orbit and the thermal control requirements of the low-temperature optical infrared remote sensing camera, the thermal control design is carried out, and the high-stability temperature control of the fully transmissive low-temperature optical system is realized by means of attitude adjustment & beveled cut hood & indirect radiation refrigeration. through the integrated design of satellite payload. The satellite was launched in 2021 and has been in stable operation in orbit for 3 years. The camera thermal control system has performed well, and the temperature of the optical and mechanical body and detector and other structures have met the imaging needs.

2. Camera Overview

The infrared camera which has a transmissive optical system works in geosynchronous orbit. The main structure of the camera includes a lens hood assembly, an optical system, a focal plane assembly, a video circuit box and a load-bearing structure. The camera inlet is in the +Z side direction for ground imaging. The schematic diagram of the camera structure is shown in figure 1. The lens hood is located at the front of the optical system, the focal plane assembly is located at the end of the optical system, the video circuit box is loaded near the focal plane assembly because the distance between them cannot be greater than 200mm. The focal plane assembly consists of a detector, a refrigerator, a Dewar and a Dewar window in order to satisfy ground testing request.

To meet the imaging requirements of the camera, the detector needs to be cooled to below 100K, and the optical system needs to be cooled to below 220K. Additionally, there should be no high-temperature structures within the entire optical path, and the visible structures within the optical path must be maintained at temperatures below 240K. The temperature specifications for various camera components are shown in table 1.

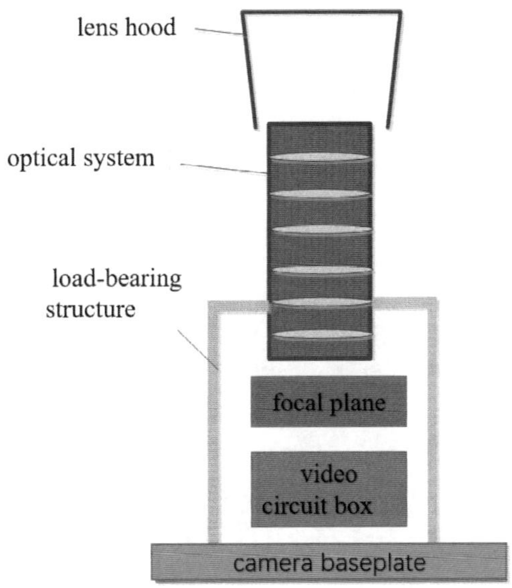

Figure 1. Schematic diagram of the main structure of the camera.

Table 1. Temperature requirements for each component.

Components		Temperature requirment
Cryogenic optical system		210±10K Temperature gradient≤5K Temperature fluctuations≤±0.3K
Detector		100±2K Temperature fluctuations≤±0.3K
Refrigerator	Hot end	210-250K
	Compressor	253-293K
Dewar window		≤240K
Video Box		258-313K

3. Thermal control design

3.1. Thermal control design of low-temperature optical system

The low-temperature optical system needs to be cooled to about 210K, and has high stability requirements. The camera is a fully transmissive optical system, and the optical system is directly facing the cold space. However, the camera operates on the geosynchronous orbit. Without evasion, the cryogenic optics will be directly irradiated by the sun. The front-end optical system will not be able to reduce to low temperatures, and the temperature of the mirror will fluctuate by more than 20°C in the case of a sharp change in heat flow. Therefore, in order to achieve high stability control and cooling of the fully transmissive cryogenic optical system, it is necessary to take necessary avoidance strategies to shield the disturbance of the temperature of the optical system by external heat flow.

Usually, the methods of lengthening the hood and avoiding the sun for a short period of time around midnight can be used to avoid the direct sunlight. The length of the hood should be designed according to the size of the aperture and the acceptable angle of avoidance. But even in this case, the inside of the hood will still be exposed to the sun for a long time. In order to reduce the temperature of the hood, on the one hand, we can use the material with high thermal conductivity to obtain the uniform temperature which can reduce the high temperature of the locally irradiated area, and on the other hand, a low-absorption-emission ratio coating is sprayed on the outside of the hood to reduce the temperature of the body by dissipating heat from the outside. The temperature of the optical system and the hood are shown in the figure 2.

Although the above scheme avoids the direct irradiation of the low-temperature optical system, but the temperature of the optical system still fluctuates greatly and the temperature of the front-end optical system is obviously high due to the high temperature and large temperature fluctuation of the hood. In order to further reduce the temperature of the optical system and temperature fluctuations, it is necessary to reduce the heat absorbed by the hood and reduce the temperature of the hood.

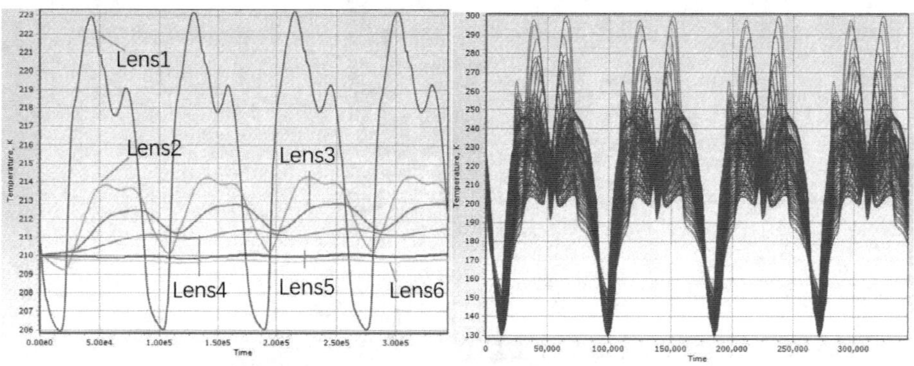

(a) (b)

Figure 2. Temperature results of the optical system (a) and the hood (b) using conventional scheme.

Scheme 1: Reflective ellipsoid light bar

In order to reduce the effect of stray light, the inside of the traditional lens hood is generally ultra-dark. The solar absorption rate of the inside is very high which resulting in a high temperature. The reflective ellipsoid light bar [9] uses the reflection principle of the ellipsoid, when the light passes through one focal point of the ellipsoid, the reflected light at any point on the ellipsoid will be reflected through another focal point. Then the heat load of the light inlet can be reduced significantly and the structure also has a good ability to eliminate stray light.

In order to reduce the effect of stray light, the inside of the traditional lens hood is generally ultra-dark. The solar absorption rate of the inside is very high which resulting in a high temperature. The reflective ellipsoid light bar [9] uses the reflection principle of the ellipsoid, when the light passes through one focal point of the ellipsoid, the reflected light at any point on the ellipsoid will be reflected through another focal point. Then the heat load of the light inlet can be reduced significantly and the structure also has a good ability to eliminate stray light.

Scheme 2: Attitude adjustment & beveled cut hood

Although the solar radiation heat flux of geosynchronous orbit satellites varies drastically, it is highly regular [8]. From 0:00 to 12:00, sunlight is incident from the +X side, and the -X side is long-term shade, from 12 :00 to 24:00, the sunlight is incident from the -X side, and the +X side is always shaded. If we make a turn at 12 o'clock and 0 o'clock, the sun will always shine from the +X side of the satellite. We can set an oblique incision hood to block the sun on +X side to avoid the sun irradiating the internal.

For the short time around 0:00 midnight, satellite attitude avoidance can be used to avoid the sun. The inside of the hood will can't be illuminated all the time, and the outside will always be illuminated. Then the multi-layer insulation component can be used as a lens hood material which is highly insulated to reduce the effect of external heat flow.

Figure 3 and figure 4 are the temperature results of the lens and the hood under the above two schemes. The two optimization schemes effectively reduce the temperature of the low-temperature optical system and improve the temperature stability. Compared the two schemes, the second scheme avoids the sunlight entering the light inlet completely, so the temperature of the optical system is lower and more stable.

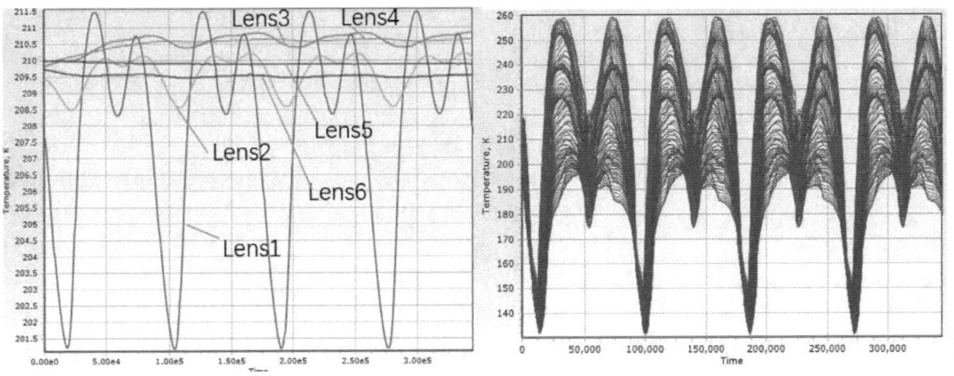

(a) (b)

Figure 3. Temperature results of the optical system (a) and the hood (b) using Scheme 1.

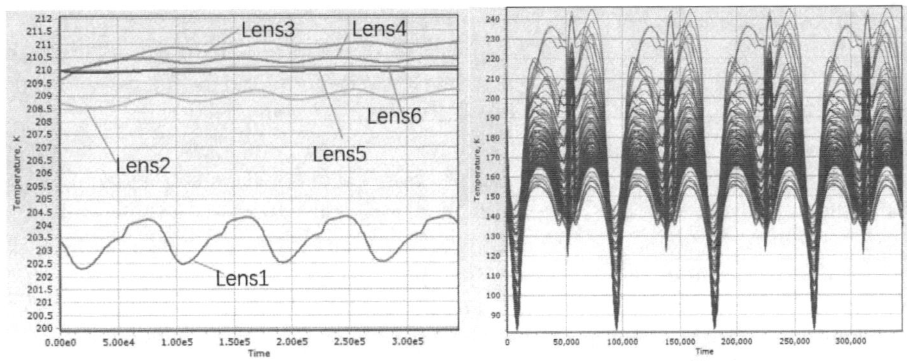

(a) (b)

Figure 4. Temperature results of the optical system (a) and the hood (b) using Scheme 2.

However, the temperature fluctuation of the first lens of the two schemes still cannot meet the requirements of ±0.3K. From the above analysis results, the temperature fluctuation of the second lens relative to the first lens, and the third lens relative to the second lens can be reduced by an order of magnitude, so an infrared window lens with high transparency can be arranged at the front of the optical system to achieve thermal insulation and ensure the temperature stability of the internal optical system.

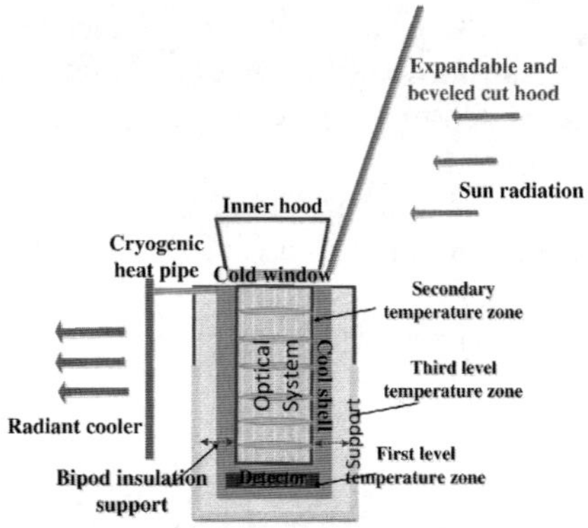

Figure 5. Schematic diagram of the thermal control scheme of the low-temperature optical system.

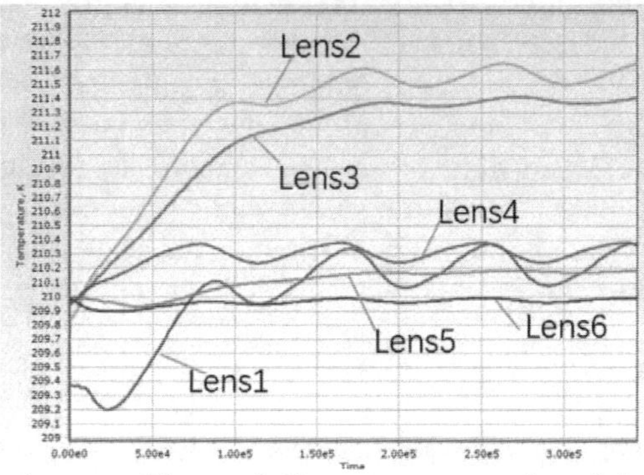

Figure 6. Temperature results of the optical system.

Combined with the tasks of this project and the attitude that can be achieved by the satellite, the optical system of this model adopts the method of scheme 2 and additionally adding a cold window at the front to achieve high stability temperature control of the optical system. The final temperature control results are as follows.

3.2. *Thermal control design of detector components*

The detector is spliced by 4 pieces of 2K*2K. It's cooled to below 100K by a vascular tube refrigerator. In order to meet the need of ground testing, the detector is sealed through the Dewar structure. The Dewar window is used to meet sealing and imaging needs in the direction of the optical path. Because the Dewar window is located within the optical path, the temperature of the Dewar window is required to be reduced to low temperature.

The structure of the detector assembly is shown in the figure 7. In order to meet the sealing requirements, efficient thermal insulation between the Dewar window and the hot end of the refrigerator was not possible. Therefore, the Dewar window needs to be refrigerated.

It is too complicated to set up a separate refrigerator for the window. Therefore, it is considered that the Dewar window borrows the cryogenic optical system for cooling, as shown in the figure 7. The cryogenic optical cold cylinder shell is lengthened and the Dewar window and some of the Dewar shell are placed in the low-temperature optical cold cylinder shell. The temperature of the hot end of the refrigerator is reduced to ensure that the temperature at the window is lower than 240K.

The lower temperature of the Dewar window and the hot end of the refrigerator also can reduce the static heat leakage of the detector. Compared with the normal temperature of 20°C the static heat leakage can be reduced by about 3W.

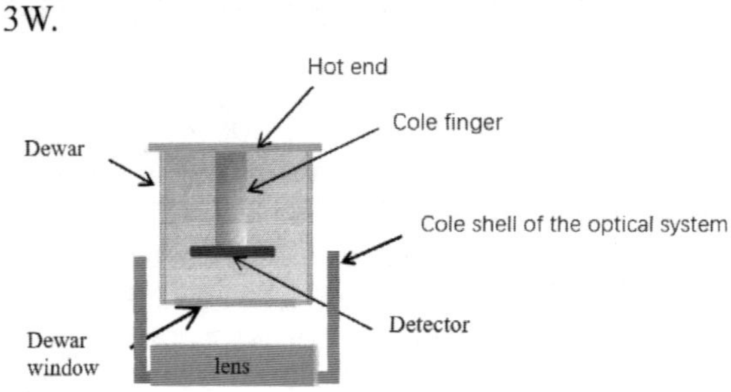

Figure 7. Schematic diagram of the thermal control for the detector assembly.

3.3. *Other thermal control designs*

The satellite provides a long-term shady side too while avoiding irradiation of the optical system. The north and south sides are periodically irradiated with seasonal changes. The shady side and coupling of the north and south sides can all be used for heat dissipation of the internal heat source. The heat sources that need to dissipate heat include the hot end and compressor of refrigerator, the video box, etc. The requirements of the heat dissipation surface are as follows according to the temperature control

Journal of Physics: Conference Series **3041** (2025) 012019 doi:10.1088/1742-6596/3041/1/012019

requirements.

The lower the temperature, the worse the efficiency of radiation heat dissipation. In order to satisfy the low temperature of the Dewar window, the temperature of the hot end needs to be reduced to below 250K. So the demand for heat dissipation surfaces on the north and south sides is much larger than that on the shade side, so the hot end needs to use the shade to measure the heat dissipation surface as much as possible. The difference is much smaller for the compressor and the video box, so when the area of the shady side is insufficient, they can also consider the coupling of the north and south side to dissipate the heat.

Table 2. Heat dissipation surface requirements for different directions.

Heat Source	Shady side	Coupling of the north and south sides
Hot end（60W）	0.5 m²	1.6 m²
Compressor（30W）	0.12 m²	0.27 m²
Video box etc. （80W）	0.2 m²	0.33 m²

4. On-orbit temperature data and analysis

The satellite was launched in 2021 and has operated stably for three years. It has experienced the test of various extreme working conditions in the synchronous orbit. Figure 8 gives the temperature curve of the cryogenic optical system. From the on-orbit data, we can see the temperature of the cold window is between 206K-209K, the temperatures of lens 1 to lens 6 are maintained between 209K-212K, the axial temperature difference of the six lenses is within 3K. The front-end optical system is directly facing the external space and the external environment, the temperature fluctuation is relatively large. The temperature fluctuation of the cold window is 2.5K, the temperature fluctuation of lens 1 is ±0.21K, which is better than the target value ±0.3K. From the results, we can see the temperature fluctuation of lens 1 is significantly reduced. The design of the cold window has a significant thermal shielding effect on the temperature stability of the internal optical system.

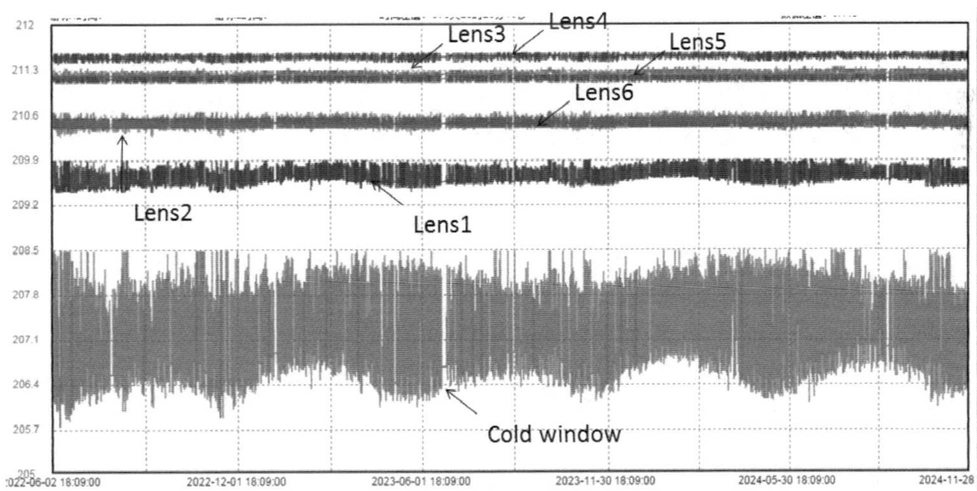

Figure 8. On-orbit temperature curves of cryogenic optics.

Figure 9 is the on-orbit temperature curve of the compressor, hot end, video box and DCDC electronics. The area of the shady side of the satellite is sufficient, so the heat dissipation surfaces such as the hot end, the compressor and the video box are all arranged on the shady side. The temperature is relatively stable because the external heat flow fluctuation is small. The temperature only rise when changing to imaging mode from standby mode.

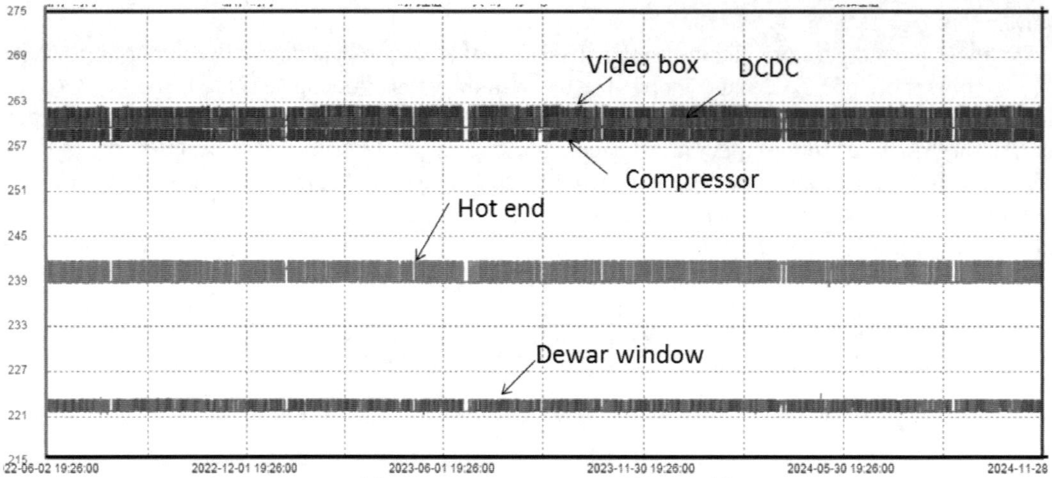

Figure 9. On-orbit temperature curves of internal heat sources.

Figure 10 shows the temperature value of the detector (the displayed value is the decimal source code value, which converts to a temperature is 100±0.3K). The temperature of the detector is controlled by a closed-loop refrigerator and the temperature of the compressor and the hot end is stable, so the detector temperature is stabilized at 100±0.3K in the case of random switching between standby and imaging modes.

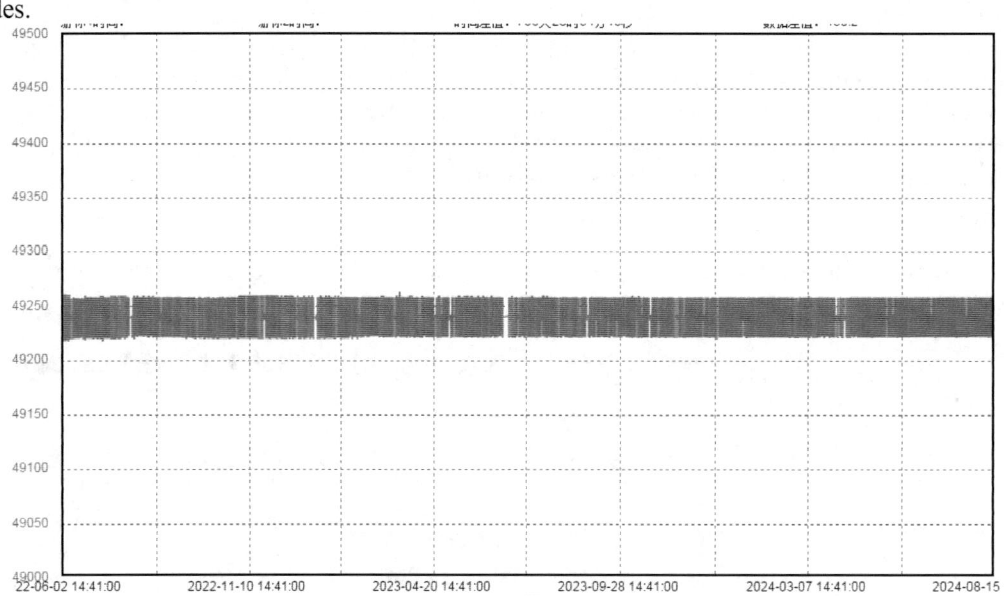

Figure 10. On-orbit temperature curve of the detector.

4. Suggestions for thermal design optimization of cryogenic cameras

4.1. Application of high thermal conductivity graphite material

The low-temperature optical system obtains uniform temperature through the external cold cylinder shell structure. The cold cylinder shell structure generally only arranges one or two groups of heat pipes[10] along the axial direction to transfer heat. Limited by the complexity of the structural layout, the number of heat pipes cannot be too much, so the temperature uniformity of the cold cylinder shell also needs its own high thermal conductivity material to guarantee. This camera uses aluminum alloy material with good thermal conductivity, and in order to meet the temperature uniformity, the thickness of aluminum alloy is designed to be 4mm. Although the requirements are met, the weight is heavier. The thermal conductivity of graphite materials can reach about 1500W/m*K [11~13], which is about 10 times thermal conductivity of aluminum alloy, but the weight is only about 70%. Figure11 shows the temperature contour of a 4mm aluminum alloy cold cylinder shell and a 1mm graphite composite (0.4mm graphite + 0.6mm carbon fiber skin) cold cylinder shell. The temperature uniformities are comparable, but the weight of the aluminum cold cylinder shell is 5 times that of the graphite composite.

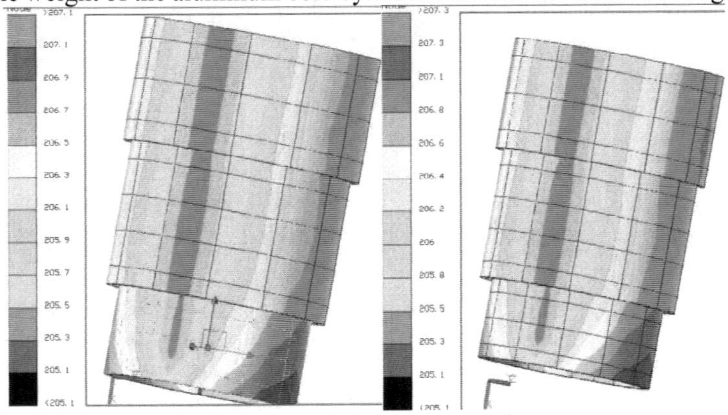

(a) (b)

Figure 11. Comparison of aluminum (a) and graphite composite (b) cold cylinder

4.2 Application of infrared high reflection coating layer

Infrared lenses generally need to be coated to ensure high transparency in the working spectrum [14], but the coating layer generally has a high absorption rate for the infrared energy of other spectral bands, so the infrared lenes are generally set to a structure with high emissivity in the thermal analysis model. The large infrared radiation heat change with the external environment is one of the main reasons why it is difficult to control the temperature of infrared lens. If the infrared emissivity on the outer side can be reduced, a coating with high permeability in the working bands but high reflection in other spectral bands can be designed, the difficulty of temperature control of the infrared optical system can be greatly reduced and the thermal design can be greatly optimized.

Figure 12 shows the temperature curves of the lenses under the various schemes described in Section 1.2 when the emissivity of the optical lens is reduced to 0.1. The conventional scheme still cannot meet the requirements of high stability, but scheme 2 and scheme 3 can all meet the requirements. And there is no need to add a lens at the front end of the camera, which enhances the adaptability of the satellite and saves on-board resources.

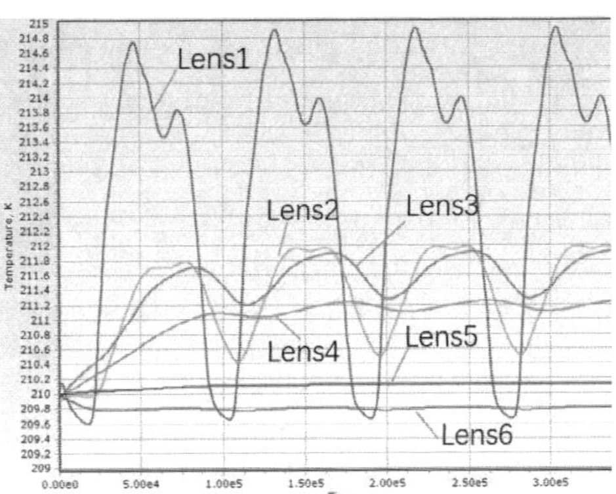

(a)

Journal of Physics: Conference Series **3041** (2025) 012019 doi:10.1088/1742-6596/3041/1/012019

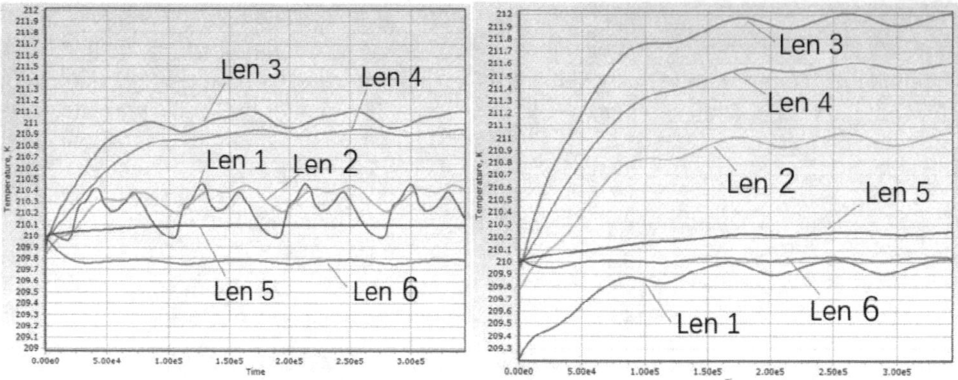

(b) (c)

Figure 12. Temperature control results of conventional scheme (a), reflective ellipsoid light bar scheme (b) and attitude adjustment & beveled cut hood scheme (c)

5. Conclusion

The thermal control design difficulties of the low-temperature optical system are analyzed according to the thermal environment characteristics and temperature control requirements of a synchronous orbit low-temperature infrared camera in this paper. The design results of different schemes are compared, and finally the scheme of attitude adjustment & beveled cut hood & indirect radiation cooling is selected to achieve the high stable temperature control requirements of the low-temperature optical system. The three-year on-orbit data is analyzed. Furthermore, combined with the thermal design of the camera, the design ideas of high thermal conductivity graphite cold cylinder shell and special coating layer for the outer side of the first lens were proposed, which will optimize the thermal control design and save on-board resources. Then the high temperature requirements can be satisfied with small energy demands. It also provides support for the subsequent high-orbit transmission infrared low-temperature optical high-precision thermal control design.

References

[1] Zhang Y, Zhou F, Ruan NJ and Wu LM. Overview of cryogenic refrigeration technology in space infrared astronomical telescopes. *Spacecraft Recovery & Remote Sensing*. **2013,34(5):**46-55

[2] Li MX,Wang SX,Liu Q, Miao JY, Wang ZS, He X. Design and verification of the space-cold optical long-wave infrared camera. *Infrared.* **2022 43(7)**:21-28

[3] Hudyma RM. Thermal MWIR Objectices.*SPIE*,**1995,2540**:119-235

[4] Yu F, Xu NN and Zhao ZM. On-orbit temperature analysis and thermal design optimization for camera on GF-4 satellite. *Journal of Beijing University of Aeronautics and Astronautics*. **2021 47(1)**:177-186.

[5] Wang DZ and He HY. Observation Capability and Application Prospect of GF-4 Satellite. *Spacecraft Recovery & Remote Sensing*. **2017 38(1)**:98-106

[6] Wang GQ, Shen X, Wang WC, Wang BY and Hu BT. On board adaptation of temperature of the FY-4 meteorological satellite radiation imager. *Journal of Infrared and Millimeter Waves*. **2018 37(5)**:540-544

[7] Xu NN, Yu F and Zhou ZH. Thermal design and validation of a geosynchronous orbit infrared camera. I*nfrared And Laser Engineering*, **2021,50(9)**:1-8

[8] Yu F, Xu NN, Zhao Y, Xu XF and Feng YG. Thermal Design and Test for Space Camera on GF-4 Satellite. *Spacecraft Recovery & Remote Sensing*. **2016,37(4)**:72-78

[9] Yang T, Li CL and Meng QL. Application of Reflective Ellipsoid Grating System in Space Optical Remote Sensor. *Spacecraft Recovery & Remote Sensing*, **2016,37(2)**:74-81.

[10] Zhang bingqiang, Application of thermal switch heat pipe in lunar exploration optical equipment [J]. Chinese Space Science and Technology,2017,37(06)

[11] Wang SY. Preparation and Thermal Conductive Properties of Modified Graphene/Copper Alloy Thermal Conductivity Materials by Electrophoretic Deposition. Guangxi Normal University.

2018
[12] Tian C. Preparation and application of heat dissipation of graphite film / Aluminium thermal conductive subsrates, Harbin Institute of Technology. 2020
[13] Peng Ren. High thermal conductivity structural design and thermophysical properties of graphite flake/Aluminum composites. Jiangsu University. 2022.
[14] Yu TY, Zhu FR, Liu DQ and Zhang FengShan. Design and deposition of broadband IR antireflection coatings on ZnS lenses. *Acta Optic Sinica*, **2005,25(2)**:270-273.

Numerical simulation study of porous media heat exchangers based on the porous media substitution model

Jun Zhang, Zhaoda Zhang, Yu Liu*, Mingrui Sun, Kangjie Liu and Guanghan Yan

Key Laboratory of Ocean Energy Utilization and Energy Conservation of Ministry of Education, Dalian University of Technology, Dalian, China

xznl@mail.dlut.edu.cn, zhangzhaoda@mail.dlut.edu.cn, liuyu@dlut.edu.cn*, MingruiSun@dlut.edu.cn, lkj@mail.dlut.edu.cn, yanguanghan@mail.dlut.edu.cn

Abstract. This study investigates the performance of porous media heat exchangers by developing a numerical simulation methodology utilizing a porous media substitution model. Based on relevant literature, elongated ellipsoidal Kelvin cells (EKC), Kelvin cells (KC), and body-centered cubic (BCC) structures are selected as porous inserts within the flow channel. A conventional plate-fin heat exchanger (FHE) serves as a benchmark for comparison. Flow and heat transfer simulations are conducted using FloEFD to assess the thermal performance and pressure drop characteristics of the EKC heat exchanger (EKCHE), KC heat exchanger (KCHE), and BCC heat exchanger (BCCHE). The results indicate that the KCHE exhibits superior heat transfer performance, achieving a 40.56% enhancement over the FHE. Conversely, the EKCHE demonstrates a reduced pressure drop, increasing by only 121.9% compared to the FHE, and is 47.7% and 19.84% lower than the pressure drops of KCHE and BCCHE, respectively. Furthermore, the EKCHE showcases an improved integrated heat transfer capability, with the area goodness factor j/f increasing by 24.11% and 61.48% relative to KCHE and BCCHE, respectively. Therefore, selecting EKC as the porous insert is recommended for optimal performance under identical structural conditions.

1. Introduction

Porous media structures offer exceptional properties, including high specific surface area, effective heat transfer, tunable porosity, high specific strength, efficient damping, and controlled energy absorption[1]. These thermal and mechanical advantages enable diverse applications across thermodynamically related fields, with compact heat exchangers for aerospace[2] being a particularly promising area of development.

Porous media play a crucial role in enhancing the overall heat transfer performance of various types of heat exchangers. T'Joen et al.[3] optimized the geometric structure and arrangement of open-cell foam to achieve low air-side pressure drops. They concluded that when foam is effectively bonded to the tube, smaller-sized, high-specific-surface-area, and compact porous media exhibit substantial advantages over spiral fin tubes at high flow rates. Yuan et al.[4] indicated that physical simulation is often necessary in practical heat exchanger studies due to implementation challenges and costs. They combined FloEFD with SIMULINK to validate the reliability of their simulation results. Anantha et al.[5] utilized FloEFD to examine the effects of spiral baffles on double-pipe heat exchanger performance, while Sheng et al.[6] investigated factors affecting the performance of cooling plates in lithium batteries and proposed a novel thermal management system for temperature control. Moreover,

Journal of Physics: Conference Series **3041** (2025) 012020 doi:10.1088/1742-6596/3041/1/012020

FloEFD can be integrated into NX software, providing a comprehensive platform for modeling, simulation, and post-processing, which significantly saves computational resources and time, playing an important role in practical engineering applications. Research on various porous media continues to grow, with KC and BCC emerging as promising options for compact heat exchangers. KC exhibit excellent heat transfer and mechanical performance[7,9]. Calati et al.[7] discussed the effects of the macroscopic geometry of Kelvin cells on thermal and hydraulic behavior in forced convection air systems. Additionally, Sun et al.[8] found that EKC exhibit superior overall heat transfer performance compared to standard KC when stretched along the flow direction. They further investigated the influence of aspect ratio on heat transfer, recommending a length-to-diameter ratio of 2 for optimal results[9]. Shi et al.[10] explored the heat transfer characteristics of various compact heat exchanger configurations, concluding that BCC structures can effectively enhance heat exchanger efficiency.

In summary, This study benchmarks the plate-fin heat exchanger and employs KC, EKC, and BCC as porous inserts to replace traditional fins. By utilizing the porous media substitution model in FloEFD, we analyze the heat transfer performance and overall efficiency of KCHE, EKCHEand BCCHE, offering valuable insights for the engineering design of porous media heat exchangers.

2. Modeling and simulation

2.1. Core model

This study investigates the overall performance of the heat exchanger, focusing on the core section while the shell design is tailored to installation requirements. The heat exchanger core, illustrated in figure 1(a), measures 120 mm × 100 mm × 30 mm and comprises 5 layers of cold and 5 layers of hot flow channels, each 2.5 mm in height. The fin arrangement within these channels is depicted in figure 1(b). The complete heat exchanger structure was modeled using NX12.0.

The cold-side channels feature standard straight fins (figure 1(d)), while the hot-side channels incorporate serrated fins. In contrast, porous medium heat exchangers, including Kelvin cells (KCHE), elongated Kelvin cells (EKCHE), and body-centered cubic cells (BCCHE), employ distinct structures to replace the serrated fins, as shown in figure 1(c). The fins or cells are aligned with the flow direction. Notably, Sun et al.[9] indicate that the long axis of EKC cells should align with the flow direction to effectively reduce pressure drop.

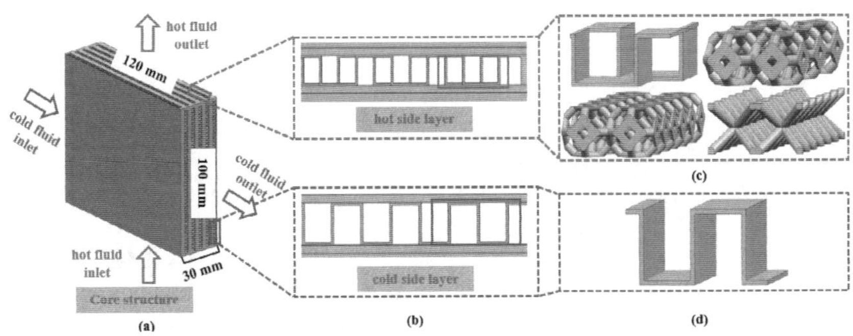

Figure 1. (a) Heat exchanger core structure size and fluid flow mode. (b) Arrangement of heat exchanger elements for cold-side and hot-side runners. (c) Details of hot-side runner sawtooth fins and porous media. (d) Detail of cold side runner through fins.

2.2. Numerical simulation

2.2.1. FloEFD porous media substitution model. Due to the significant size disparity between individual cells and the heat exchanger, modeling the flow cells according to the real porous medium structure directly could result in reaching 9000 cells, complicating mesh generation and simulation. Thus, we directly utilize the software's built-in porous media *substitution* model to represent the metal foam instead of its actual structure.

According to the software user guide, relevant settings were configured. In FloEFD, the volumetric heat transfer coefficient h_V is used to describe heat transfer in porous media, providing a comprehensive characterization of its properties[11]. The porosity is set at 0.85[12]. Due to the unidirectional flow of air in the channels and significant resistance in the other two directions, the permeability type is defined as isotropic. A cell unit model is created based on the channel height in the core. Following the simulation methods reported by Zhang et al.[13], resulting in velocity-pressure drop and velocity-volumetric heat transfer coefficient characteristic curves.

The type of resistance calculation is pressure drop-velocity, which produces a flow resistance coefficient k to describe the flow resistance. k is given by equation (1).

$$k = \frac{\Delta P}{uL\rho} \tag{1}$$

Where ρ denotes fluid density, L is the flow length, and u and ΔP represent flow velocity and calculated pressure difference. The software defines GH3625 as an isotropic porous matrix and defines the thermal conductivity, with heat transfer characterized by a velocity-volume heat transfer coefficient[8] curve. Physical properties of GH3625 are listed in Table 1.

$$h_V = h\frac{S_V}{V} \tag{2}$$

S_V refers to the specific surface area, with the heat transfer coefficient h [8]defined by equation (14).

$$h = \frac{q}{T_W - T_f} \tag{3}$$

Where q is the density of the heat flow received by the actual fluid and solid, and T_W and T_f are the temperatures of the heating surface and fluid respectively.

Table 1. Physical property parameters of GH3625.

Material	ρ(kg/m³)	C_P(kJ/kg·K)	k_s(W/m·K)
GH3625	8440	430	12.1

2.2.2. Physical modeling and boundary conditions. Heat transfer simulations of the complete heat exchanger were conducted using FloEFD. The software employs a Cartesian coordinate system with an adaptive hexahedral mesh for discretization, facilitating flow and heat transfer analysis[6]. The mesh level is set to 7, indicating finer mesh resolution, as shown in figure 2(c).

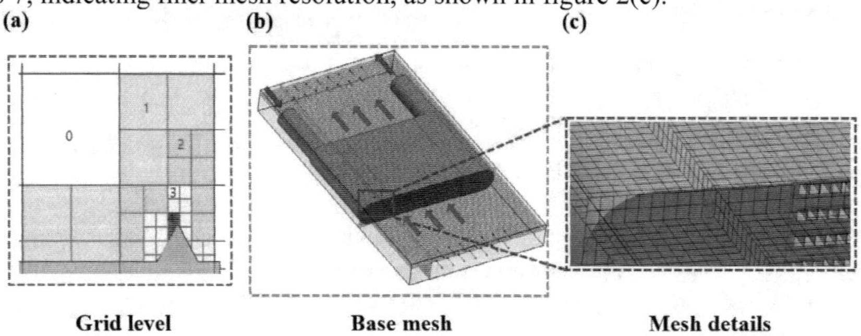

Figure 2. (a) FloEFD grid level. (b) Base mesh of the heat exchanger. (c) Details of the 7-level global mesh refinement.

In addition, the FloEFD software is based on the fundamental conservation equations of mass, energy, and momentum. For the heat exchanger model, the following simplifying assumptions are made: 1) steady-state flow; 2) exclusion of cavities without flow; 3) consideration of fluid flow rate relative to conductivity; 4) neglect of radiative heat transfer; and 5) disregard of gravity, rotation, and free surface effects[14]. Detailed equations and related symbols are provided in reference[14].

Given the pressure inlet and mass flow outlet, the specific boundary condition settings are shown in table 2. $m_{l,hot}$ is the hot-side flow rate, $m_{l,cold}$ is the cold-side flow rate, P_{hot} is the hot-side pressure, P_{cold} is the cold-side pressure, $T_{hot,in}$ is the inlet temperature of the hot side, and $T_{cold,in}$ is the inlet temperature of the cold side. Both the cold and hot sides consist of air, with the cold side being ambient air and the hot side comprising high-temperature gas from the compressor[15]. The focus is on the heat exchanger's enhanced cooling effect on the hot side air.

Table 2. The boundary conditions for heat exchangers.

Parameter		value
	$m_{l,hot}$ (kg/s)	0.12
Hot side	$T_{hot,in}$ (K)	460
	P_{hot} (kPa)	460
	$m_{l,cold}$ (kg/s)	0.24
Cold side	$T_{cold,in}$ (K)	330
	P_{cold} (kPa)	190

3. Results and discussions

3.1. Data reduction

This study focuses on the heat transfer and flow characteristics on the hot side of the heat exchanger. Using the traditional fin-and-tube heat exchanger (FHE) as a benchmark, we evaluate the heat transfer capabilities of KCHE, EKCHE, and BCCHE. The heat transfer rate for the hot side air is presented in equation (4).

$$Q = cm\Delta T \tag{4}$$

In certain operational conditions, heat exchangers may face stringent pressure drop requirements. Therefore, it is essential to evaluate both the heat transfer performance and the associated pressure drop. The overall performance is characterized by the area goodness factor j/f[13], as described by equations[13] (5) to (9). The j/f factor is closely related to the microstructure of porous media, with performance assessments limited to porous media heat exchangers.

$$\mathrm{Re} = \frac{\rho_f u d_h}{\mu} \tag{5}$$

$$Nu = \frac{h_V d_h}{\lambda} \tag{6}$$

$$j = \frac{Nu}{\mathrm{Re}\,\mathrm{Pr}^{1/3}} \tag{7}$$

$$f = \frac{\Delta P}{\Delta L}\frac{d_h}{\rho_f u^2 / 2} \tag{8}$$

$$j / f = \frac{Nu}{\mathrm{Re}\,\mathrm{Pr}^{1/3} f} \tag{9}$$

Where, ΔT represents the temperature difference on the hot side of the heat exchanger, while m denotes the mass flow rate, ρ_f is the fluid density. Fluid properties are closely linked to the porous region's characteristics. For detailed symbol definitions, refer to Zhang et al[13].

In this study, The j/f factor focuses on the core structure to evaluate microstructural performance of porous media, enabling selection of optimal configurations. Operational considerations include: 1) Validated for fully developed turbulent flow (Re>3000) with required corrections in transitional regimes; 2) Clogging propensity: Porous media exhibit higher clogging risks than traditional fins. This metric provides targeted guidance for porous media optimization while emphasizing scenario-specific constraints.

3.2. Heat transfer capacity and overall performance of heat exchangers

As shown in figures 3(a) and (b), porous media heat exchangers outperform the conventional plate-fin FHE. The KCHE demonstrates the highest thermal performance, achieving a 40.56% enhancement in heat transfer on the hot side compared to FHE. Cloud diagram analysis (figure 4(a), green solid-line box) reveals that KCHE exhibits a more pronounced temperature drop and distinct thermal interface, indicating superior heat transfer efficiency. This is attributed to the high specific surface area of KC and enhanced turbulent mixing induced by complex flow channels. The EKCHE and BCCHE show heat transfer improvements of 34.09% and 31.24%, respectively.

As depicted in the cloud diagrams, the structural complexity of KC units generates higher flow resistance and vortices under identical shell configurations. Sparse vortex streamlines (figure 4(b)) further amplify hydraulic losses, correlating directly with unit geometry. This study examines structural variations among porous media and their effects on flow resistance and vortex generation. Shell-side influences were not addressed. Results confirm that heat exchanger reliability depends on the strength properties of porous media, while operational efficiency correlates with their overall performance.

In contrast, EKCHE reduces pressure drop by 47.7% and 19.8% relative to KCHE and BCCHE, respectively, owing to its elongated Kelvin cell (EKC) design. By stretching KC units along the flow direction (aspect ratio=2, figure 1c)[9], EKCHE minimizes transverse resistance while retaining high specific surface area. This optimization yields smoother streamlines (figure 4(b)), reducing turbulence dissipation.

Although EKCHE exhibits a 4.61% lower heat transfer than KCHE, its j/f factor (integrated thermal-hydraulic performance) surpasses KCHE and BCCHE by 24.1% and 61.5%, respectively, within the 20–40 m/s flow velocity range.

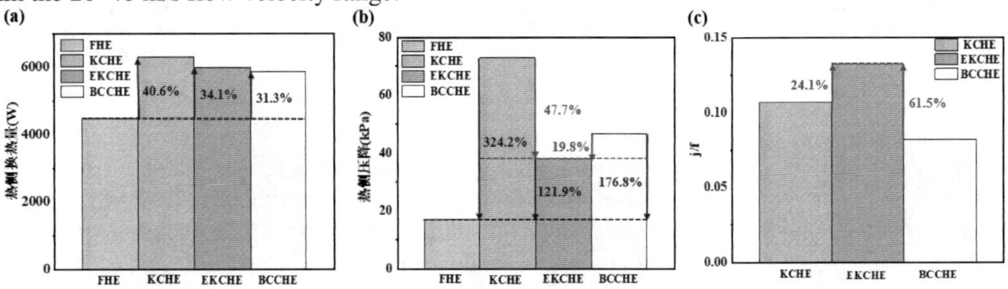

Figure 3. (a) Heat transfer rates of four types of heat exchangers. (b) Pressure drops across four types of heat exchangers. (c) Comprehensive performance of porous medium heat exchangers.

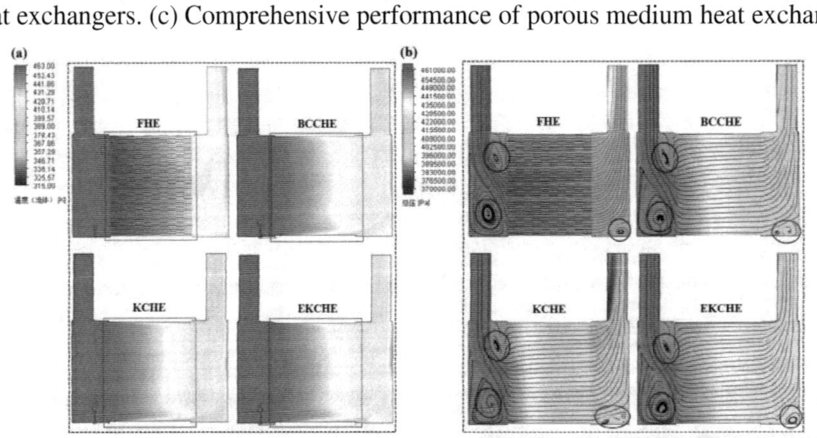

Figure 4. (a) Temperature distribution in hot-side channels of four heat exchangers. (b) Pressure distribution and flow streamlines of four heat exchangers.

4. Conclusions

This study validates the feasibility of simulating heat exchangers using the porous media substitution model in FloEFD. Key findings include:

KCHE achieves a 40.56% heat transfer enhancement over FHE but incurs a 324.2% pressure drop increase, making it suitable for high-heat-flux, pressure-tolerant scenarios (e.g., aerospace transient thermal management[16]) when paired with high-strength materials like GH3625.

EKCHE balances performance with a 34.09% heat transfer improvement and a 121.9% pressure drop increase relative to FHE. Its j/f factor exceeds KCHE and BCCHE by 24.1% and 61.5%, respectively, demonstrating superior overall efficiency. Structural elongation (aspect ratio=2) in EKCHE reduces manufacturing complexity to a non-critical factor while maintaining thermal efficacy.

Thus, EKC is recommended as the optimal porous insert for heat exchangers under identical structural conditions. This work corroborates prior studies on cellular structure performance and provides novel insights for porous media heat exchanger design

References

[1] Fiedler T and Movahedi N 2023 Compact aluminium foam heat exchangers *Metals* **13(8)** 1440

[2] Mahalle A M and Jajoo B N 2013 An approach for modeling and simulation of sintered bronze foam sink for high heat dissipation *Applied Thermal Engineering* **51(1-2)** 899-907

[3] T'Joen C, De Jaeger P, Huisseune H, De Paepe M and Asme 2010 *PROCEEDINGS OF THE ASME INTERNATIONAL HEAT TRANSFER CONFERENCE - 2010, VOL 4: HEAT TRANSFER MEASUREMENT TECHNIQUES, HEAT TRANSFER EQUIPMENT, THERMOELECTRICS* (14th International Heat Transfer Conference: pp 361-370

[4] Yuan W G and Wang Y B 2014 *MECHANICAL SCIENCE AND ENGINEERING IV* vol 472, ed Y Wang (4th International Conference on Mechanical Science and Technology (ICMSE 2014): pp 389-393

[5] Anantha S, Gnanamani S, Mahendran V, Rathinavelu V, Rajagopal R, Tafesse D and Nadarajan P 2022 A cfd investigation and heat transfer augmentation of double pipe heat exchanger by employing helical baffles on shell and tube side *Thermal Science* **26(2)** 991-998

[6] Sheng L, Su L, Zhang H, Li K, Fang Y D, Ye W and Fang Y 2019 Numerical investigation on a lithium ion battery thermal management utilizing a serpentine-channel liquid cooling plate exchanger *International Journal of Heat and Mass Transfer* **141** 658-668

[7] Calati M, De Monte E and Mancin S 2021 Numerical analysis of the effects of the structure shape and orientation of kelvin cell porous structures during air forced convection *Applied Sciences* **11(13)** 6189

[8] Sun M, Zhang L, Hu C, Zhao J, Tang D and Song Y 2022 Forced convective heat transfer in optimized kelvin cells to enhance overall performance *Energy* **242** 122995

[9] Sun M R, Yang L, Hu C Z, Zhao J F, Tang D W and Song Y C 2021 Simulation of forced convective heat transfer in kelvin cells with optimized skeletons *International Journal of Heat and Mass Transfer* **165** 120637

[10] Shi X H, Wang L X, Chen W, Li Z C, Chai X M and Chyu M K 2023 Effect of induced vortex and configuration layout on heat transfer enhancement of helium-xenon mixture *Applied Thermal Engineering* **225** 120168

[11] Iasiello M, Bianco N, Chiu W and Naso V 2020 Anisotropic convective heat transfer in open-cell metal foams: assessment and correlations *International Journal of Heat and Mass Transfer* **154** 119682

[12] Sabet S, Barisik M, Buonomo B and Manca O 2022 Thermal and hydrodynamic behavior of forced convection gaseous slip flow in a kelvin cell metal foam *International Communications in Heat and Mass Transfer* **131**

[13] Zhang Z D, Yan G H, Sun M R, Yan H, Zhao J F, Song Y C and Liu Y 2023 Pore-scale simulation of forced convection heat transfer in metal foams with uniform and gradient structures *Applied Thermal Engineering* **225** 120074

[14] JIN L T 2019 Experimental Study on Flow and Heat Transfer of Truss Structure Heat Exchanger (Dalian: Dalian University of Technology Library) chapter 4 pp 44-46

Journal of Physics: Conference Series **3041** (2025) 012020 doi:10.1088/1742-6596/3041/1/012020

[15] Pandey S, Kallath H, Choi H Y, Sibilli T, Min J K, Chetwynd-Chatwin J and Ha M Y 2022 Numerical study of the effect of flow nonuniformities on the low-pressure side of a cooled cooling air heat exchanger *Applied Thermal Engineering* **217.**
[16] Wang C Eri Q, Wang, Y, & Ding, W 2023 Flow and heat transfer characteristics of intake-precooler system for hypersonic precooled aero-engine *Applied thermal engineering* **229** 120596.

Journal of Physics: Conference Series **3041** (2025) 012021 doi:10.1088/1742-6596/3041/1/012021

Heat insulation and deformation resistance assessment of a novel variable-diameter lattice structure

Jian-Xiong Zheng[1], Chen Li[1], Ling Yang[2], Ai-Fang Chao[2], Qi-Chen Ren[3], Cheng-Wei Fei[1*]

[1] Department of Aeronautics and Astronautics, Fudan University, Shanghai 200466, P.R. China;
[2] AECC Hunan Power Machinery Research Institute, Zhuzhou 412002, China
[3] Shanghai Aviation Industrial （Group） Co., Ltd, Shanghai 200232, China

*Email address: cwfei@fudan.edu.cn

Abstract: High speed, large load capacity and long range are design requirements of future aircraft, which results the increase of aerodynamic load and temperature of aircraft surface. For requirements of high heat insulation performance and high load-bearing capacity of the skin structure of high-performance aircraft, an improved body-centered cubic lattice with Z-reinforcing variable-diameter beam body-centered cubic (ZV-BCC) structure is designed combined with Z-reinforcing body-centered cubic (Z-BCC) and variable-diameter beam. By simulation, it is verified the relative density of the ZV-BCC structure is reduced by 32.7% compared with the Z-BCC structure. This study is to propose an improved ZV-BCC lattice cell for heat insulation structure of aircraft, which provides an insight to support the high-performance and lightweight design of advanced aircraft.

1. Introduction

As the velocity of aircraft increases, the working condition becomes complex, which leads to significant pressure on the thermal protection system[1]. Heat insulation sandwich lattice structure including truss, minimal surface, topology-optimized, and 3D plate lattices[2,3], is applied in thermal protection system.

Truss lattice structures have received significant attention due to its simple unit cell configurations, and excellent load-bearing performance[4]. Existing lattice cell designs typically use small volumes design and uniform diameter beams. However, for single-layer thermal protection sandwich structures, using larger cell size and uniform diameter beams leads to excessive weight. Therefore, this paper presents a cell beam dividing and variable diameter design strategy for the structural loading conditions.

2. Theory and methods

2.1. Mechanical properties analysis methods

Lattice structures are composed of repeating unit cells arranged in a specific pattern[5]. Among them, the BCC structure has attracted considerable attention due to its simple geometry and straightforward, effective design methodology. The lattice of BCC structure is located in the center of the cube and the eight vertices, which is shown in Figure 1.

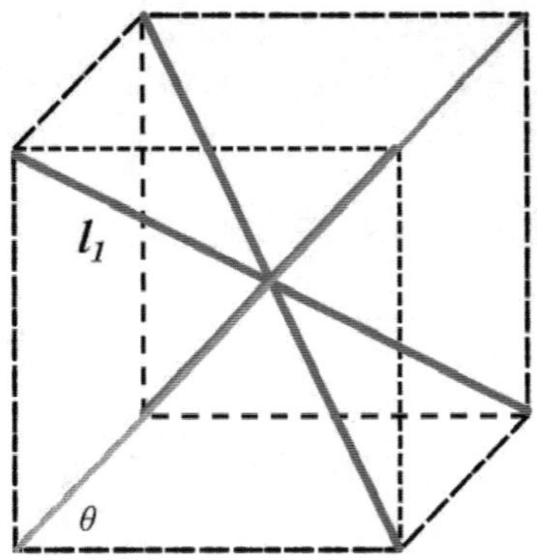

Figure 1. Schematic model of the BCC

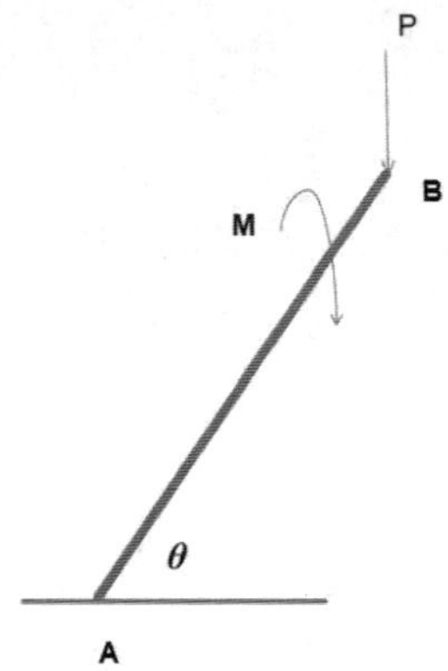

Figure 2. Force diagram of single bar structure

where l_1 is the length of the non-Z-axis reinforced members; l is the length of the unit cell; and θ is the angle between the member and the horizontal plane.

The relative density is used to describe the degree of lightweight structure design, which affects the mechanical properties of the lattice structure. The relative density is shown in equation (1).

$$\bar{\rho} = \frac{8\pi \cdot r_1^2 \cdot l_1}{l^3} \tag{1}$$

where ρ is the relative density; r_1 indicates the radius of the inclined beam structure.

When the lattice structure is subjected to a vertical load, with each unit cell experiencing a pressure of P_1, the forces experienced by the five struts are respectively $P=P_1/4$. Because the lattice structure is symmetrical, the force model of the lattice structure can be simplified as figure 2, the deformation of point B is

$$h_z = \frac{P l_1^3 \cos^2\theta}{12EI} = \frac{P l_1 l^2}{12EI} \tag{2}$$

The results indicate that the value of h_z is significantly related to l. Therefore, in order to ensure the structural stability, it is necessary to incorporate reinforced beams in the z-direction.

2.2 Thermal properties analysis methods

Using the homogenization method to study the heat transfer characteristics of a single unit cell. Based on Fourier's law and the law of conservation of energy, the three-dimensional unsteady heat conduction differential equation for constant properties and no internal heat source is

$$\frac{\partial T}{\partial t} = a\left(\frac{\partial^2 T}{\partial x^2} + \frac{\partial^2 T}{\partial y^2} + \frac{\partial^2 T}{\partial z^2}\right) \qquad a = \frac{\rho}{\lambda c} \tag{3}$$

Where T is temperature; t indicates time; x, y, z are the spatial coordinates; a is the thermal diffusivity; ρ represents density; c is specific heat capacity; and λ demonstrates thermal conductivity.

As shown in equation (3), heat conduction depends on the density, specific heat capacity, and thermal conductivity of the material. Thermal conductivity can be calculated by using Fourier's law of heat conduction:

$$\Phi = -\lambda A \frac{\partial T}{\partial x} \tag{4}$$

where Φ is the heat flux; A indicates the heat transfer area; x represents the coordinate along the direction of heat conduction. The equation (4) indicates that Φ is related to A. Therefore, to reduce the heat flux, it is necessary to decrease the cross-sectional area of the beam.

3. Structural design of variable diameter lattice cells

3.1 Design of cellular

Based on the Z-reinforcing body-centered cubic (Z-BCC) structures, a modified body-centered cubic lattice with Z-reinforcing variable-diameter beam body-centered cubic (ZV-BCC) unit cell structure is proposed. The unit cell model is shown in figure 3 and the specific data is listed in table 1.

Figure 3. Design and Modeling of ZV-BCC

Table 1. Lattice structure parameters

Parameters	Description / Numerical
Overall size	4*3*3
Diameter of cell truss (non-reinforced beam)(mm)	0.6
Cell truss diameter (reinforced beam)(mm)	0.4
Truss angle(°)	45
Width(mm)	13
Height(mm)	14
Length(mm)	17
Material	Titanium alloy

To verify the performance of the ZV-BCC structure, a 4x3x3 stacked unit cell test piece is established here and compared with the BCC, Z-BCC and SC-BCC through static and thermal analysis, considering deformation resistance, thermal conductivity and lightweight properties. To validate the lattice unit cell model, a finite element model of the unit cell is established, as shown in Figure 4:

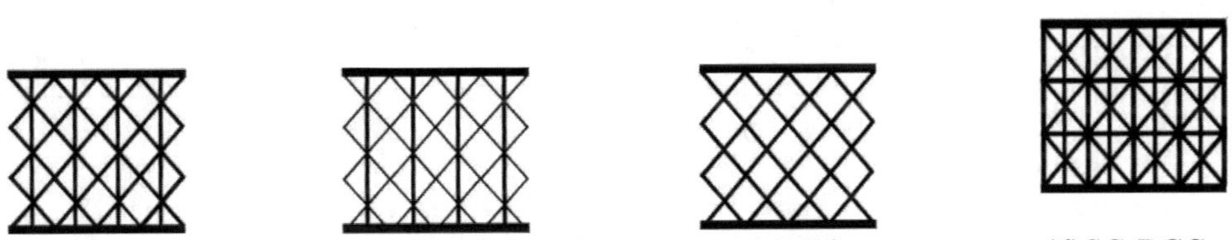

(a)Z-BCC (b) ZV-BCC (Vary) (c) BCC (d)SC-BCC

Figure 4. Different lattice structure configurations

3.2 Simulation model

The structural performance of the four different unit cells is analyzed by applying the following conditions to the model, shown in Table 2.

Table 2. Boundary conditions

Parameters	Description / Numerical	Unit
Pressure	20	MPa
Temperature of hot end	140	°C
Incoming temperature	80	°C
Thermal convection coefficient	450	W/m°C

Substituting the relevant parameters for calculation, the mechanical and thermodynamic simulation results are shown in Fig. 5. Detailed simulation results are shown in the Table 3.

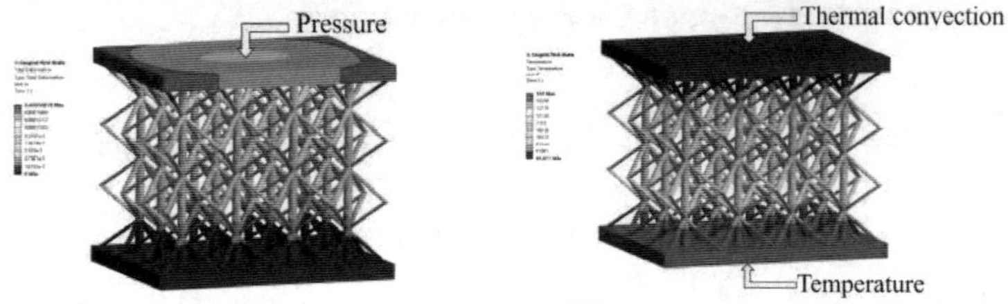

(a) Deformation cloud map (b) Temperature distribution cloud map

Figure 5. Mechanical and temperature analysis results of ZV-BCC

Table 3. Structural performance of lattice structures

Model	Maximum deformation/**mm**	Relative density	Maximum temperature difference/**K**
BCC	1.8081	0.089	55.595
Z-BCC	0.1319	0.104	53.176
ZV-BCC	0.1688	0.070	55.123
SC-BCC	0.0873	0.152	50.645

As shown in Figure 5 and Table 3, under the same loading conditions, the relative density of the ZV-BCC structure is reduced by 32.7 % compared to the Z-BCC structure. At the same time, the deformation of the ZV-BCC structure is somewhat higher compared to the Z-BCC structure. In terms of thermodynamic analysis, under the same thermal boundary conditions, the ZV-BCC structure exhibits a higher temperature difference then Z-BCC, suggesting that its thermal conductivity lower than Z-BCC. Therefore, using the ZV-BCC lattice structure as a heat insulation filler can effectively improve heat-insulating property of the structure.

4. Conclusion

This paper designed a novel lattice structure to improve the heat insulation and load-bearing performance of heat insulation layers. The Z-reinforcing variable-diameter beam body-centered cubic (ZV-BCC) structure was proposed to enhance the structural performance of traditional BCC structures. From this study, some main conclusions are summarized as follows:

(1) With the same conditions, the deformation of ZV-BCC structure is reduced by ~90.7% than other structures.

(2) Compare with Z-BCC and SC-BCC, the relative density of ZV-BCC are reduced by ~32.7%.

(3) Considering about the deformation and temperature difference of BCC, Z-BCC, and SC-BCC, ZV-BCC present better comprehensive performance.

Theoretical calculation and simulation are applied to prove the superb performance of ZV-BCC, which provides a new direction and theoretical support for the future development of heat insulation structure in aeroengine.

References

[1] Fei C W, Li C, Lin J Y, et al. Structural design of aeroengine radiators: State of the art and perspectives[J]. *Propulsion and Power Research*, 2024,13(3):319-334.

[2] Duan Shengyu, Wang Panding, Liu Chang, et al. Research progress on design, optimization and performance characterization of additive manufactured 3D Lattice structures[J]. *Aeronautical Manufacturing* Technology, 2022, 65(14): 36-48, 57. (in Chinese)

[3] Ankun H, Yaojie W E N, Baicheng Z. Research progress on properties of metal lattice structures in additive manufacturing[J]. *Aeronautical Manufacturing Technology*, 2023, 66(11): 90-101.

[4] Aziz A R, Zhou J, Thorne D, et al. Geometrical scaling effects in the mechanical properties of 3d-printed body-cantered cubic (BCC) lattice structures[J]. *Polymers*, 2021, 13(22): 3967.

Journal of Physics: Conference Series **3041** (2025) 012021 doi:10.1088/1742-6596/3041/1/012021

[5] Wang X M, Su Y D, Wu B, et al. Application for additive manufacturing of lattice materials on integrated aircraft structures and functions[J]. *Aeronaut. Manuf. Technol*, 2018, 61: 16-25.

Acknowledgments
This paper is co-supported by National Natural Science Foundation of China (Grant No. 52375237) and National Science and Technology Major Project (Grant J2022-IV-0012). The authors would like to thank them.

Prediction of the two stage turbine cavities thermal state with calculating unsteady processes of hot gas inflow

G Popov[1], V Zubanov[1], S Melnikov[1] and A Scherban[1]

[1]Department of Aircraft Engine Theory, Samara National Research University, Samara, 34 Moskovskoye Highway, 443086, Russia

zubanov.vm@ssau.ru

Abstract. The paper presents an assessment of the influence of unsteady processes on determining the temperature of the surfaces of the hub cavities and the degree of hot gas inflow by the example of a two-stage turbine of a gas turbine engine. The distinctive feature of the performed research is the use of CFD-model including simultaneously domains of four blade rows and three hub cavities. The modelling was completed in both steady and unsteady formulations in ANSYS CFX. The requirements necessary to perform unsteady modelling are explicitly described and fulfilled. It is revealed that it is not enough to determine the thermal state of the cavity surfaces at a particular moment of time of unsteady calculation. First, it is necessary to reach a steady state, and then average the parameters over time, particularly the surface temperatures, for example, for one rotor revolution. The visualisation of the effect of hot gas inflow caused by the static pressure gradient behind the stator nozzle blades is shown. This effect is one of the main causes of hot gas flow into the hub cavities, but at the same time it cannot be predicted by numerical modelling in a steady-state setting.

1. Introduction

The gas turbine engine (GTE) integrates cutting-edge advancements in various fields, including materials science, manufacturing techniques, thermo-mechanics, combustion processes, heat transfer, solid mechanics, reliability, and electronics. Contemporary GTEs are tasked with enhancing fuel efficiency, which is measured by specific fuel consumption. To achieve a reduction in specific fuel consumption, it is necessary to elevate the parameters of the GTE thermo-gas dynamic cycle. This enhancement has resulted in an increase in gas temperature prior to the turbine, rising from 1000-1150 K in the first generation of GTEs to levels between 1850 K and 1950 K in the fifth generation. Additionally, the lifespan requirements for engines have escalated from several hundred hours to between 50,000 and 70,000 hours [1]. A critical element of a GTE is the turbine, which functions under extreme conditions characterized by high temperatures and significant centrifugal forces. Modern GTE turbines feature cooled blades, and cooling air is directed through the inter-disk cavities. The durability and longevity of the turbine blades are crucial for the overall reliability and lifespan of the GTE; notably, the cost of replacing cooled turbine blades can account for as much as 50% of the total repair expenses for the engine [2, 3]. However, the air utilized for turbine cooling can represent 10-15% of the total air intake at the gas generator compressor's inlet. Consequently, the efficiency of the GTE is influenced by the volume of air extracted for cooling purposes. It is essential to minimize this airflow to the lowest necessary level while still maintaining the operational integrity of both the turbine and the GTE. In the design phase of the turbine, significant emphasis is placed on accurately assessing the thermal state of its components. Misjudging the thermal conditions of turbine elements during design or final calculations can result in alterations to radial clearances around working blades and labyrinth seals. In severe instances, elevated disc temperatures may lead to catastrophic failures, posing a critical risk to engine integrity [4]. Therefore, forecasting the thermal state of the turbine and its components is a vital undertaking.

2. Background

Processes in rotor-stator cavities have been studied for a long time [5]. A considerable number of works are devoted to the description of the working body flow and heat exchange in rotating cavities. In turbines, two types of turbine hub cavities are identified:

- cavities due to the axial clearance between the stator and rotor, which are inflated with cooling air to prevent hot gas from entering and overheating the rotor disk;
- cavities between the LPT vanes and rotor, which have a guaranteed hot gas inflow.

Works [6] and [7] indicate that the circumferential non-uniformity is formed under the action of counterpressure from the side of the leading edge of the row standing after the cavity: in the region of the leading edge there are increased pressure values, in the region of the flow part of blade channel - decreased. In [8, 9] it is stated that the circumferential non-uniformity of pressure over the cavity is caused by the blade row in front of the cavity. In [9], it is stated that the non-uniform pressure field above the cavity is influenced only by the previous blade row, which contradicts the conclusions drawn in [6] and [7]. In [10], the cause of the circumferential non-uniformity of the flow is conditioned by the interaction between the rotor and stator blades: regions of increased pressure are formed where the rotor blades pass the flap traces. The flow of hot gas into the cavity is also influenced by the design of the tract seal. If an axial gap seal is used, a strong vortex forms in the axial gap of the seal, interacting with the flows inside the cavity. The presence of a seal with a simple radial gap leads to vortex localization [8].

The hot gas inflow and the workflow inside the cavity are determined by two factors [10, 11]:

- rotor pumping effect. Centrifugal forces in the boundary layer on the disc surface push the cold air outwards. If the flow rate through the turbine's rim sealing is greater than the cooling air flow rate entering the hub cavity, the difference in flow rate is compensated for by the inflow of hot gas from the turbine's flow path. The incoming hot gas descends along the stator surfaces and heats them, and remains circulating in the cavity;
- static pressure gradient behind the stator nozzle blades. Interaction of rotating rotor blades and stator nozzle blades leads to periodic occurrence of areas of increased static pressure, which affects the hot gas flow near the rim sealing. In the areas with reduced static pressure behind the nozzle blades on the suction side, hot gas flows (injects) into the pre-hub cavity.

Conducting field experiments to determine the thermal state of a turbine is extremely difficult and expensive. The alternative to field experimentation is numerical investigation. In 90% of the studies reviewed, CFD was used as the primary research tool and in 45% of the studies CFD supplemented the experimental data.

In the stationary formulation CFD models had one interblade channel each, in part of the studies in the non-stationary formulation domains had several interblade channels to coordinate the angular periodicity of the rotor and stator domains. Full-round models were rarely used [10]. In most studies the working process was considered for only one hub cavity that reliably had cooling air blowdown [5, 13-15]. The authors in work [14] obtained a difference between steady and unsteady simulations of 5.6% in the cooling depth parameter (degree of hot gas inflow). According to the results of unsteady modelling, the authors predict the hot gas flow into the hub cavity from the turbine flow path. According to the results of the unsteady calculation the authors predicted the inflow of hot gas, which led to an excess of the allowable temperature in the deflector and the turbine disc by 277 C.

The work [13] conclude that the k-ω SST turbulence model provides a good compromise between accuracy and computational cost and is suitable for modelling working processes in rotor-stator cavities of turbines.

Not all works devoted to the modelling of the working process in the turbine hub cavities fulfilled the conditions (or provided a description) to ensure the Courant number less than 10 and the area ratio in the interface of about 1.0. There was no information about the output of the working process on the steady-state mode in the works [15].

To correctly predict parameters in the interblade channels and hub cavities of the turbine, grid models must satisfy the following conditions:

- cell growth ratio must be less than ER<1.25;
- value of dimensionless parameters of the first cell height must be $y+\leqslant 1.0$;
- number of the prismatic layers of the wall layer in the gas-air regions must be more than 25;
- number of elements in the axial gap between the blade rows in the cavity inflow area must be more then 20 (excluding prismatic layers);
- volume ratio of the first cell of the tetrahedral element to the last cell of the prismatic layer is less than $V_2/V_1\leqslant 1.2$.

Based on the results of the information analysis, the aim of the work was formulated: to evaluate the influence of unsteady processes in the turbine's hub cavities on its thermal state while explicitly fulfilling all recommendations for unsteady modelling.

A two-stage turbine of a GTE core engine was chosen as the subject of the study (Figure 1). The turbine has four blade rows and three hub cavities. Cavities №1 and №3 have an explicit blowdown with cooling air, which is necessary to prevent hot gas flowing in. Cavity №2 has a clear hot gas flow: hot gas after 1st rotor blade (RB1) flows into cavity №2, mixes with air cooling RB1 root attachment, passes through the labyrinth sealing under 2nd nozzle blade (NB2) and is discharged into the turbine flow path before RB2. A correct determination of the thermal state is required for cavity №2.

3. Turbine geometry configuration

Unsteady gas dynamic calculations with 'rotor-stator' interfaces should be performed with surfaces area ratio in the interfaces $F_2/F_1\approx 1$. For this reason, several blades in each row were considered (Figure 2). The ratio of areas of contacting surfaces at interfaces F_2/F_1 was provided equal to $F_2/F_1=0.982...1.018$.

Mesh volumetric models of the hub cavities were created in ANSYS ICEM CFD, mesh volumetric models of the nozzle and rotor blade passages were created in Numeca AutoGrid5. The created mesh model consisted of 63.7 mln. elements.

Gas dynamic modelling of the working process in the turbine under study was performed in ANSYS CFX. Boundary conditions corresponding to the normal operating mode were set.

It is worth noting that using the developed CFD model it is possible to study the influence of unsteady effects only on gas-dynamic parameters, but it is impossible to determine the temperature stresses (without appropriate modification to consider FSI – fluid solid interaction).

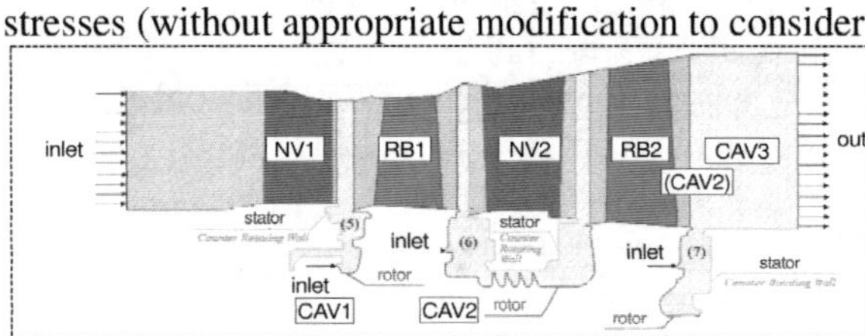

Figure 1. Meridional section of the investigated turbine with bushing cavities.

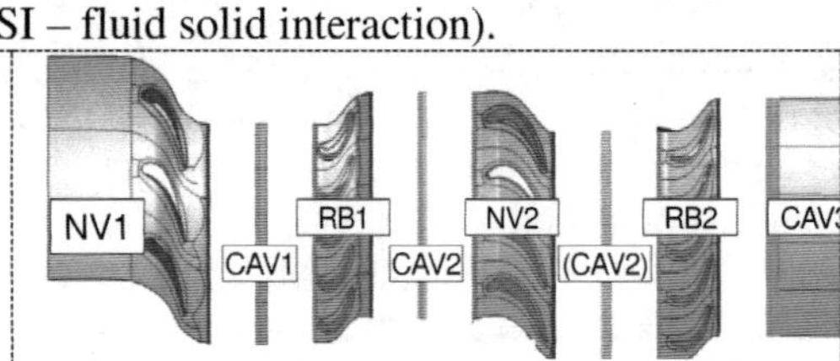

Figure 2. Adopted turbine geometry configuration for CFD-modelling.

4. Results

4.1. Convergence checking

The verification of convergence was established through the identification of minimal mathematical residuals and the stability of integral parameters for a fully converged solution. Initially, the convergence of mass and momentum was validated. Subsequently, the consistency of all integral parameters across iterations was confirmed as an additional criterion. In total, over 400 integral parameters (including those within the cavities) were utilized to assess the convergence verification.

4.2. Reaching the Converged Mode During Unsteady Modeling

Gas-dynamic modeling of the operational process in the turbine under investigation was conducted using both steady-state and unsteady formulations within ANSYS CFX. The outcomes from the steady-state analysis served as the initial solution for initializing the unsteady calculations. The fluctuations in parameters for the converged steady-state analysis were found to be under 1%. In the unsteady simulation, a minimum of 9 turbine revolutions was necessary to stabilize the working processes within the hub cavities to a steady-state condition.

The temperature variations of certain surfaces after a number of rotor revolutions, following initialization from the steady-state calculation, are depicted in Figure 3. To achieve steady-state thermal conditions in the turbine cavities, the following was observed:

- cavity No.1 - more than 4 turns. The average integral value of surface temperature for the last turn of unsteady simulation is higher by 38°C than the value in steady-state simulation;
- cavity No.2 - more than 3 turns. The average integral value of surface temperature for the last turn of unsteady calculation is 25°C higher than the value of steady-state simulation;
- cavity No.3 - more than 8 turns. The average integral value of surface temperature for the last turn of unsteady calculation is higher by 120°C than the value in steady-state simulation.

In cavity No. 2, which experiences a distinct inflow, surface temperatures are elevated compared to other cavities; however, the disparity between steady-state and unsteady modeling is minimal. Cavity No. 3 exhibited the longest duration to achieve a steady-state thermal condition, likely due to having significantly less cooling air passing through it—specifically, 2.8 times less cooling air flows through cavity No. 3 compared to cavity No. 1.

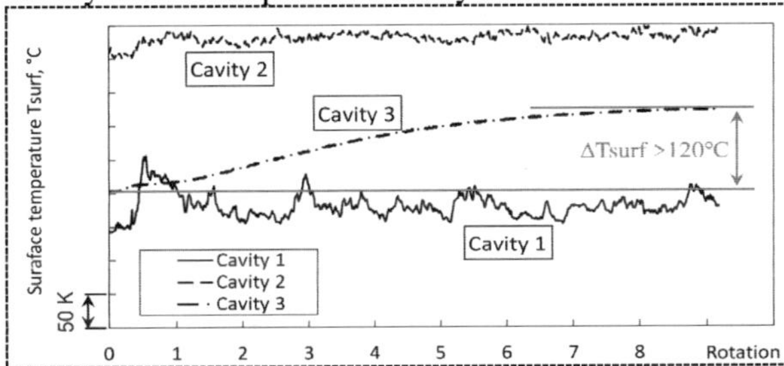

Figure 3. Surface temperature depending on the rotor revolutions number.

4.3. Comparison of Steady and Unsteady Modeling Results

The degree of hot gas inflow was estimated by the formula $\varepsilon_{S.R}=(T^*_{out\_5\%}-T^*_{mix\_t})/(T^*_{out\_5\%}-T^*_{cool\_air})$, where $T^*_{out\_5\%}$ is a total gas temperature in absolute motion at the outlet from the blade domain, located to the left in relation to the cavity under consideration, at a path height of 5%, K; $T^*_{cool\_air}$ is a total temperature of cooling air in absolute motion at the inlet to the hub cavity (minimum cavity temperature), K; $T^*_{mix\_t}$ is a total temperature of the mixture in absolute motion. The value $\varepsilon_{S.R}=1$ means that there is no hot gas in the mixture, and $\varepsilon_{S.R}=0$ – the volume contains only hot gas.

The relative error of the cooling depth $\varepsilon_{S.R}$ between the steady-state and unsteady staging was as follows: cavity No. 1 was 40.1%, cavity No. 2 was 65.9%, and cavity No. 3 was 58.2%.

Figure 4 illustrates the relationship between static pressure and circumferential position at a height of 5% in the cross-section located just before the entrance of hub cavity No. 1. Meanwhile, Figure 5 presents the static pressure distribution within the blade passages NB1 and RB1 at the same height (a), as well as the pressure variation along the height in the midsection of hub cavity No. 1. The averaged integral static pressure obtained from unsteady modelling was found to be 1.5% lower than that from steady-state modelling. In the unsteady simulations, parameters were transmitted without averaging across the interface located behind NB1. The static pressure gradient observed behind the stator nozzle blades during the interaction between stator and rotor blades facilitates the flow of hot gas into the hub cavity from the suction side of the stator blades. Consequently, Figure 5 provides a visualization of the aforementioned influence of hot gas flow.

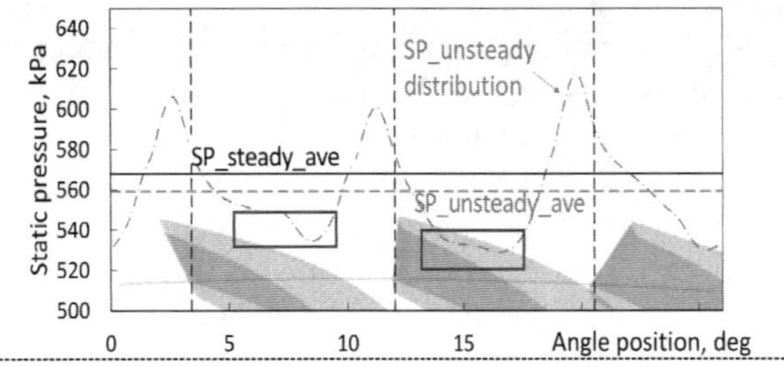

Figure 4. Dependences of static pressure on circumferential position in the section before the beginning of hub cavity No. 1 at a height of 5 %.

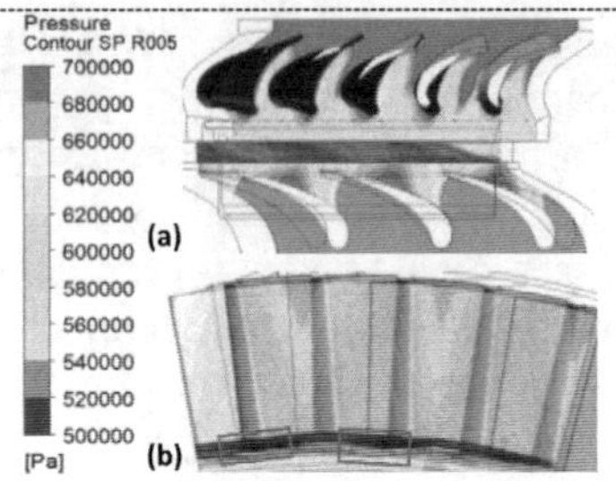

Figure 5. Static pressure distribution in the blade passages NB1 and RB1 at a height of 5 % (a) and by height in the section in the middle of the hub cavity No. 1 (b).

4.4. Parameters Change Assessment During One Rotor Turn in Converged Unsteady Modeling

The unsteady modelling was continued: one more rotor turn to the existing of 9 was calculated. It was determined that the temperature of hot gas flowing into the hub cavities varies within the range $\Delta T^{*}_{0\text{-}5\%}=1\text{-}2\ K$ and does not significantly affect the variation of the hot gas flow rate parameter $\varepsilon_{S.R}$. The characteristic surfaces R01 and R02 for all cavities depending on the rotor turn have been estimated. Figure 6 shows the temperature variation of surfaces R01 and R02 for cavity No. 1. Typical positions of surfaces R01 and R02 are shown in Figure 7 for cavity No. 1 along with the distribution of the hot gas inflow degree parameter for two time steps (rotor revolution positions).

The temperature changes of the surfaces within one rotor turn are as follows: cavity No.1 - $\Delta T_{R01.cav1}$=57°C and $\Delta T_{R02.cav1}$=41°C; cavity No.2 - $\Delta T_{R01.cav2}$=71°C and $\Delta T_{R02.cav2}$=33°C; cavity No.3 - $\Delta T_{R01.cav3}$=23°C and $\Delta T_{R02.cav3}$ =5°C. In cavity No.2, the temperature of the mixture and surfaces after the first tooth of the labyrinth sealing varied within 3 °C.

The revealed significant change in temperature of characteristic surfaces affecting the working blades extension indicates that it is not enough to determine the thermal state of the surfaces of the hub cavities only by the results of unsteady calculation. It is also necessary to perform averaging of surface temperatures over time, for example, for one rotation of the rotor.

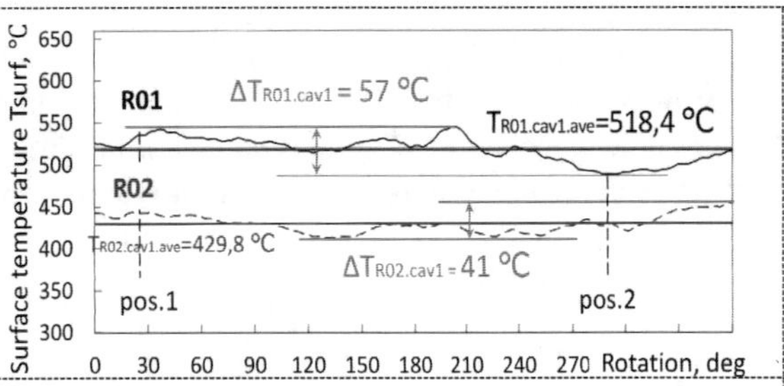

Figure 6. Temperature variation of surfaces R01 and R02 by rotor revolution.

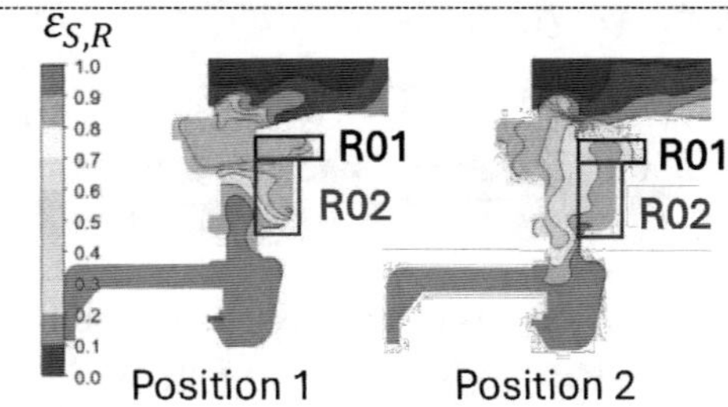

Figure 7. Distribution of the hot gas inflow rate parameter in cavity No.1 under unsteady modelling for two characteristic time moments

5. Conclusion

Correct prediction of the thermal state of the turbine and its parts is an important task in its design and finalisation. Incorrect determination of the thermal state of the turbine components can lead to changes in radial clearances above the working blades, in labyrinth sealings, which results in re-consistency of the entire pneumohydraulic scheme of the engine. In the worst case, the temperature increase of discs can lead to their destruction. The authors in this paper made an attempt to evaluate the influence of unsteady processes on the thermal state of the surfaces of the hub cavities of a two-stage turbine of a GTE core engine under explicit fulfilment of all recommendations necessary for unsteady modelling.

A distinctive feature of the performed research is the use of CFD-model including simultaneously domains of all turbine blades (NB1, RB2, NB2, RB2) and three hub cavities. In this case, cavity No. 2 has a clear inflow of hot gas, which after flowing behind RB1 mixes with cooling air, passes through the labyrinth sealing under NB2 and is blown out in front of RB2.

The results of the study revealed the insufficiency of determining the thermal state of the cavity surfaces based on the unsteady calculation data at a particular moment of time. It is necessary to perform time averaging of the surface temperatures, for example, for one rotor revolution.

The visualization of the effect of hot gas inflow caused by the static pressure gradient behind the stator nozzle blades is shown. This effect is one of the main causes of hot gas inflow into the hub cavities, but at the same time it cannot be predicted by numerical simulations in the steady-state setting.

References

[1] Skibin V A, Solonin V I and Palkin V A 2010 Works of leading aircraft engine-building companies to ensure the creation of advanced aircraft engines (Moscow: Central Institute of Aviation Engine Building)

[2] Airlines Continue the Search for Improved Reliability 2003 MRO Management

[3] Petukhov A N and Kiselev F D 2018 Critical zones of working turbine blades of aviation GTE turbines and recommendations to ensure their operational reliability and service life Aviation Industry Vol pp 58-62

[4] Kutyrev VV 2006 Destruction of turbomachine discs and justification of the allowable strength reserves Conversion in mechanical engineering Vol 6 pp 14-16

[5] Owen J M and Rogers R H 1995 Flow and Heat Transfer in Rotating Disc Systems Volume 2: Rotating Cavities (Taunton, U. K.: Research Studies Press) p 319

[6] Fiore M, Gourdain N, Boussuge J and Lippinois E 2018 Numerical Study of a Linear Cascade With Upstream Cavity Using Various Rim-Seal Geometries and Purge Rates ASME Turbo Expo GT2018-76615

[7] Chilla M, Hodson H and Newman D 2012 Unsteady Interaction Between Annulus and Turbine Rim Seal Flows ASME Turbo Expo: Power for Land, Sea, and Air Vol 8 Turbomachinery pp 1359-1370

[8] O'Mahoney T D, Hills N J, Chew J W and Scanlon T 2010 Large-Eddy Simulation of Rim Seal Ingestion ASME Turbo Expo: Power for Land, Sea, and Air, Volume 4: Heat Transfer, Parts A and B pp 1155-1165

[9] Gentilhomme O O, Hills N J, Turner A B and Chew J W 2002 Measurement and Analysis of Ingestion Through a Turbine Rim Seal ASME Turbo Expo: Power for Land, Sea, and Air Vol 3: Turbo Expo p 925-934

[10] Tisarev A Yu and Vasilevich N M 2013 Study of the hot gas ingestion into inner cavities of gas turbines (Samara: SamSC RAS) p 987–992

[11] Johnson B V, Mack G J, Paolillo RE et al. 1994 Turbine rim seal gas path flow ingestion mechanisms AIAA paper 94

[13] Poncet S and Schiestel R 2007 Numerical modelling of heat transfer and fluid flow in rotor-stator cavities with throughflow Int. J. HeatMass., Tran p 50.

[15] Nguyen Q Q and Laskin A S 2014 Numerical modelling of the influence of axial clearance on unsteady forces in a turbine stage Global Energy, No. 2 (195)

Acknowledgments

The work was carried out with financial support from the Ministry of Education and Science of Russia within the framework of a state assignment (project number FSSS-2023-0008, "Fundamental problems of dynamics and vibroacoustics in hydraulic and gas systems of machines and power plants").

Analysis of flow pulsation characteristics of engine fuel supply system with external gear pump

L Wang[1], F Wan[2], Y R Wang[3] and Y Sang[4]

[1] Product design institute, AVIC Guizhou Honglin Aviation Power Control Technology Co., Ltd, Guiyang, China.

[2] Product design institute，AVIC Guizhou Honglin Aviation Power Control Technology Co., Ltd，Guiyang，China.

[3] Department of Mechanical Engineering，Dalian University of Technology，Dalian，China.

[4] Corresponding author，Department of Mechanical Engineering，Dalian University of Technology，Dalian，China..

agsymsu@163.com

Abstract. Gear pump is very important to the engine fuel supply system, but it is easy to produce violent instantaneous flow pulsation in the working process, which causes the vibration of the pump body and the pipeline, and affects the reliability and service life of the engine fuel system. In this paper, Fluent software, UDF function and dynamic grid technology are used to conduct transient simulation analysis on the internal flow field and flow of gear pump at different speeds, and the influence of speed on the output characteristics of gear pump is analyzed by comparing with theoretical and practical data. The numerical simulation results show that the instantaneous output flow of the gear pump fluctuates in a zigzag pattern, and the flow pulsation rate, pulsation amplitude at the mesh increase with the increase of the rotational speed. The simulation results are almost consistent with the actual measurement data, but there is a large deviation between the theoretical results and the two. The research results provide powerful theoretical guidance for parameter measurement, state detection and fault diagnosis of external gear pump of engine fuel supply system.

1. Introduction

Due to its simple and compact structure, low cost, strong self-priming ability, low internal leakage, insensitivity to oil dirt, strong resistance to impact load, and reliable operation, the external gear pump is widely used in industrial production fields [1]. As the power source of the hydraulic system, the pump's operating state directly determines the reliability of the entire system. Among them, the instantaneous output flow fluctuation of the gear pump, which is the main noise source of the hydraulic system, leads to system oscillation, not only reducing its own performance and service life, but also causing certain harm to the actuator and its driving working parts, and even reducing the efficiency of the entire hydraulic system and causing frequent faults [2]. Therefore, studying the instantaneous output flow characteristics of the external gear pump and its influencing factors is of great significance for the improvement of the engine fuel supply system.

Domestic and foreign scholars have carried out a lot of theoretical, simulation and experimental research on the flow pulsation characteristics of gear pump. An G C et al. [3] established a mathematical model of instantaneous output flow of gear pump and proposed a method to calculate the displacement of involute external gear pump with the maximum error within 5%. Wei X l et al. [4] derived the internal cavity pressure model, cavity volume model and flow pulsation model of the circular arc gear pump, and found that when the center distance error was less than 0.01mm, the outlet flow rate of the circular arc gear pump gradually increased, which had good dynamic characteristics. Liu Q [5] derived the instantaneous flow formula of the external gear pump by using the control volume method, and discussed the influence of design parameters such as gear modulus, number of teeth, pressure Angle and coefficient of tooth crest on the flow characteristics of the gear pump. Manring N D et al. [6] studied the influence of the number of main and slave gear teeth on flow pulsation of external gear pump, and found that more driving gear teeth and less slave gear teeth are conducive to reducing flow pulsation amplitude.

Domestic and foreign scholars have done a lot of mathematical model derivation and simulation calculation of the relationship between gear parameters and flow pulsation for gear pump flow pulsation, but the theoretical model often ignores some parameters and lacks comprehensive consideration of the research object, resulting in a large error compared with the simulation results. Moreover, for the engine fuel supply system, the speed of gear pump is a parameter that is easy to adjust and has a great influence on its internal flow field. Determining the influence of gear pump speed on flow pulsation is helpful to provide guidance for the stable operation of engine fuel supply system. With the help of fluid simulation software Fluent, combined with UDF programming function and dynamic grid technology, this paper conducted dynamic simulation on the internal flow field model of gear pump, simulated and calculated the instantaneous flow of gear pump at different speeds, compared the fitting degree of theoretical flow, actual flow and simulated flow, and analyzed the influence of rotational speed on the fluid flow pulsation of external gear pump.

2. Establishment of gear pump simulation model

2.1. Establishment of three-dimensional model

The structural parameters of the external gear pump of the engine fuel supply system studied in this paper are shown in Table 1. Working parameters of the gear pump are as follows: Three states of 1500r/min, 3500 r/min and 6000 r/min are selected for speed. Under 1500r/min, the working pressure of gear pump is 0.56MPa, and the working pressure of other speeds is 3MPa.

Table 1. Gear structure parameters.

Parameter name	Numerical value
Number of teeth z	10
Normal plane modulus m_n	1.5mm
Normal pressure Angle α_n	28°
Tooth width B	10mm
Coefficient of displacement x	0
Tip circle diameter D_a	18mm
Center distance a	15.05
Helix Angle β	0°
Top gap	3mm
Bottom clearance	3.6mm

According to the above gear pump structure parameters, gear tolerances $\phi 18_{-0.043}^{-0.032}$ and pump body tolerances $\phi 18_{+0.033}^{+0.041}$, UG and GearTrax are used to draw a three-dimensional model of the gear pump, as shown in Figure 1

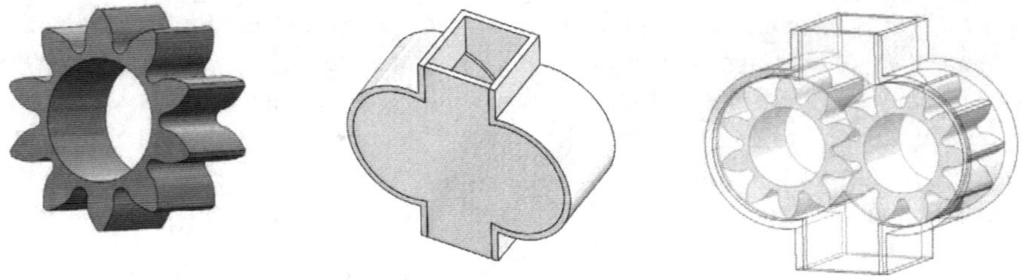

(a)Gear 3D model (b)Pump body 3D model (c)Assembly 3D model

Figure 1. Gear pump 3D model

2.2. Establishment of finite element model

In order to accurately and quickly calculate and simulate the internal flow field of gear pump at different speeds, this paper completed the fluid domain extraction in SolidWorks, and transformed the three-dimensional model into a two-dimensional model by "freezing" and "thin surface" in Design Modeler.

As shown in Figure 2, the two-dimensional external gear pump model is meshed using triangular mesh, in which local mesh encryption was carried out and the size was refined to 0.02mm. Due to the small distance between the two gears meshing and the top circle of the gear and the inner wall of the pump, the mesh changes quickly when the flow field is simulated by using the dynamic mesh technology, and negative volume is easy to occur. Therefore, the radial clearance and meshing clearance between the top of the gear and the inner wall of the pump cavity need to be refined to improve the mesh quality.

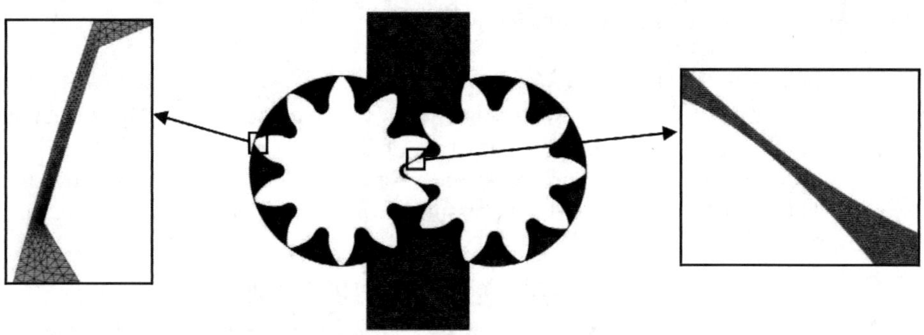

Figure 2. Gear pump finite element model

2.3. Grid independence verification

In general, the denser the mesh, the higher the accuracy of the solution results. However, with the increasing number of grids, the calculation cost will increase, and when the number of grids reaches a certain level, the improvement of calculation accuracy is no longer obvious [7]. Therefore, in order to ensure the accuracy and efficiency of fluid calculation, taking 1500r/min as an example, the number of generated grids was controlled by changing the size of grid cells. Detailed data were shown in Table 2. Simulation calculation was carried out respectively, and the calculation results of instantaneous output flow after stability were compared, as shown in Figure 3.

Table 2. Mesh parameters of different sizes.

Cell size	Number of nodes	Total number of units
2mm	81758	151801
1mm	83270	154387
0.2mm	94753	175855
0.1mm	114387	214343

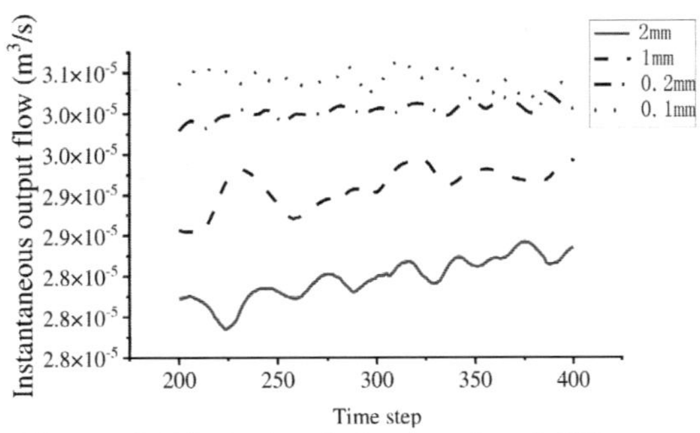

Figure 3. Mesh simulation results of different sizes

A comparative analysis of the data curve in Figure 4 shows that: With the decrease of unit size and the increase of the total number of units, the instantaneous output flow curve began to increase significantly. When the unit size was reduced from 0.2mm to 0.1mm, the curve variation amplitude began to flatten out. Considering the calculation accuracy and calculation time, 0.2mm was finally adopted as the mesh unit size for the simulation of the gear pump model.

2.4. Setting of simulation parameters

The time term is Transient, the solver is Pressure-based coupled, and the turbulence model uses the Realizable k-ε model for complex flow fields of rotating flows. A Mixture two-phase flow Model was adopted, and Schnerr and Sauer models were selected to set aviation fuel as the first phase, air as the second phase, surface tension coefficient of 0.1N/m, and saturation pressure of 6000Pa. The boundary conditions set the inlet type as the pressure inlet and the outlet type as the pressure outlet, and set the corresponding pressure value and the volume fraction of oil and gas in the inlet and outlet two-phase flow. Gear pump rotation motion using DEFINE_CG_MOTION UDF function, select the compilation method. In dynamic mesh, spring Smoothing method and local mesh Remeshing method are used to realize the updating iteration of mesh motion deformation region, the size function is opened, and the size reconstruction interval is set to 1. The simulation time step is 1e-6, the time step is 1000, and the number of iterations in each time step is 40.

3. Theoretical analysis of gear pump flow characteristics

According to literature [8-10], the instantaneous flow formula of ordinary external gear pump is as follows:

$$Q_{sh} = B\omega(R_a^2 - R^2 - R_b^2\varphi^2) \tag{1}$$

Where, B is the width of the gear, ω is the rotational angular speed of the gear, R_a is the radius of the top circle of the gear, R is the pitch circle radius of the gear, R_b is the base circle radius of the involute, and φ is the Angle of the gear.

It can be seen from the formula (1) that the instantaneous flow rate of the external gear pump is a continuity function with $2n\pi / z$ (n as a positive integer) as the period, and it is symmetric in a period. Therefore, calculating the change of the function in the period $[-\pi / z \;\; \pi / z]$ can get the function characteristic curve on the whole interval, that is, the pulsation characteristic of the output flow of the gear pump.

Take the derivative of both sides of equation (1) as:

$$Q_{sh}^{'} = -2B\omega R_b^2\varphi \tag{2}$$

Let $Q_{sh}^{'} = 0$, get $\varphi = 0$, take the derivative of both sides of equation (2)

$$Q_{sh}^{''} = -2B\omega R_b^2 < 0 \tag{3}$$

According to equation (3), the maximum value of Q_{sh} is obtained at $\varphi = 0$:

$$Q_{sh\max} = B\omega(R_a^2 - R^2) \tag{4}$$

Q_{sh} get the minimum at $\varphi = \pm\pi / z$:

$$Q_{sh\min} = B\omega(R_a^2 - R^2 - R_b^2 \frac{\pi^2}{z^2}) \tag{5}$$

Since the instantaneous output flow of gear pump is constantly changing, according to literature [8], the flow pulsation rate can be used to represent the size of flow pulsation. The formula of flow pulsation rate is:

$$\delta = \frac{q_{\max} - q_{\min}}{q} \times 100\% \tag{6}$$

Where, $q_{\max}$ is the maximum instantaneous output flow of the gear pump, $q_{\min}$ is the minimum instantaneous output flow of the gear pump, q is the average output flow of the gear pump.

The structural parameters and speed of the gear pump are respectively brought into the calculation, and the theoretical instantaneous output flow characteristics of the gear pump at each speed are obtained, as shown in Table 3:

Table 3. Theoretical instantaneous output flow characteristic results of gear pump at each speed.

Speed r/min	Maximum instantaneous output flow m^3/s	Minimum instantaneous output flow m^3/s	Average output flow m^3/s	Flow pulsation rate
1500	3.88773×10^{-5}	3.20789×10^{-5}	3.66112×10^{-5}	18.57%
3500	9.07137×10^{-5}	7.48507×10^{-5}	8.54261×10^{-5}	18.57%
6000	1.55509×10^{-4}	1.28316×10^{-4}	1.46445×10^{-4}	18.57%

4. Simulation results and analysis

The speed was set as 1500r/min, 3500 r/min and 6000 r/min respectively, and other parameters remained unchanged. The instantaneous output flow curve of the gear pump at each speed and the simulated instantaneous output flow characteristic results of the gear pump at each speed were obtained by simulation, as shown in Figure 4 and Table 4.

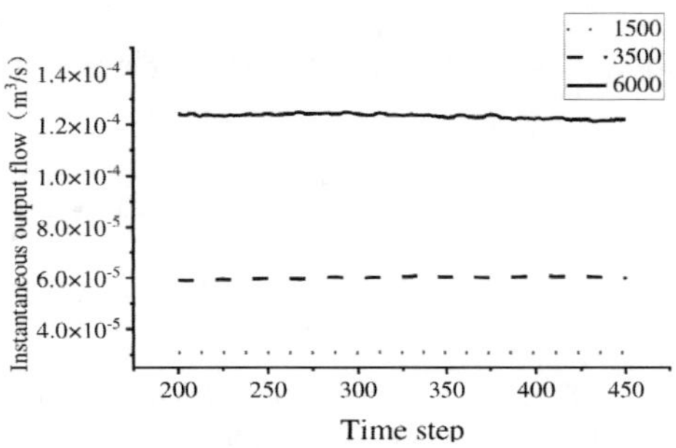

Figure 4. Instantaneous output flow curve of gear pump at each speed

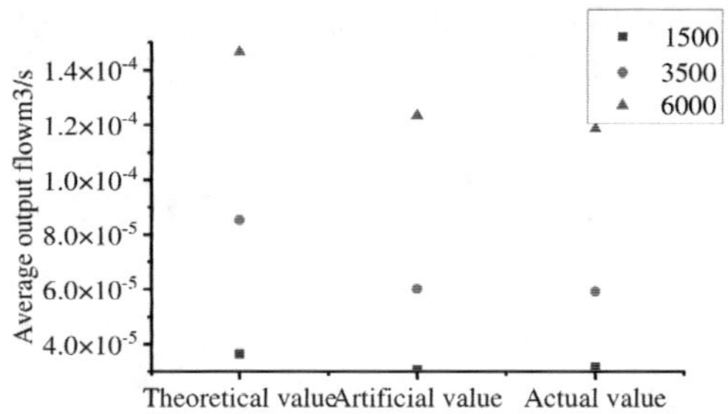

Figure 5. Comparison of theoretical, simulated and actual average output flow of gear pump

Table 4. Simulation instantaneous output flow characteristics of gear pump at each speed.

Speed r/min	Maximum instantaneous output flowm^3/s	Minimum instantaneous output flowm^3/s	Average output flow m^3/s	Flow pulsation rate
1500	3.11251×10^{-5}	3.06362×10^{-5}	3.09119×10^{-5}	1.58 %
3500	6.07848×10^{-5}	5.95581×10^{-5}	6.01554×10^{-5}	2.04%
6000	1.24931×10^{-4}	1.21222×10^{-4}	1.23375×10^{-4}	3.00%

It can be seen from Figure 4 and Table 4 that the instantaneous output flow at the outlet of the gear pump is not a fixed value, but a zigzagged fluctuation with the rotation of the gear pump; With the increase of speed, the instantaneous output flow pulsation rate of the gear pump increased from 1.58% to 3.00%, and the pulsation amplitude increased from 4.88878×10^{-7} to 3.70901×10^{-6}.

As shown in Figure 5, by comparing the theoretical average output flow of the external gear pump of the engine fuel system, the simulated average output flow and the actual average output flow, it is found that the simulation calculation results of the external gear pump of the engine fuel supply system are in good agreement with the actual measurement data, but there are large errors between the two and the theoretical calculation results. This is because the theoretical calculation process will ignore the influence of many factors, such as the leakage inside the gear pump body, the viscosity of the liquid, etc., resulting in a large deviation from the actual situation; However, the physical model and the solver setting in the simulation process will consider all kinds of situations comprehensively, so as to ensure the accuracy of the simulation calculation.

5. Conclusion

In this paper, the external gear pump of the engine fuel system is taken as the research object. According to the basic parameters and processing size of the gear pump, the three-dimensional modeling software SolidWorks is used to draw the three-dimensional model of the gear pump and extract the fluid domain. According to the working parameters and external conditions of the gear pump, the fluid dynamics calculation software Fluent was used to realize the rotating motion of the gear pump by using UDF function and dynamic grid technology. The dynamic simulation process of gear pump meshing rotating motion was comprehensively and accurately calculated and simulated. By simulating the instantaneous output flow of the gear pump at different speeds, it was found that: The instantaneous output flow rate of gear pump fluctuates in a zigzag pattern. With the increase of speed, the flow pulsation rate and pulsation amplitude increase. The theoretical average output flow, the simulated average output flow and the actual average output flow of the external gear pump of the engine fuel system are compared and analyzed, and it is found that the simulation calculation is more in line with the actual calculation.

References

[1] Mithun, M.G., Koukouvinis, P., Karathanassis, I.K. and Gavaises, M., 2019. Numerical simulation of three-phase flow in an external gear using immersed boundary approach. *Applied Mathematical Modelling*, **72(8)**, pp.682-699.

[2] Yu, H., Wang, H.M. and Zhong, H.T., 2011. Flow characteristics analysis of multipole Compound Gear Pump. *Journal of Harbin Institute of Technology*, **43(11)**, pp.54-59.

[3] An, G.C., Li, P.G. and Guo, J.Q., 2019. Displacement Calculation of External Gear Pump Based on Tooth Profile Curve. *Machine Tool & Hydraulics*, **52(12)**, pp.25-32.

[4] Wei, X.L., Feng, Y.B. and Han, X.X., 2012. Research on Flow Pulsation Characteristics of Circular Arc Gear Pump considering Center Distance Error. *Chinese Hydraulics & Pneumatics*, **46(10)**, pp.71-77.

[5] Liu, Q., 2013. Study on Instantaneous Flow and Pulsation Characteristics of External Gear Pump. *Mechanical Transmission*, **37(6)**, pp.17-21.

[6] Manring, N.D., Kasaragadda, S., 2003. The theoretical flow ripple of an external gear pump. *ASME Journal of Dynamic Systems, Measurement, and Control*, **125(3)**, pp.263-276.

[7] Yu, J.H., Xiao, J.S., Zhu, Z.B. and Ye, F., 2007. Parallel multi-grid algorithm and its performance of fluent software. *Dcabes 2007 Proceedings Vols I & II*, **14(17)**, pp. 33-36.

[8] He, C.X., 1982. Hydraulic Components *China Machine Press*.

[9] Kong, F.Y., He, Y.Y. and Zheng, D., 2019. Analysis of Influencing Factors on Flow Characteristics of External Gear Pump. *Journal of Drainage and Irrigation Machinery Engineering*, **32(2)**, pp.108-112.

[10] Houzeaux, G. and Principe, J., 2008. A variational subgrid scale model for transient incompressible flows. *International Journal of Computational Fluid Dynamics*, **22(3)**, pp. 135-152.

Journal of Physics: Conference Series **3041** (2025) 012024 doi:10.1088/1742-6596/3041/1/012024

An equal-margin control method for inlet unstart protection in the solid rocket ramjet

Zhao Wang[1,2], Xiang Tang[1,2], Meng Huang[1,2], Xiaotao Tian[1,2], Bo Zhang[1,2]

[1] National Key Laboratory of Land and Air Based Information Perception and Control, Xi'an Modern Control Technology Research Institute, Xi'an 710065, China
[2] Xi'an Modern Control Technology Research Institute, Xi'an 710065, China

3141674879@qq.com

Abstract. For the solid rocket ramjet with fixed geometry, the unstart protection control must be carried out in order to keep the inlet under the start state. Firstly, the maximum backpressure that the inlet can withstand under different flow conditions was obtained by the steady state model of the inlet. Then, the unsteady model of solid rocket ramjet was established. Finally, the double-loop TD-PID controller was adopted to adjust the gas flow to achieve equal-margin control, so that the unstart margin of the inlet can be maintained at the designated value, which provides a strong premise and guarantee for the normal operation of the engine.

1. Introduction

For a solid rocket ramjet with fixed geometry, the inlet determines the best performance that the engine can achieve through downstream parameter adjustment [1]. The inlet unstart restricts the improvement of the performance of the whole propulsion system, and even makes the engine unable to generate thrust, which is easy to cause dynamic load, resulting in structural damage, and makes the aircraft more difficult to control [2]. Thus, there are many control methods, such as mass bleeding[3-4], gas flow regulation[2], vortex generation[5] to inhibit the occurrence of unstart.

Generally speaking, there are two types of inlet unstart modes: low Mach number unstart and backpressure unstart. The backpressure unstart is formed due to the larger backpressure or the lower static pressure of freestream [6]. This paper focuses on the research of backpressure unstart. For the supersonic inlet, due to the influence of the backpressure of the combustion chamber, the normal shock wave will move forward and backward in the expansion section of inlet [7]. Therefore, backpressure unstart protection control is designed for the optimal performance of the inlet under the start state. The backpressure unstart protection control mainly includes three parts: (a) unstart boundary identification, that is, to obtain the maximum backpressure that the inlet can withstand under different flow conditions. There has been much work devoted to the identification of backpressure unstart boundary. The method based on instantaneous or steady state information is adopted to detect whether the inlet can start [8]. The former mainly includes cumulative sum algorithm [9], generalized likelihood ratio algorithm [9], derivative-based detection [10], and recursive Fourier transform [11]. As far as the latter concerned, there are three ways to describe the unstart state: shock position [12], pressure distribution [13-14], and backpressure[15-16]. Among the three ways, it is more convenient to determine the unstart boundary by backpressure. For the sake of simplicity, the inlet unstart boundary by backpressure was estimated by the steady state model of the inlet. (b) establish the unsteady model of solid rocket ramjet to obtain the corresponding relationship between backpressure at the inlet outlet and incoming flow disturbance as well as the change of gas generator throat area. The unsteady model based on lumped parameter method is too complex with many coefficients, and the response of pressure disturbance and shock wave position is simplified into inertia link [2]. For the normal shock capturing, a large number of literature is based on the idea of small deviation linearization model [17-21], but due to the strong nonlinear characteristics of the dynamic position of the shock wave, these models cannot represent the dynamic behavior of the shock position in the process of large-scale change. Therefore, the shock position is calculated by directly solving the one-dimensional unsteady flow equation [4,22-23] or by two-dimensional numerical simulation [24]. According to the calculation results of [4], it is meaningful to adopt the one-dimensional equation to calculate the inlet flow field. In addition, zero-dimensional conservation equation was used to solve the parameter evolution law of the secondary combustion chamber. For the model of inlet and secondary combustion chamber, the key point of coupling between them is the calculation method of backpressure at the inlet outlet. Referring to the sudden expansion coefficient in [25], the scale factor between backpressure at the inlet outlet and the pressure in the secondary combustion chamber was introduced, and the scale factor was approximated by the ratio relationship between the two pressures under steady state. (c) unstart control method. The equal-margin control strategy of inlet unstart in [6] was introduced. The unstart margin represents the distance between the current pressure ratio of the inlet and the maximum pressure ratio it can withstand. Chai realized the control effect of rapid and no-overshoot for the nonlinear controlled plant by introducing the tracking differentiator (TD) transient process into the conventional PID controller [26]. The double-loop TD-PID controller was designed to control the unstart margin of the inlet, so that the inlet can play its optimal performance under the premise of starting.

As described above, the unstart boundary was firstly identified by the inlet steady-state model, the flow field in the inlet was calculated by quasi-one-dimensional Euler equation, and the flow field in the secondary combustion chamber was calculated by zero-dimensional conservation equation, and the backpressure at the inlet outlet under unsteady condition was approximated by the scale factor under the steady-state condition. Finally, the double-loop TD-PID controller was adopted to adjust the gas flow to achieve equal-margin control, so that the inlet unstart margin can be adjusted at the designated value, and the control effect was tested by numerical simulation.

2. Control principle and model

2.1 Control principle

An equal-margin double-loop controller is used to realize the inlet unstart protection, the principle of which is shown in Figure 1. First, for the outer loop control, the inlet unstart margin ζ_i is given, and then it and the current unstart margin ζ are input into the unstart margin controller to control the gas generator pressure of the inner loop. For the inner loop control, the target gas pressure p_{ri} and the current gas pressure p_r are input to the controller to control the motor, and the actuator is controlled by the motor, so that the pressure of the gas generator can be adjusted, and the outlet pressure of the inlet can be further affected, so that a new margin can be obtained, and the cycle is repeated until the given unstart margin is reached, and the equal-margin control of the inlet can be achieved. It should be noted that the unstart margin of the inlet is defined as follows:

$$\zeta = \frac{\chi_u - \chi_b}{\chi_u} \tag{1}$$

where χ_b is the ratio of the current outlet pressure to the entrance pressure of the inlet, and χ_u is the maximum pressure ratio that the inlet can tolerate under the condition of incoming flow.

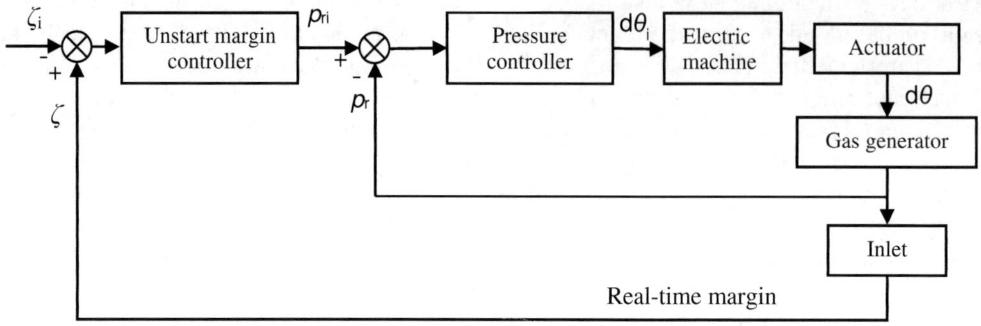

Figure 1. Structure diagram of inlet unstart margin control system

2.2 Unstart boundary recognition

As shown in Figure 2, air is led in the secondary combustion chamber through two 2-D inlets to mix with the primary gas injected by the gas generator, and secondary combustion occurs in the secondary combustion chamber. Sections 0, 1, c, t, gt, e represent the inlet lip, inlet outlet, secondary combustion chamber outlet, gas generator throat and nozzle throat, respectively. For simplicity, the steady-state performance model in [27] was used to estimate the inlet unstart boundary.

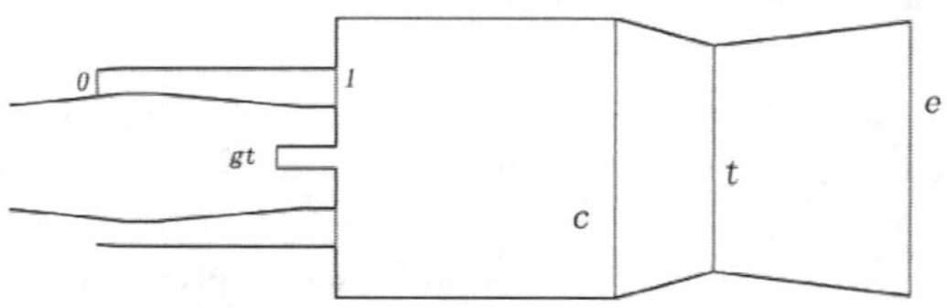

Figure 2. Diagram of geometric model for solid rocket ramjet

2.3 Modeling of solid rocket ramjet control system

2.3.1 Gas flow regulation system. The system model of gas flow regulation is seen in [28].

2.3.2 Inlet. The expression of the quasi-one-dimensional Euler equation of conservation type is [29]:

$$\frac{\partial \boldsymbol{U}}{\partial t}+\frac{\partial \boldsymbol{F}(\boldsymbol{U})}{\partial x}=\boldsymbol{G}(\boldsymbol{U}) \tag{2}$$

where, $\boldsymbol{U}=[\rho A \quad \rho uA \quad \rho EA]^T$, $\boldsymbol{F}(\boldsymbol{U})=\left[\rho uA \quad (\rho u^2+p)A \quad \rho HuA\right]^T$

$\boldsymbol{G}(\boldsymbol{U})=\left[0 \quad p\frac{\partial A}{\partial x} \quad 0\right]^T$, $E=e+\frac{1}{2}u^2, H=h+\frac{1}{2}u^2$, $e=\frac{p}{(\gamma-1)\rho}$, $h=e+\frac{p}{\rho}$.

The van Leer scheme [30] is used to solve the above formula.

2.3.3 Hot zone. The hot zone in this paper refers to the control body enclosed by section gt, section 1, the head section of the secondary combustion chamber and section t. For the hot zone, the mass conservation equation is as follows:

$$V_{\mathrm{h}}\frac{\mathrm{d}\rho_{\mathrm{h}}}{\mathrm{d}t}=2m_1+m_{\mathrm{g}}-m_{\mathrm{t}} \tag{3}$$

where m_1 refers to the air flow of a single inlet outlet, m_t refers to the gas flow of the nozzle throat, and V_h refers to the volume of the hot zone.

The momentum conservation equation in the hot zone is as follows:

$$V_h \frac{d\rho_h u_h}{dt} = m_g u_{gt} + p_{gt} A_{gt} + p_1 \left(A_c - A_{gt} - 2A_1 \right) + 2\left(m_1 u_1 + p_1 A_1 \right) - \left(m_t u_t + p_t A_t \right) + \left(\frac{p_h + p_t}{2} \right) \left(A_t - A_c \right) \tag{4}$$

3. Controller design

PID controller is widely used in process control and industrial automation. Considering that the input required by the actuator is angle increment, incremental TD-PID controller for the gas pressure controller of the inner loop is adopted, that is [26, 31]:

$$d\theta_i(t) = d\theta_i(t-1) + k_p\left(e_r(t) - e_r(t-1)\right) + k_i e_r(t) T + k_d\left(e_c(t) - e_c(t-1)\right) \tag{5}$$

For the equal-margin controller of the outer loop, the incremental PID controller is adopted, that is [31]:

$$dp_{ri}(t_o) = dp_{ri}(t_o - 1) + k_{po}\left(e_{ro}(t_o) - e_{ro}(t_o - 1)\right) + k_{io} e_{ro}(t_o) T_o + k_{do}\left[e_{ro}(t_o) - 2e_{ro}(t_o - 1) + e_{ro}(t_o - 2)\right] \tag{6}$$

4. Numerical simulation

4.1 Initialization

The unsteady model is used to obtain the steady state results of incoming flow height 12km, Mach number 2.5, and gas generator slide valve position 0° (i.e., gas mass flow 0.304kg/s). In addition, the initial scale factor ψ is 0.954 and the unstart margin ζ is 0.159. In order to improve the performance of the inlet, the unstart margin can be appropriately reduced.

4.2 Control results and analysis

The target value of the unstart margin ζ_i is set to 0.14, and the corresponding scale factor is 0.9612. It should be noted that, for simplicity, the scale factor is set to vary linearly from 0.954 to 0.9612 over a period of 0.6s.Figure 3 and Figure 4 show the change of inlet outlet parameters and unstart margin with time, respectively. As can be seen from Figure 5, due to the closed-loop control, the steady-state solution of the unstart margin is 0.1404, and the relative error between it and the target value 0.14 is only 0.29%, which verifies the validation of the control method. In addition, due to the artificial viscosity added to the calculation of van Leer scheme, the mass of inlet entrance and exit is not conserved. As shown in Figure 3, there is a relative error of 2.1% between the inlet exit flow of 1.175kg/s and the inlet entrance flow of 1.2kg/s.

5.Conclusion

An equal-margin control method for inlet unstart protection in the solid rocket ramjet was established in this paper. Specifically, the unstart boundary was firstly identified by the steady-state model of the inlet. Then, the inlet unstart control system was modeled. Finally, a double-loop TD-PID controller was designed to control the unstart margin. The results show that the relative error between the unstart margin and the target value is only 0.29%, which fully verifies the effectiveness of the control method.

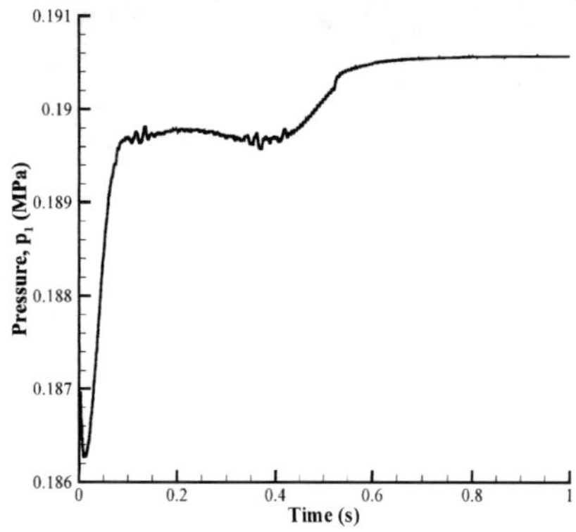

(a) pressure

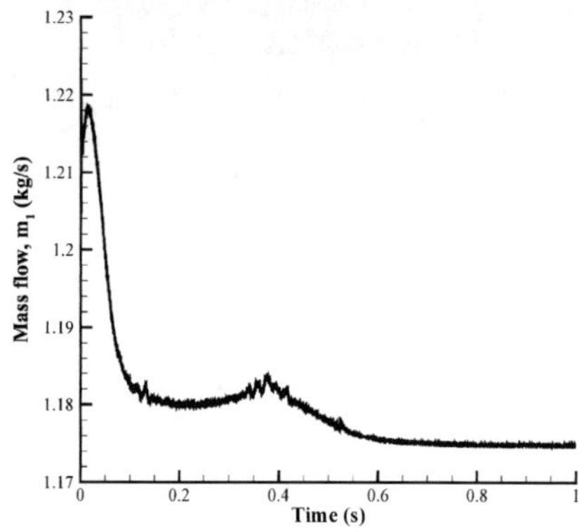

(b) mass flow

Figure 3. Variation of inlet outlet parameters with time

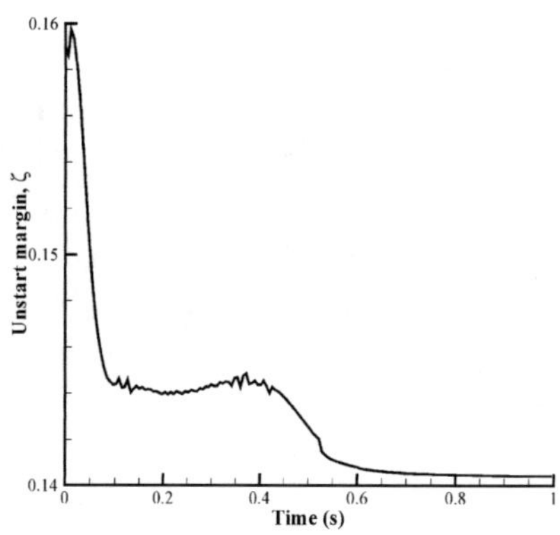

Figure 4. Variation of unstart margin with time

References

[1] M.Shao. Study on Control Method and Integrated Flight/Propulsion Control Method of Compound Adjustable Ducted Rocket, *Northwestern Polytechnical Univesity*, Xi,an, 2016 (in Chinese).

[2] W.Niu. Study on Control Methods for Variable Flow Ducted Rockets, *Harbin Institute of Technology*, Harbin, 2009 (in Chinese).

[3] J.T. Chang, D.R. Yu, W. Bao, Y. Fan, Y. Shen, Effects of boundary-layers bleeding on unstart/restart characteristics of hypersonic inlets, *Aeronaut. J.* 113 (1143)(2009) 319–327.

[4] Y. DU, H. ZHANG, Research on dynamic identification and control method of normal shock position of supersonic inlet, *Journal of Propulsion Technology* 31(6)(2018)888–897 (in Chinese).

[5] Valdivia A，Yuceil K B，Wagner J L，et al. Dolling active control of supersonic inlet unstart using vortex generator jets. *AIAA paper* 2009-4022，2009.

[6] J. Chang, Y. Fan, W. Bao, D. Yu, and Y. Shen. Unstart Margin Control of Hypersonic Inlets, *Acta Astronaut*, 66 (1), 78–87, 2010.

[7] K. George, W. C. Joseph, E. P. Daniel, and J. W. Kyle. Quasi 1D modeling of mixed compression supersonic inlets, *AIAA-2012-0775*, 1-15, 2012.

[8] Chang, J., Li, N., Xu, K., Bao, W., and Yu, D., "Recent Research Progress on Unstart Mechanism, Detection and Control of Hypersonic Inlet," *Progress in Aerospace Sciences*, Vol. 89, Feb. 2017, pp. 1–22.

[9] S. Trapier, S. Deck, P. Duveau, Time-frequency analysis and detection of supersonic inlet buzz, *AIAA J.* 45 (9) (2007) 2273–2284.
[10] J.T. Chang, R.S. Zheng, L. Wang, W. Bao, D.R. Yu, Backpressure unstart detection for a scramjet inlet based on information fusion, *Acta Astronaut.* 95 (2014) 1–14.
[11] J.T. Chang, L. Wang, B. Qin, W. Bao, D.R. Yu, Real-time unstart prediction and detection of hypersonic inlet based on recursive Fourier transform, Proc. Inst. Mech. Eng. G: *J. Aerosp. Eng.* 229 (4) (2015) 772–778.
[12] Kopasakis G, Connolly J W. Shock Positioning Controls Design for a Supersonic Inlet [R]. *AIAA 2009-5117.*
[13] D.R.Yu, J.T. Chang, W. Bao, Z. Xie, Optimal classification criterions of hypersonic inlet start/unstart, *J. Propuls. Power* 23 (2) (2007) 310–316.
[14] J.T. Chang, Q.H. Hu, D.R. Yu, W. Bao, Classifier utility modeling and analysis of hypersonic inlet start/unstart considering training data costs, *Acta Astronaut.* 69(9) (2011) 841–847.
[15] J.Chang, D.Yu, W.Bao, L.Qu, Dimensionless analysis of the unstart boundary for 2D mixed hypersonic inlets, *The Aeronaut Journal* 112(1135) (2008)547–555.
[16] Juntao Chang, Daren Yu , Wen Bao , Liang Qu, Influence Factors of Unstart Boundary for Hypersonic Inlets. *AIAA paper 2008-4586*, 2008.
[17] R. G. Willoh. A mathematics analysis of supersonic inlet dynamic, *NASA TN D-4969*, 1-19, 1979.
[18] B. Qian and Q. Zeng. Linear Dynamics of Supersonic Ramjet, *Journal of Aerospace Power*, 5(2), 179-181, 1990 (in Chinese).
[19] F. E. C. Culick and T. Rogers. The response of normal shocks in diffusers, *AIAA Journal*, 21(10), 1382-1390, 1983.
[20] G. Douglas, MacMartin. Dynamics and control of shock motion in a near-isentropic inlet, *AIAA 2002-2943*,1-33, 2002.
[21] Umair Ahsun, Ali Merchant, Paduano James D, and Mark Drela. Design of a near-isentropic supersonic inlet using active control, *Journal of Propulsion and Power*, 21(2), 292-299, 2005.
[22] J. Chang, W. Bao, D. Yu, and L. Qu. Dynamics Simulations of Shock-Motion of One-Dimensional Supersonic Inlets, *Missile and Space Vehicle*, 290(4), 48-51, 2007 (in Chinese).
[23] Z. Yao, B. Wu, and J. Liang. One-dimensional analysis of shock movement caused by combustion in ramjet, *Journal of Rocket Propusion*, 39(2), 52-58, 2013 (in Chinese).
[24] Y. Du, H. Zhang, and M. Liu. Real-time simulation modeling for integrated inlet and engine system based on normal shock position calculation, *Journal of Aerospace Power*, 33(1), 247-256, 2018 (in Chinese).
[25] D. Fang, Z. Xia, W. Zhang, C. Jiang, Q. Li, J. Cao, and etc. The Adaptive Control Characteristics of Unchocked Solid Rocket Ramjet Engine, *Journal of Ballistics*, 12(4), 31-34, 2000 (in Chinese).
[26] J. Chai, X. Chen, K. He, J. Zhou, and X. Li. Research on Tracking Differentiator for Variable Flow Gas Generator, *Journal of Propulsion Technology*, 40(5), 1153-1161, 2019 (in Chinese).
[27] Q. Ji. Research on the Performance of Solid Fuel Ramjet for missile, *Nanjing Univesity of Science and Technology*, Nanjing, 2006 (in Chinese).
[28] J. Chai. Research on Control Algorithm and Experiment of Gas Flow Regulation System for Variable Flow Ducted Rocket, *Nanjing Univesity of Science and Technology*, Nanjing, 2018 (in Chinese).
[29] Z. Yao and Q. Zhou. Introduction to Computational Fluid Dynamics, *Tsinghua University Press*, Beijing, 2010 (in Chinese) [John D. Anderson, JR. Computational Fluid Dynamics: the Basics with Application, McGraw-Hill, New York, 1995].
[30] C. Yan. Methods and Applications of Computational Fluid Dynamics, *Beihang University Press*, Beijing, 2006 (in Chinese).
[31] J. Liu. MATLAB Simulation of Advanced PID Control (4th edition), *Publishing House of Electronics Industry*, Beijing, 2016 (in Chinese).

Journal of Physics: Conference Series **3041** (2025) 012025 doi:10.1088/1742-6596/3041/1/012025

Study on ventilation and compression performance of honeycomb-based double-wall panels for auxiliary fuel tank

Jun Chen, Dongguang Xu[*] and Wenjia Wu

Shanghai Aircraft Design and Research Institute, Commercial Aircraft Corporation of China, Ltd., Shanghai 201210, China

[*]Corresponding author's e-mail: xudongguang@comac.cc

Abstract. Configuring auxiliary fuel tanks (AFTs) is the most effective way to extend the range of commercial aircrafts. AFTs are constructed from honeycomb-based double-wall panels, which must be equipped with ventilation and drainage functions. In this study, six distinct structures of honeycomb-based double-wall panels are designed and investigated. Computational fluid dynamics (CFD) simulations and Finite Element Analysis (FEA) were conducted to analyze the influence on ventilation performance and compression performance of different internal structures of double-wall panels, respectively. The results of the simulation revealed that excellent ventilation performance and superior compression property are inherently contradictory, necessitating a careful balance and selection based on specific circumstances. These findings can also guide the structural design of honeycomb-based double-wall panels, thereby reducing the experimental cost and improving the efficiency of AFTs design.

1. Introduction

In order to meet the demand for increasing range of the commercial civil aircraft, the fuel capacity of the aircraft should be improved [1,2]. Auxiliary fuel system is an optimized choose to improve the fuel capacity of civil aircraft. During flight, the system utilizes pneumatic pressure to transfer extra-carry fuel from the auxiliary fuel tanks (AFTs), which are typically installed in the fuselage, to the center-wing tank, as shown in Figure 1. The areas surrounding AFTs are identified as Flammable Fluid Leakage Zone (FFLZ). CCAR25.863 of civil aircraft airworthiness authority requires that "The areas of flammable fluid leakage zones should be determine, and in each area where flammable fluids or vapors might escape by leakage of a fluid system, there must be means to minimize the probability of ignition of the fluids and vapors, and the resultant hazards if ignition does occur" [3].

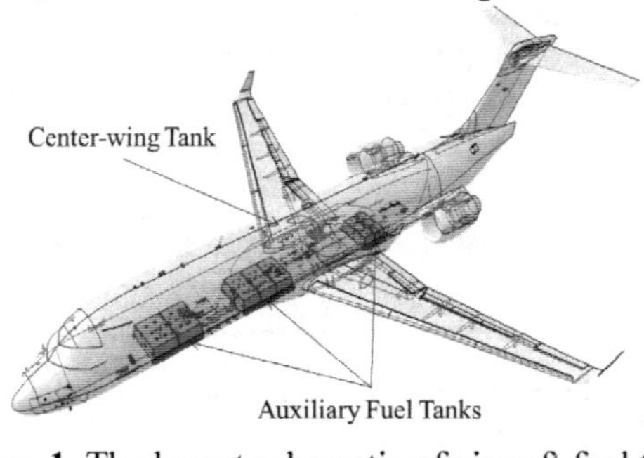

Figure 1. The layout schematic of aircraft fuel tanks.

The specific AFTs are constructed from double-wall structure panels, which comprise the inner wall for storing the fuel and the outer wall for blocking the leakage. The area between inner and outer wall is classified as FFLZ. The schematic diagram illustrating the ventilation and drainage systems of AFTs is presented in Figure 2. In the case of fuel leakage through the inner skin wall to the FFLZ, the vented fresh air from cabin can evacuate the leaked fuel and vapor away from the FFLZ to outside the aircraft, which can reduce the risks of combustion explosion. However, the influence of airflow channel design on the ventilation and compression performance of double-wall structural panels has not been clearly investigated.

In this study, the ventilation and compression performance of double-wall panels with different internal structure are investigated. Computational fluid dynamics (CFD) calculations are conducted for each panel to compare with the ventilation performance numerically. Concurrently, Finite Element Analysis (FEA) is implemented to each panel to assess the compression performance numerically.

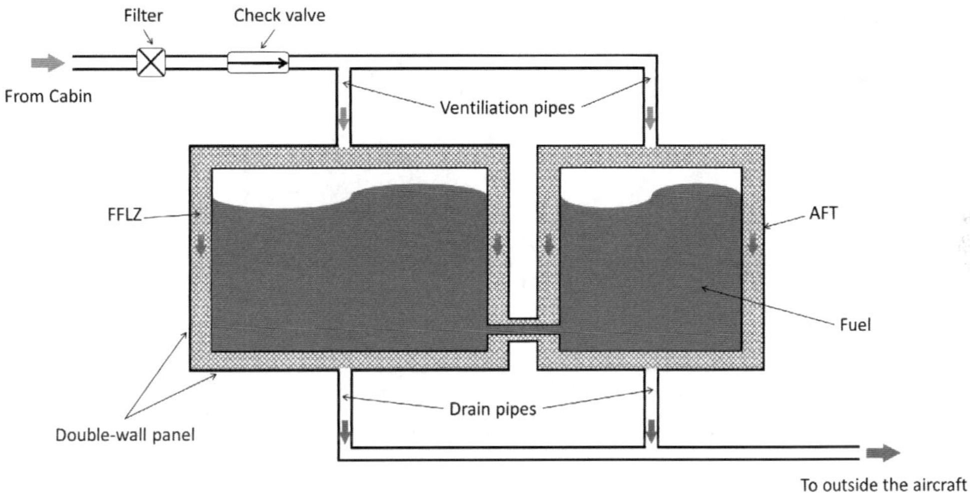

Figure 2. The schematic diagram of ventilation and drain of AFTs.

2. The structure of double-wall panels

The structure of a well-known commercial aircraft's AFTs is shown in Figure 3. The AFTs are composed of six double-wall structure panels (Top panel, Bottom panel, Front panel, Rear panel, Left panel, and Right panel), and the double-wall structure panels consist of two aluminium skins and an aluminium honeycomb core. The inner skin of the panel serves as the primary fuel barrier and the outer skin serve as a leakage barrier. The honeycomb core is treated with opening grooves at the end or opening holes in the center to ensure continuous ventilation with cabin air. As illustrated in Figure 4, six honeycomb cores with distinct structures have been designed according to different groove or hole sizes, shapes and directions. The width of the honeycomb wall is 3.7 mm, and the height of the honeycomb is 15 mm. The actual structural parameters are presented in Figure 5 and Table 1.

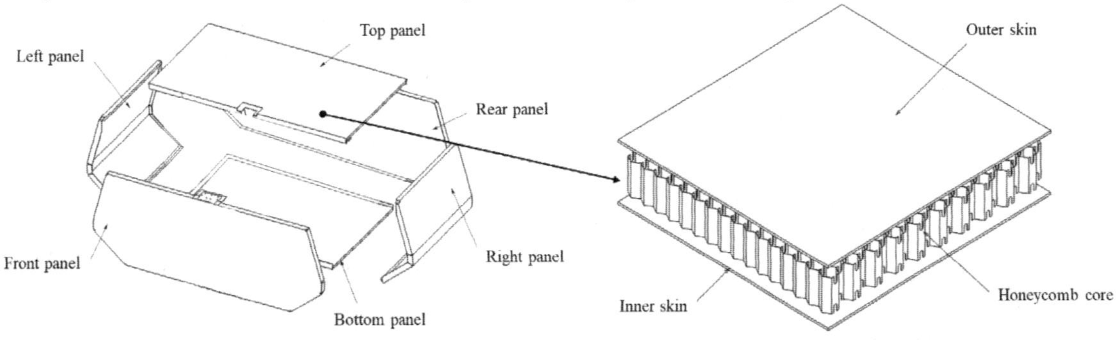

Figure 3. The structure of a commercial aircraft's AFTs and its double-wall structure panel.

Journal of Physics: Conference Series **3041** (2025) 012025 doi:10.1088/1742-6596/3041/1/012025

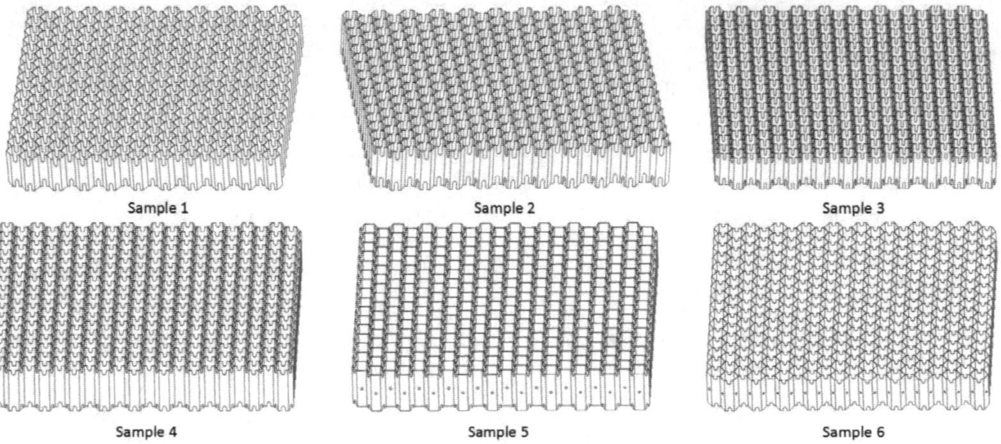

Figure 4. The six different structures of double-wall panels.

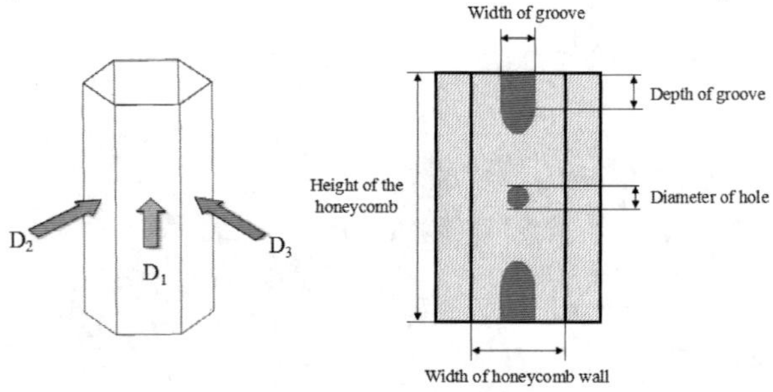

Figure 5. The schematic diagram of single honeycomb.

Table 1. The shape, direction and size of groove and hole.

Sample	Structure of honeycomb core (Shape-Direction)	Depth of "U" groove	Width of "U" groove	Diameter of hole
1	"U" groove-D_1	2.54 mm	1.5 mm	-
2	"U" groove-D_1&D_2	2.54 mm		-
3	"U" groove-D_1&D_2&D_3	2.54 mm		-
4	"U" groove-D_1&D_2&D_3	1 mm		-
5	"U" groove-D_2&D_3+hole-D_1	1 mm		1 mm
6	"U" groove-D_1+hole-D_2&D_3	1 mm		1 mm

3. Simulation of ventilation and Compression

3.1. CFD simulation

CFD is a branch of science which can predict the fluid flow, mass transfer and any other related physical phenomenon. In this study, CFD modelling was performed by using Star CCM+, which is verified effective in the field of fluid computing in the aviation industry [4].

Mesh generation is a crucial part in CFD calculation, which affects the accuracy and efficiency of calculation. Based on the dimensions of the honeycomb samples, the mesh generation is conducted with basic size of 2 mm and refined to 0.2 mm in the vicinity of the honeycomb walls. Through the sensitive analysis, it is found that further refinement of the mesh does not significantly change the calculation results. Figure 6 shows the mesh of honeycomb and fluid domain of double-wall panel. In order to ensure simulation stability, the fluid domain was extended to avoid reverse flow.

The Reynolds-Average Navier-Stokes (RANS) equation, as well as mass conversion and turbulent viscosity models, govern the flow [5]. And the fluid flow is considered to be in-compressible, turbulent, and three dimensions. In order to simplify calculation, the standard k-ε turbulence model is applied in this work. The other boundary conditions were also listed in Table 2.

Journal of Physics: Conference Series **3041** (2025) 012025 doi:10.1088/1742-6596/3041/1/012025

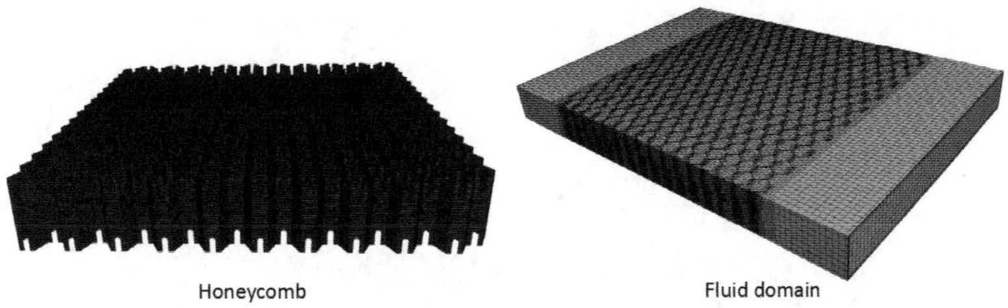

Figure 6. The mesh of honeycomb and fluid domain of double-wall panel.

Table 2. Boundary conditions of CFD simulations.

Boundary Condition	Setting
Turbulence model	standard k-ε
Flow	Ideal gas
Wall roughness	Smooth wall
Shear condition	No slip
Inlet condition	Total pressure (2000 Pa)
Outlet condition	Pressure outlet (0 Pa)
Convergence criterion	10^{-5}

3.2. FEA simulation
The finite element model was established in ABAQUS [6,7]. The material properties of 5052 aluminum alloy materials properties were used to build the honeycomb model. The model was meshed by sweeping grid with the continuous shell element C3D10 with the mesh size of 1 mm. The support platen and the loading platen were defined as rigid bodies and the C3D10 element with the mesh of 2 mm was used to build their models. The models were cut to a quarter of its original size to reduce computation time and improve efficiency. Through the sensitive analysis, it is found that further refinement of the mesh does not significantly change the calculation results. The loading of 7 Mpa is adopted in this study.

4. Results and discussion

4.1. Ventilation performance
In order to capture the gas flow within the fluid domain of double-wall panel, derivative planes, including top-section, middle-section, and cross-section, were established within the fluid domain, as shown in Figure 7. Figure 8 and 9 show the gas velocity distribution at top-section and middle-section of the fluid domain. It is observed that the gas primarily flows along the direction of the grooves or holes and seldom changes direction when the orientation of the grooves or holes aligns with the ventilation direction. Compared Sample 1, Sample 2, and Sample 3, it could be inferred that the uniformity of gas velocity distribution increases with increasing number of grooves. Sample 3 and Sample 4 are the honeycomb structure of grooving in D_1, D_2, and D_3 directions with the groove depth of 2.54 mm and 1 mm, which can be clearly seen in the velocity distribution that the gas velocity distribution is more uniform. Compared with Sample 3 and Sample 4, it could be inferred that the ventilation gas flows more difficult into the honeycomb core as the depth of the groove decreases. By converting the groove in the D_1 direction of Sample 4 to a central hole with a diameter of 1 mm, Sample 5 is obtained. Similarly, by changing the grooves in the D_2 and D_3 directions of Sample 4 to central holes with a diameter of 1 mm, Sample 6 is obtained. Figure 10 shows gas velocity vector distribution at cross-section of the fluid domain. Comparing Sample 4, Sample 5, and Sample 6, it could be inferred that introducing holes at the center of the honeycomb core, aligned with the ventilation flow direction, can significantly enhance the gas velocity and flow rate of the ventilation flow gas within the honeycomb core.

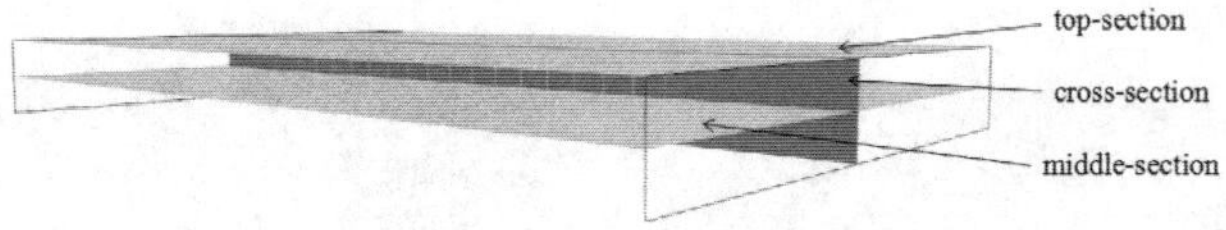

Figure 7. Derivative planes are established in fluid domain of double-wall panel.

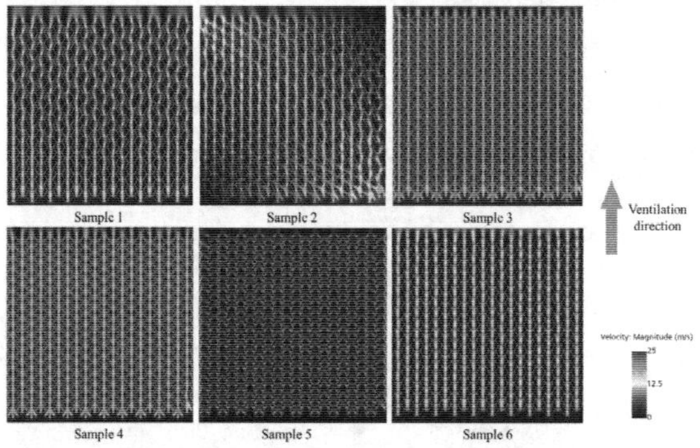

Figure 8. Gas velocity distribution at top-section of fluid domain of double-wall panel.

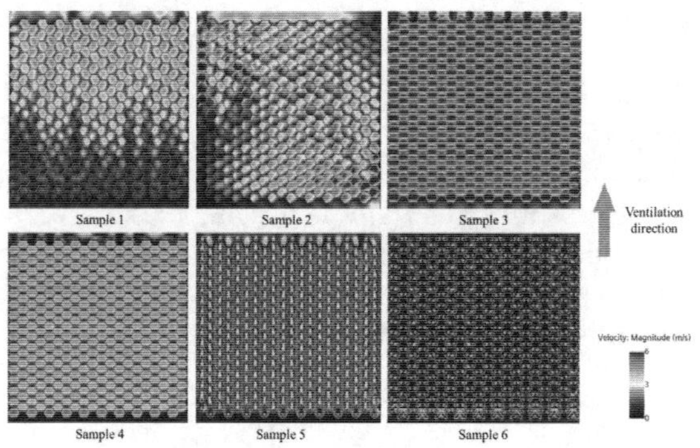

Figure 9. Gas velocity distribution at middle-section of fluid domain of double-wall panel.

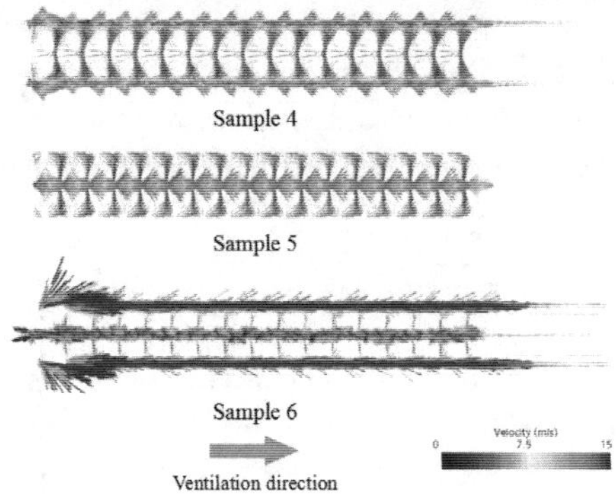

Figure 10. Gas velocity vector distribution at cross-section of Sample 4, Sample 5, and Sample 6.

Figure 11 and Table 3 present the gas average velocity and the gas flow rate at the outlet of double-wall planes. The gas average velocity at outlet of Sample 1, Sample 2, Sample 3, Sample 4, Sample 5, and Sample 6 are 5.62 m/s, 4.35 m/s, 3.61 m/s, 3.17 m/s, 0.73 m/s, and 4.21 m/s, respectively. The gas flow rate at the outlet of Sample 1, Sample 2, Sample 3, Sample 4, Sample 5, and Sample 6 are 571.6 L/min, 442.4 L/min, 367.2 L/min, 322.4 L/min, 74.2 L/min, and 427.9 L/min, respectively. The gas average velocity and gas flow rate at the outlet decreases with the increasing number of grooves, which suggests that the direction of gas flow becomes more dispersed as the number of grooves increases, which can expand the gas flow throughout the whole fluid domain instead of flowing

Journal of Physics: Conference Series **3041** (2025) 012025 doi:10.1088/1742-6596/3041/1/012025

directly from the inlet to the outlet, thereby increasing the uniformity of the gas flow. The gas average velocity and gas flow rate at outlet decreases as the decrease of depth of groove, which means that it has a large flow resistance when the gas flows through a shallower groove, which is not conducive to the flow and diffusion of the gas in the fluid domain. After replacing the groove in the D_1 direction with hole, the gas average velocity and gas flow rate decrease sharply. Conversely, after replacing the grooves in the D_2 and D_3 directions with holes, the gas average velocity and gas flow rate increase slightly. It implies that the gas prefers to flow through the grooves rather than the holes, suggesting that holes present a greater flow resistance compared to grooves.

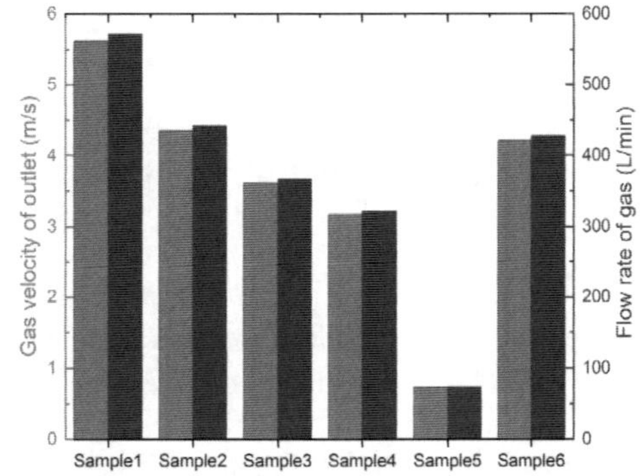

Figure 11. The gas average velocity and the gas flow rate at outlet of double-wall planes.

Table 3. The gas average velocity and the gas flow rate at outlet of double-wall planes.

Sample	Gas average velocity	Gas flow rate
1	5.62 m/s	571.6 L/min
2	4.35 m/s	442.4 L/min
3	3.61 m/s	367.2 L/min
4	3.17 m/s	322.4 L/min
5	0.73 m/s	74.2 L/min
6	4.21 m/s	427.9 L/min

4.2. Compression performance

The amount of deformation of honeycomb in load direction is shown in the Figure 12 and Figure 13. The maximum amount of deformation of Sample 1, Sample 2, Sample 3, Sample 4, Sample 5, and Sample 6 are 0.2358 mm, 0.6182 mm, 1.587 mm, 0.6873 mm, 0.335 mm, and 0.179 mm, respectively. It is noticed that the maximum amount of deformation increases as the number of grooves increases, while decreases as the depth of the groove decreases at the same load. Increasing the number of grooves and the depth of grooves is not conducive to the compressive strength of honeycomb. After replacing the grooves at the end of the honeycomb with holes in the center of the honeycomb, the maximum amount of deformation decreases sharply and large deformation also occurs around the holes, which means that opening holes in the center of the honeycomb reduces the stability of the honeycomb, while the weakest mechanical properties of the honeycomb core exist in the end of the honeycomb rather than the center.

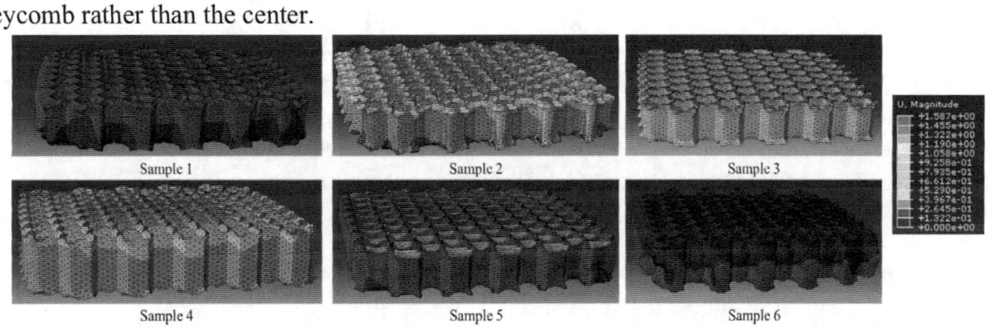

Figure 12. The amount of deformation of honeycomb in load direction.

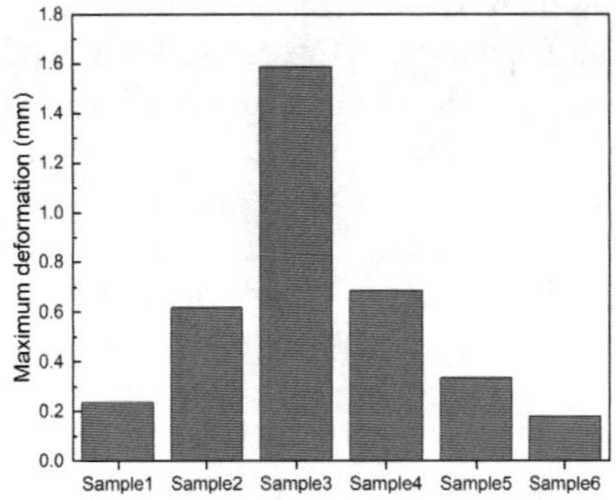

Figure 13. The maximum amount of deformation of honeycomb at the same load.

5. Conclusion

The ventilation and compression performance of double-wall panels with Six different internal structures are investigated using computational fluid dynamics simulation and finite element analysis in this research. Based on the ventilation simulations and compression analyses, the following conclusions are drawn:

- The uniformity of gas flow increases with increasing number of flow directions. However, the compression strength of the honeycomb core decreases as the number of grooves increases.
- The ventilation gas flows into the honeycomb core more easily, while the compression strength of the honeycomb core is significantly reduced as the depth of the groove increases.
- Opening holes in the center of the honeycomb core can greatly improve the flow rate of the ventilation flow inside the honeycomb core, but reduce the stability of the honeycomb core.
- The gas is more willing to flow through the grooves rather than the holes in the honeycomb core, indicating that the holes have greater flow resistance compared to grooves.

This study provides design and performance simulation methods for honeycomb-based double-wall panels, which hold significant potential in manufacturing auxiliary fuel tanks. The results of the simulation calculation can guide the structural design of honeycomb-based double-wall panels and reduce the experimental cost, which can improve the efficiency of auxiliary fuel tanks design.

6. References

[1] Jaidev Vyas J, Balamurugan G and Rakesh M 2015 SIMULATION OF TYPICAL CIVIL AIRCRAFT FUEL TRANSFER PROCESS. 2015 5th Nirma University International Conference on Engineering (NUiCONE).

[2] A. Filippone, Advanced Aircraft Flight Performance, Cambridge University Press, New York, 2012.

[3] Auxiliary fuel system installations: AC 25-8, Federal Aviation Administration, 2006.

[4] T. M. Wang, H. Wang, G. J. Tang, Theory, Methodology, Tools and Applications for Modeling and Simulation of Complex System, 2016, Volume 645.

[5] F. K. Osman, J. Zhang, L. Lai, Effect of Turbulence Model on Flow Characteristics of a Vertical Fire Pump. Journal of Applied Fluid Mechanics, 15, 1661-1674, 2022.

[6] H. Feng, L. Liu, Qi. Zhao, Experimental and numerical investigation of the effect of entrapped air on the mechanical response of Nomex honeycomb under flatwise compression. Compos. Struct. 182, 617–627, 2017.

[7] L. H. Xu, C. Ni, G. Liu, J. L. Li, Analysis of Honeycomb and Sandwich Structure Compression Property and the Damage of Double Cell Wall. Fibers and Polymers. 24, 3245–3260, 2023.

Acknowledgments

This research was funded by the Shanghai Post-doctoral Excellence Program.

Effects of mainstream incidence angle on the turbine blade leading edge film cooling of cylindrical and laidback holes

Li Hongye[1,2,3,4], Xu Qingzong [*1,2,3,4], Du Qiang[1,2,3,4], Xu Guangyao[1,2,3,4], Liu Haoyang[1,2,3,4], Xu Jiawei[1,2,3,4]

1 Advanced Gas Turbine Laboratory, Institute of Engineering Thermophysics, Chinese Academy of Sciences, Beijing 100190, China
2 University of Chinese Academy of Sciences, Beijing 100049, China
3 National Key Laboratory of Science and Technology on Advanced Light-duty Gas-turbine, China
4 Key Laboratory of Advanced Energy and Power, Institute of Engineering Thermophysics, Chinese Academy of Sciences, Beijing,100190, China
5 Qingdao Institute of Aeronautical Technology, China

*xuqingzong@iet.cn

Abstract. The blade leading edge is directly exposed to hot gas, and its cooling performance is highly sensitive to the mainstream incidence angle. This study numerically investigates the effects of the mainstream incidence angle on the film cooling effectiveness of the blade leading edge. It compares the cooling performance of cylindrical and laidback holes at varying mainstream incidence angles of -2°, 0°, and +2° under blowing ratios of 0.75, 1.5, and 2.25. Film cooling effectiveness for both types of holes increases with the increasing incidence angle under blowing ratios of 0.75 and 1.5, while the opposite occurs under the blowing ratio of 2.25. This is due to the fact that the effect of the incidence angle on the mixing of coolant and mainstream varies with the blowing ratios. The area-averaged film cooling effectiveness for laidback holes remains at least 26% higher than that for cylindrical holes at varying incidence angle under each blowing ratio. Moreover, the cooling performance stability of laidback holes is better than that of cylindrical holes when the mainstream incidence angle changes from design to off-design conditions.

1. Introduction

The turbine blade is a key component in gas turbines, facing the highest thermal stress and temperatures, particularly at the leading edge. Film cooling is a common technique to prevent erosion from hot gases. Enhancing film cooling performance can improve both the thermal efficiency and durability of turbine blades.

The most common type of hole applied on the blade leading edge is the cylindrical hole. Ou et al.[1] explored the relative location of rows. Results showed that the row with an injection angle of ±15° is more sensitive to the adverse effects of mainstream than the row with ±40°. The relative location was also explored by Xie et al.[2] on the blade leading edge, and they proposed that the impacts varied on the suction and pressure side. Azzi and Ali [3] explored inclination angles of 25°, 30°, 35° and 60°. They found that the cooling performance peaked at an inclination angle of 25°. Liu et al.[4] also studied the inclination angle, and their results indicated that an increase in the angle reduced the coolant coverage area. Garg et al.[5] examined the lateral hole spacing and suggested that cooling performance improves as the hole spacing decreases. Rozati et al.[6, 7] performed LES simulations to explore the influence of blowing ratios. Their findings showed that the cooling performance reduced as the blowing ratio increased within the range of 0.4 to 1.2. However, according to Ou and Rivir [8], at a turbulence of 20%, the cooling performance increased as the blowing ratio increased from 1.0 to 2.0.

To enhance the film cooling effectiveness of turbine blade leading edge, laidback holes have been introduced as a solution. Reiss et al.[9] compared cylindrical, laidback and laterally expended holes. Their results revealed that LBF provided superior cooling performance compared to the other two types. Kim et al.[10] explored a variety of hole geometries and found that laidback holes offered a more uniform cooling performance distribution. Liu et al.[11, 12] investigated cylindrical and laidback holes with different inclination angles and hole spacing. Their findings also showed that laidback holes performed better than cylindrical holes.

As the engine operating conditions continuously change, the turbine also operates under off-design mainstream incidence angles. Numerous studies have been undertaken to explore how the incidence angle influences the flow field and cooling performance at the leading edge. Albert et al.[13] investigated the incidence angles of 0°, -3° and +3° on a semi-cylindrical model, discovering that the position of the mainstream stagnation line, which shifts with the incidence angle, greatly influences film cooling performance. Benabed et al.[14] studied the effects of incidence angle on an asymmetrical turbine blade. Results showed that cooling performance was strongly affected by the incidence angle at low blowing ratios but diminished at higher values. Chaves et al.[15] and Liu et al.[16] experimentally studied the cooling performance at off-design incidence angles in linear cascade environment. They found that effects of incidence angle showed different on the suction and pressure side.

In summary, film cooling performance for laidback holes is superior to that for cylindrical holes. Multiple investigations have been carried out to examine film cooling effectiveness for cylindrical holes at off-design incidence angles. Nevertheless, to the authors' knowledge, there are few studies concerning the effect of mainstream incidence angle on the film cooling effectiveness for laidback holes. In the present study, a comparison of the film cooling effectiveness for cylindrical and laidback holes will be conducted at varying mainstream incidence angles under different blowing ratios. To explore the underlying mechanisms, flow characteristics will also be presented.

2. Numerical method

2.1. Geometrical details

Figure. 1 shows the calculating blade, which is enlarged from the mid-section of an actual turbine blade by a factor of 6.7, as shown in Fig. 1. A comparison of the actual turbine blade and the scaled model can be seen in the study of Li et al.[17]. There are three rows of film cooling holes on the blade leading edge. F2 is located at the blade stagnation line where $S/d_f = 0$,while F1 and F3 are positioned at ±40° on each side of the blade stagnation line. Under the design incidence angle, $i = 0$, the mainstream stagnation line forms an angle of 18° with F2. Under off-design conditions, the mainstream stagnation line moves to suction side (SS) at positive incidence angle, while it moves to pressure side (PS) at negative incidence angle.

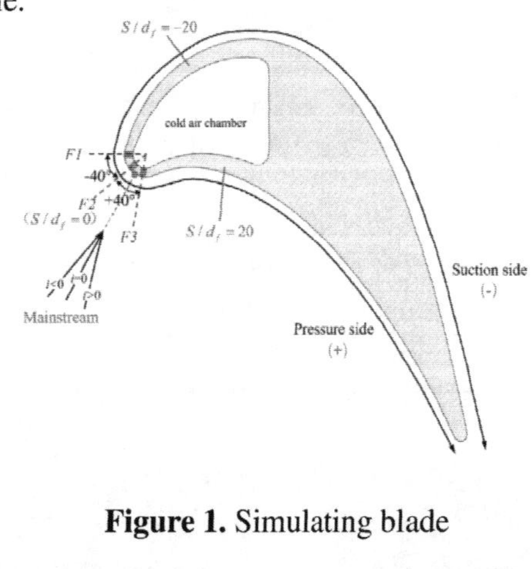

Figure 1. Simulating blade

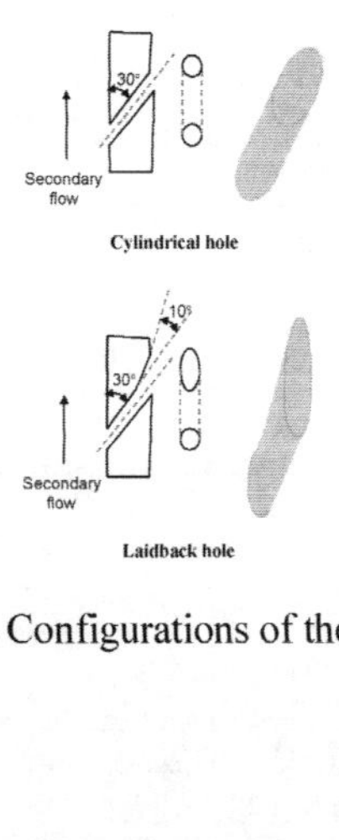

Figure 2. Configurations of the film holes

The current paper presents a comparative study of CYF and LBF. Fig. 2 presents the hole configurations. The LBF have a 10-degree backward expansion relative to the hole axis. The expansion section length to the hole length ratio of the LBF is $1/2$. The measuring section of the LBF has an identical diameter d_f of 3mm as the CYF. The inclined angle of each hole is 30° from the surface, and the compound angle is 90° from the mainline flow. The film holes are spaced in three staggered rows, with a hole-to-hole pitch p of 12mm, i.e., $p/d_f = 4$.

2.2. Computational methods and boundary conditions

Commercial CFD software ANSYS CFX 19.0 is adopted to solve the steady Reynolds Averaged Navier-Stocks (RANS) equations, which include continuity, energy and momentum equations. The SST k-ω turbulence model with automatic wall treatment is chosen for simulations.. Equations are discretized using the finite volume method, and a high-resolution scheme is applied for convective variables. The convergence criterion is satisfied when the momentum and enthalpy residuals drop to 10^(-6) with negligible variation.

Figure 3 shows the computational domain and boundary conditions. On both sides of the mainstream, periodic conditions are established to simulate the cascade environment. The mainstream inlet velocity and turbulence intensity are 25 m/s and 5%, respectively. To investigate the flow and heat transfer characteristics of leading edge film cooling under off-design mainstream inlet conditions, incidence angles of -2°, 0° and +2° are simulated. The mainstream total temperature is 298K and the coolant temperature is 288K. The coolant inlet is controlled by mass flow rate, which is given according to blowing ratios of 0.75, 1.5 and 2.25. Blowing ratio M is defined as:

$$M = \frac{\rho_c U_c}{\rho_g U_g} \tag{1}$$

where, ρ_c and U_c represent the density and velocity of the coolant at the cold air chamber, respectively, while, ρ_g and U_g denote the density and velocity of the mainstream inlet, respectively. The outlet condition is set as atmosphere pressure.

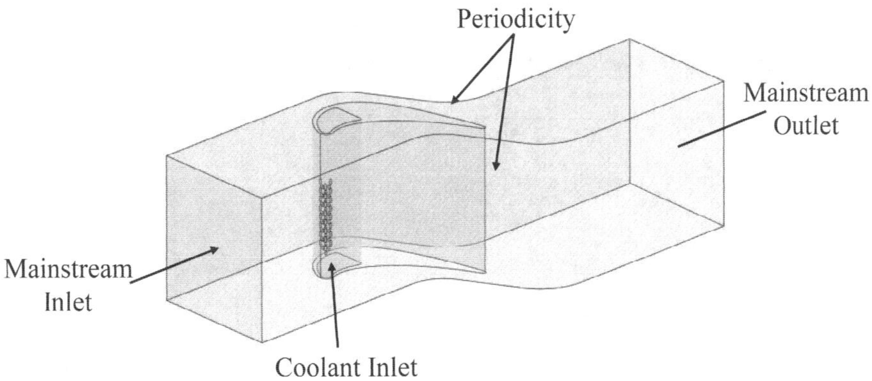

Figure 3. Computational domain and boundary condition

Table. 1 Boundary conditions for the CFD model

Dimension	Value
Mainstream inlet velocity (m/s)	25
Mainstream inlet temperature (K)	298
Mainstream inlet turbulence (%)	5
Mainstream inlet incidence angle (°)	-2, 0, +2
Coolant temperature (K)	288
Blowing ratio	0.75, 1.5, 2.25

2.3. Mesh generation and grid independence analysis

Unstructured mesh for the simulating model is created using ANSYS ICEM19.0 package, as illustrated in Fig. 4. The simulation domain is discretized using tetra/mixed mesh, incorporating 20 prism layers in the near-wall areas to precisely capture the flow field and cooling performance. The first layer height and the mesh growth rate in the boundary layers are set to 0.001mm and 1.2, respectively, ensuring that the near wall $Y+$ is less than 1 to fit the turbulence model.

A grid convergence study is carried out to confirm that the computational outcomes are unaffected by the grids number. In this analysis, three various mesh numbers at 9, 13 and 22 million are compared. The effect of grid number is illustrated using the predicted spanwise-averaged film cooling effectiveness on the leading edge external wall. The film cooling effectiveness is defined as:

$$\eta = \frac{T_g - T_{aw}}{T_g - T_c} \tag{2}$$

where, T_g and T_c represent the mainstream inlet temperature and coolant temperature, respectively, and T_{aw} denotes the adiabatic wall temperature. Fig. 5 illustrates the comparison of the spanwise-averaged η across three different grid resolutions. It is evident that the numerical results exhibit only minor changes when the grids number exceeds 13 million. To balance accuracy of simulation and the cost of numerical computing, the final grid elements number for the calculating domain is chosen to be 13 million.

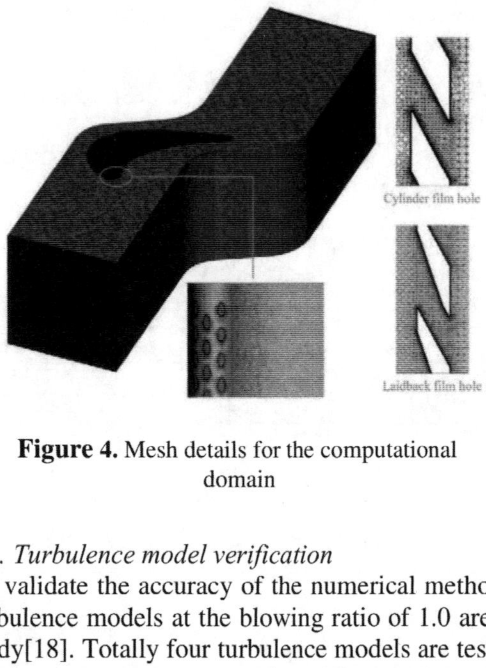

Figure 4. Mesh details for the computational domain

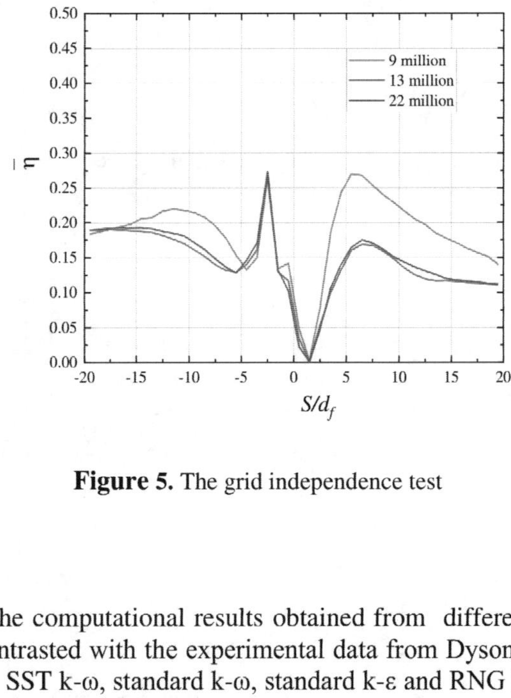

Figure 5. The grid independence test

2.4. Turbulence model verification

To validate the accuracy of the numerical method, the computational results obtained from different turbulence models at the blowing ratio of 1.0 are contrasted with the experimental data from Dyson's study[18]. Totally four turbulence models are tested: SST k-ω, standard k-ω, standard k-ε and RNG k-ε. Fig. 6 shows the comparison of spanwise-averaged $\bar{\eta}$ on the leading edge external wall. All four turbulence models show similar trends in $\bar{\eta}$ along flow direction compared to the experimental results.. Regarding the values, the results from the standard k-ε model show the greatest deviation. Within the interval of $4 < X/d_f < 10$, the results from the remaining three turbulence models aligns well with the experimental results. However, the RNG k-ε overestimates $\eta\eta$ within the range of $0 < X/d_f < 2$, while the standard k-ω severely underestimates it in the region of $2 < X/d_f < 3$. From the comparison of $\bar{\eta}$, the SST k-ω turbulence model aligns better with experimental data in terms of trends and numerical outputs.

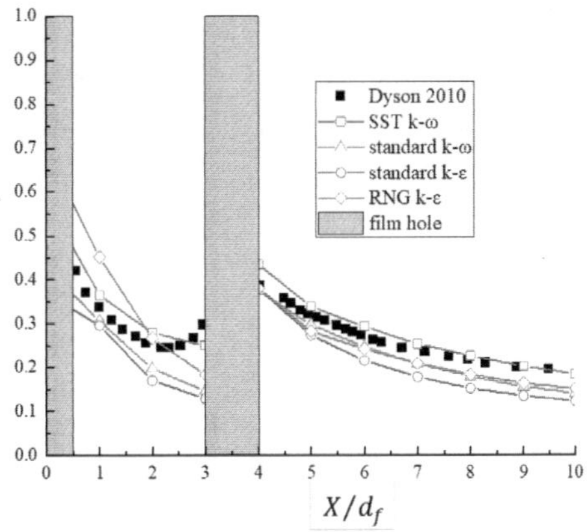

Figure 6. Comparison of spanwise-averaged adiabatic cooling effectiveness between numerical results and experimental data [18]

Fig. 7 presents a comparison of η distributions on the external wall contrasting experimental findings with results from the four turbulence models. Obviously, the coolant coverage in the numerical results from the standard k-ω and standard k-ε turbulence models is much worse than in the experimental results. Although the RNG k-ε turbulence model has similar coolant coverage as the SST k-ω turbulence model, η predicted by the SST k-ω turbulence model aligns more closely with experimental results.

All in all, the simulation results from the SST k-ω turbulence model align more closely with the experimental results. Also, it is commonly used in the open literature to simulate the cooling performance of turbine blades[19-22]. Therefore, in this study, the SST k-ω turbulence model is employed to investigate the flow field and cooling performance of the blade leading edge film cooling.

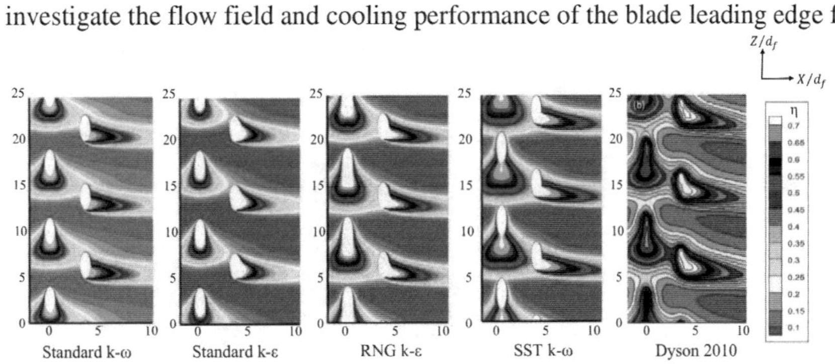

Figure 7. Comparison of adiabatic cooling effectiveness distribution on the leading edge external wall between numerical results and experimental data[18]

3. Results and discussion

3.1. Film cooling effectiveness

The contours of film cooling effectiveness distribution on the blade leading edge external wall for different cases are shown in Fig. 8. At the blowing ratio M of 0.75, it can be seen in Fig. 8(a) that η distribution of both film hole types is periodic along the spanwise direction. Due to film superposition characteristics, coolant coverage on suction side is superior to pressure side. For each case, η diminishes along the flow direction. As the mainstream incidence angle i increases, η rises noticeably on both sides for the CYF. For the LBF, η rises on the PS wall while changes slightly on the SS wall.

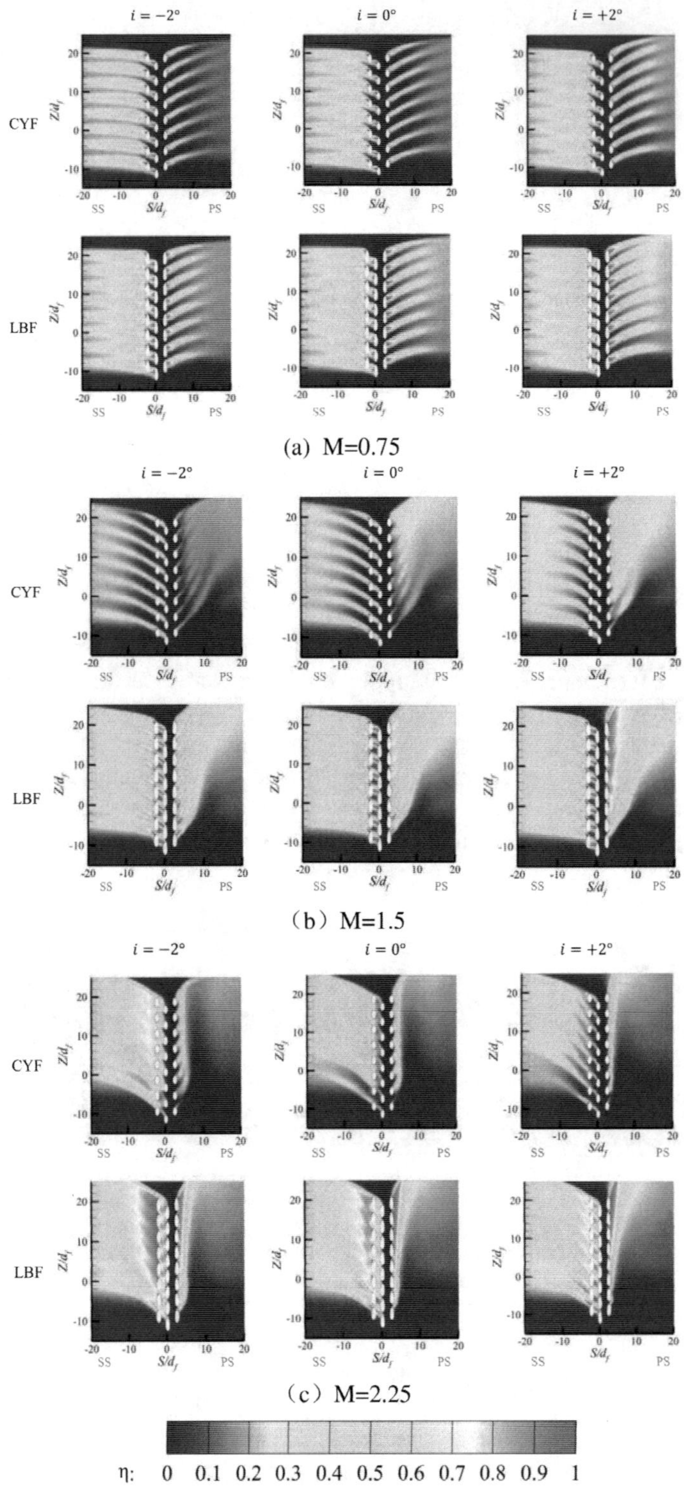

Figure 8. Distribution of the film cooling effectiveness

Fig. 8(b) shows η distribution at M of 1.5. For the CYF, coolant traces from hole row F2 and F1 overlap, and the downstream cooling coverage expands. As mainstream incidence angle increases, η on the suction side improves significantly. Moreover, coolant from hole row F3 lifts off from the wall surface at the hole exits and reattaches at the downstream region, converging together near the blade tip region. As the mainstream incidence angle increases, the mainstream stagnation line moves to the PS side, which suppresses the coolant blowing off from the PS wall. Consequently, η on the PS wall increases with rising incidence angle. For the LBF, coolant from hole row F3 also converges together near the blade tip region and η improves as the mainstream incidence angle increases. Differently, coolant from hole row F3 for the LBF can attach the external wall well, resulting in much higher η on the PS wall in comparation to the CYF. In addition, the coolant coverage on the suction side is excellent, with variations in i having negligible influence on η in this region.

When M reaches 2.25, as shown in Fig. 8(c), the phenomenon becomes noticeably different from that at relatively lower blowing ratios. Both types of film holes have a decreasing η on the SS wall when the mainstream incidence angle increases. For the CYF, coolant can cover the suction side wall effectively at negative incidence angle of -2°, and the coverage slightly declines when the incidence angle turns to designed condition. As the incidence angle increases to +2°, coolant lifts off from the suction side and η becomes lower. Although coolant from F1 and F2 for the LBF remains attached to the suction side wall, it is noticeable that η decreases with decreasing incidence angle. Coolant from F3 for both types of film holes is blown significantly away from the blade leading edge pressure side wall, which reduces η. Moreover, η on the PS wall increases only marginally with rising incidence angle for both types of film holes.

Figure 9 shows the spanwise-averaged film cooling effectiveness $\bar{\eta}$ for cylindrical and LBF. At M of 0.75, $\bar{\eta}$ for both types of holes follow the same trend along the flow direction. A peak is observable at the exit of each row, and the peak for F1 is much higher than the other two peaks due to superposition characteristics of F1 and F2. Therefore, for both types of film holes, $\bar{\eta}$ on the SS wall is higher than on the PS wall, while for the LBF, the variation between the SS side and PS side is smaller compared to CYF. As i increases, $\bar{\eta}$ increases on the PS side, while it changes minimally on the SS side.

As shown in Fig 9 (b), the peak of $\bar{\eta}$ at the exit of hole row F2 almost disappears, because of the overlap of coolant traces from hole row F2 and F1. For the CYF, $\bar{\eta}$ rises with mainstream incidence angle, as a larger incidence angle facilitates better coolant attachment to the suction and pressure side walls. For the LBF, $\bar{\eta}$ on the suction side shows little variation, while it increases with mainstream incidence angle on the pressure side.

As presented in Fig. 9(c), $\bar{\eta}$ decreases dramatically on the SS wall for the CYF, and it also decreases in the same region for the LBF. At the exits of hole row F3, $\bar{\eta}$ for the CYF grows with increasing mainstream incidence angle, while it first increases and then decreases for the LBF. Downstream of F3, on the region of $S/d_f > 5$, $\bar{\eta}$ for both types of film holes is hardly affected by the incidence angle.

Moreover, at M of 1.5 and 2.25, a low peak can be observed around the position of $S/d_f = -5$ on the SS side for the CYF. The same phenomenon has also been noticed by Liu et al[16]. The region of $-10 < S/d_f < 0$ is a concave surface, where flow is likely to separate. As i shifts to positive value, the mainstream stagnation point moves to the SS side. Therefore, at the condition of $i \geq 0°$, the separation becomes more severe, which reduces the low peak value. When M increases to 2.25, the low peak can only be observed when $i \geq 0°$, because the increased coolant mass flow rate partly compensate for the drawback of flow separation at $i < 0°$ .However, there is no such low peak for LBF, as the coolant attachment of LBF is much better.

CYF

LBF

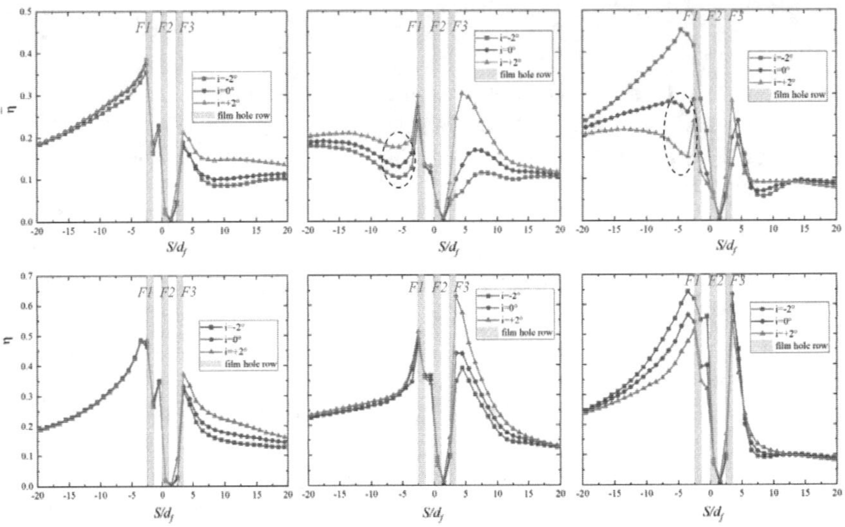

(a) $M = 0.75$ (b) $M = 1.5$ (c) $M = 2.25$

Figure 9. Spanwise-averaged film cooling effectiveness for cylindrical and LBF at various incidence angle under different blowing ratios.

Figure.10 shows the area-averaged film cooling effectiveness for cylindrical and LBF on the region of $-20 < S/d_f < 20$. At relatively lower M of 0.75 and 1.5, area-averaged η rises with mainstream incidence angle for both types of film holes. For the CYF, as $i$changes from -2° to +2°, the area-averaged η increases by 15.8% and 48.7% at M of 0.75 and 1.5, respectively. In contrast, for the LBF, the area-averaged η increases by 11.7% and 15.9% at the same blowing ratios as the incidence angle shifts from -2° to +2°. However, at M of 2.25, area-averaged η decreases with increasing incidence angle. It reduces by 30.8% and 15.1% for CYF and LBF when incidence angle changes from -2° to +2°, respectively. It is obvious that when $i$changes from design condition to off-design conditions, the ability to maintain the cooling performance is much better for the LBF than for the CYF. Moreover, at the same mainstream incidence angle under the same blowing ratio, the area-averaged η for the LBF is higher than that for the CYF by up to 95% and by at least 26%.

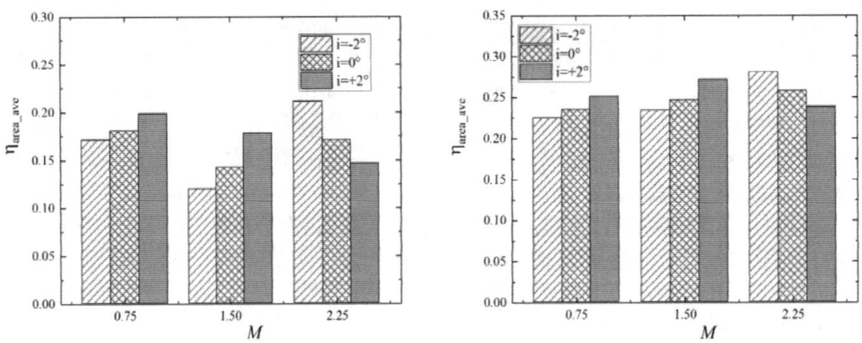

CYF LBF

Figure 10. Area-averaged film cooling effectiveness for the cylindrical and LBF ($-20 < S/d_f < 20$)

3.2. Thermal field and flow characteristics

In order to explore mechanisms by which the mainstream incidence angle affects the film cooling effectiveness for cylindrical and LBF under different blowing ratios, a detailed examination of thermal field and flow characteristics is conducted in this section. The non-dimensional thermal field distribution downstream of F1 and F3 is showed to analyze the mixing of coolant and mainstream on SS and PS wall. Midplane velocity contours are used to reveal the effects of mainstream incidence angle on the local mainstream velocity and in-hole flow structure. Local blowing ratios are presented to analyze the coolant attachment to the external wall. Evaluating the discharge coefficient is intended to investigate the flow resistance of cylindrical and LBF.

Figure 11 shows the dimensionless thermal fields spanwise and perpendicular to the blade surface directly downstream of F1 and F3 at $S/d_f = -5$ and $S/d_f = 5$. In order to better show the flow details, only parts of the selected sections are shown.

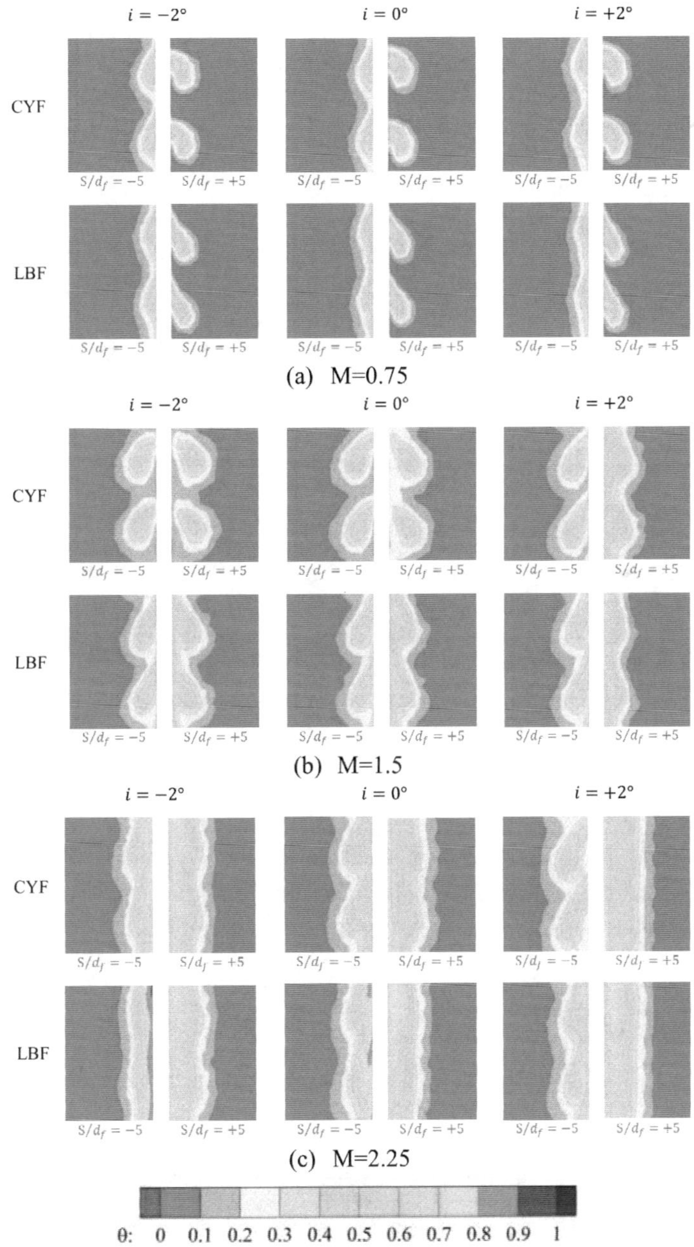

Figure 11. Dimensionless temperature distributions of the downstream section and pressure side of the holes

Dimensionless temperature is defined as follows:

$$\theta = \frac{T_g - T}{T_g - T_c} \tag{3}$$

At M of 0.75, it is clear that the coolant attaches the wall better on the SS side than on the PS side. Significant separation can be seen on the PS side for both cylindrical and LBF. The separation is more pronounced at smaller mainstream incidence angle. It can explain the worse η on the PS wall in Fig. 8(a). However, the separation for the LBF is less severe than for the CYF, thus η for the LBF is better than for the CYF.

As M increases to 1.5, the attachment of coolant to the leading edge pressure side wall improves a lot, although separation can still be observed at incidence angle of 0° and -2° for the CYF. Correspondingly, η is improved on the PS wall for both types of holes. Nevertheless, separation occurs on the SS side for CYF. This explains the low peak in Fig 9(b).

When M reaches 2.25, the coolant covers the leading edge pressure side wall very well for both types of film holes at different incidence angles. Although coolant blows away severely from the PS wall, the concave surface effectively gathers the coolant at the exit of hole row F3. As shown in Fig. 9(c), $\bar{\eta}$ at $S/d_f = 5$ shows a relatively high value. For the suction side, the high momentum coolant from F1 flows along the inclined direction of film holes and covers the upward spanwise wall well. However, as the mainstream stagnation point moves to the SS side with increasing incidence angle, local mainstream velocity at the exit of F1 and F2 increases. Consequently, the coolant is constrained to flow in its original direction and mixes severely with the mainstream, which makes η decreases with increasing incidence angle.

4. Conclusion and future work

A numerical simulation has been conducted to investigate the effect of mainstream incidence angle on the cooling performance and flow fields for cylindrical and laidback film holes on the turbine blade leading edge under different blowing ratios. Film cooling effectiveness distributions, spanwise-averaged and area-averaged film cooling effectiveness have been compared. Flow fields of mid-hole planes spanwise and perpendicular to the blade's surface have been presented to reveal the flow mechanisms. The major results are summarized below:

(1) The mainstream incidence angle notably affects the film cooling effectiveness because of the shifting mainstream stagnation point. At low blowing ratios, film cooling effectiveness increases with incidence angle, while it shows the opposite at high blowing ratios. However, coolant coverage for the laidback holes remains better than cylindrical holes under different conditions.

(2) Laidback holes show at least 26% higher area-averaged η than cylindrical holes. When the mainstream incidence angle shifts from design to off-design conditions, the variation in film cooling effectiveness for laidback holes is significantly smaller than that for cylindrical holes. This demonstrates that laidback holes exhibit higher cooling effectiveness and better performance stability than cylindrical holes.

This study numerically studied the film cooling effectiveness of cylindrical and laidback holes under different mainstream incidence angles. The findings provide valuable insights for the design of turbine blade leading edge cooling structures and the allocation of coolant under off design conditions. Future work will focus on experimental investigations of cooling performance under varying mainstream incidence angles.

5. References

[1] S. Ou, J.C. Han, Leading edge film cooling heat transfer through one row of inclined film slots and holes including mainstream turbulence effects, ASME. J. Heat Transfer, 116(3) (1994) 561-569.

[2] G. Xie, Z. Tao, Z.-y. Zhou, R.-q. You, S.-z. Xia, H.-w. Li, Hole arrangement effect to film cooling performance on leading edge region of rotating blade, Int. J. Therm. Sci., 169 (2021) 107034.

[3] A. Azzi, B. Ali Jubran, Influence of leading-edge lateral injection angles on the film cooling effectiveness of a gas turbine blade, Heat Mass Transf., 40 (2004) 501-508.

[4] Z. Liu, F. Li, Z.X. Zhang, Z.P. Feng, T.W. Simon, Conjugate heat transfer predictions on combined impingement and film cooling of a blade leading edge model, Heat Transf. Eng., 42(16) (2021) 1363-1380.
[5] V.K. Garg, R.E. Gaugler, Leading edge film-cooling effects on turbine blade heat transfer, Numer. Heat Transf. A, 30(2) (1996) 165-187.
[6] A. Rozati, D.K. Tafti, Large eddy simulation of leading edge film cooling—Part II: Heat transfer and effect of blowing ratio, ASME. J. Turbomach, 130(4) (2008).
[7] A. Rozati, D.K. Tafti, Effect of coolant–mainstream blowing ratio on leading edge film cooling flow and heat transfer–LES investigation, Int. J. Heat Fluid Flow, 29(4) (2008) 857-873.
[8] S. Ou, R.B. Rivir, Leading edge film cooling heat transfer with high free stream turbulence using a transient liquid crystal image method, Int. J. Heat Fluid Flow, 22(6) (2001) 614-623.
[9] H. Reiss, A. Bölcs, Experimental study of showerhead cooling on a cylinder comparing several configurations using cylindrical and shaped holes, ASME. J. Turbomach, 122(1) (2000) 161-169.
[10] Y.J. Kim, S.-M. Kim, Influence of shaped injection holes on turbine blade leading edge film cooling, Int. J. Heat Mass Transf., 47(2) (2004) 245-256.
[11] C.L. Liu, H.R. Zhu, X. Zhang, D.C. Xu, Z.W. Zhang, Experimental investigation on the leading edge film cooling of cylindrical and laid-back holes with different radial angles, Int. J. Heat Mass Transf., 71 (2014) 615-625.
[12] C.L. Liu, H.R. Zhu, Z.W. Zhang, Experimental investigation on the leading edge film cooling of cylindrical and laid-back holes with different hole pitches, in: Proc. ASME Turbo Expo: Power for Land, Sea, and Air, GT2012-68027, 2012
[13] J.E. Albert, D.G. Bogard, F. Cunha, Adiabatic and overall effectiveness for a film cooled blade, in: Proc. ASME Turbo Expo: Power for Land, Sea, and Air, GT2004-53998, 2004
[14] M. Benabed, A. Azzi, B. Jubran, Numerical investigation of the influence of incidence angle on asymmetrical turbine blade model showerhead film cooling effectiveness, Heat Mass Transf., 46 (2010) 811-819.
[15] K. Chavez, T.N. Slavens, D. Bogard, Experimentally measured effects of incidence angle on the adiabatic and overall effectiveness of a fully cooled turbine airfoil with shaped showerhead holes, ASME. J. Turbomach, 139(9) (2017) 091007.
[16] C. Liu, H.R. Zhu, Z.Y. Fu, R.H. Xu, The effects of inlet Reynolds number, exit Mach number and incidence angle on leading edge film cooling effectiveness of a turbine blade in a linear transonic cascade, in: Proc. ASME Turbo Expo: Power for Land, Sea, and Air, GT2015-42888, 2015
[17] H.Y. Li, Q. Du, Q.Z. Xu, G.Y. Xu, H.Y. Liu, D.W. Chen, Conjugate Heat Transfer Characteristics of a Film-Cooled Turbine Blade Leading Edge With Staggered-Oblique Impinging Jets, ASME. J. Turbomach, 145(11)(11) (2023) 111008.
[18] T.E. Dyson, Effects of Hole Pitch Variation on Overall and Internal Effectiveness in the Leading Edge Region of a Simulated Turbine Blade with Heat Flux Measurements, 2010.
[19] Z. Liu, Y. Lv, C.Y. Wang, Z.P. Feng, Numerical simulation on impingement and film composite cooling of blade leading edge model for gas turbine, Appl. Therm. Eng., 73(2) (2014) 1432-1443.
[20] Y. Jiang, Q. Zheng, P. Dong, J. Yao, H. Zhang, J. Gao, Conjugate heat transfer analysis of leading edge and downstream mist–air film cooling on turbine vane, Int. J. Heat Mass Transf., 90 (2015) 613-626.
[21] H. Fawzy, Q. Zheng, Y.T. Jiang, Impingement cooling using different arrangements of conical nozzles in a film cooled blade leading edge, Int. Commun. Heat Mass Transf., 112 (2020) 104506.
[22] G. Chen, Y. Liu, Y. Rao, J. He, Y. Qu, Numerical investigation on conjugate heat transfer of impingement/effusion double-wall cooling with different crossflow schemes, Appl. Therm. Eng., 155 (2019) 515-524.
[23] J.D. Moore, M.J. Horner, D.G. Bogard, Adiabatic Effectiveness and Thermal Field Measurements of a Shaped Hole in the Showerhead of a Model Turbine Blade, ASME. J. Turbomach, 143(9) (2021) 091017.

Acknowledgments
We wish to acknowledge the financial support of the National Natural Science Foundation Outstanding Youth Foundation (Grant No. 52122603), the National Natural Science Foundation of China (Grant No. 52106064), the National Science and Technology Major Project (J2019-II-0022-0043), the Taishan Scholars Program and the cloud computing supported by the Beijing Super Cloud Computing Center.

Journal of Physics: Conference Series **3041** (2025) 012027 doi:10.1088/1742-6596/3041/1/012027

Numerical Study on a Medium-pressure Centrifugal Fan Near Stall Condition with Different Computational Fluid Dynamics Methods

Liren Wang[1], Wei Lyu[1, 2]and Yan Wang[1, *]

[1]College of Aerospace Engineering, Nanjing University of Aeronautics and Astronautics, Nanjing, 210016, China
[2] HENGCHI ENVIRONMENTAL EQUIPMENT(NANJING) CO., LTD, Nanjing, 211212, China

*aerowangy@nuaa.edu.cn

Abstract. Centrifugal fans have always played an essential role in engineering applications. To better study the flow characteristics of centrifugal fans, this paper carried out numerical studies on a medium-pressure centrifugal fan near stall condition using two different computational fluid dynamics (CFD) methods: the multiple reference frame (MRF) method and the sliding mesh method (SMM). The accuracy of these two methods under different conditions is compared. It is found that under the standard operating point, both methods can provide accurate pressure predictions. However, the SMM method is significantly better near the stall point than MRF due to its capability to capture unsteady flow characteristics. The flow details near the stall point are analyzed, and the fluid energy conversion process is explained. It is found that phenomena like flow separation are mainly concentrated on both sides of the high-pressure difference cover plates. It is observed that the transient flow field can more clearly display flow separation phenomena and better capture dynamic characteristics and local flow features, thus offering a more comprehensive description of the internal flow of the centrifugal fan.

1. Introduction

As an essential impeller machinery in the industrial field, centrifugal fans are widely used in energy, aerospace, chemical, and other industries. This type of fan generates negative pressure through the high-speed rotation of the internal impeller, which draws in the outside air through the collector and discharges the air through the impeller channel and volute to achieve gas delivery. Due to the complexity of its internal structure and the asymmetry of the vortex shell, the internal flow of the centrifugal fan is very complex, showing the characteristics of being fully three-dimensional, unsteady, non-uniform, and nonlinear. In addition, complex flow phenomena, such as the separation flow of the blade's leading edge and the secondary flow between the leaves, often occur. These phenomena usually exist in the form of various vortex systems[1], which affect fans' ventilation efficiency and safety.

As a result, investigating the internal flow characteristics of the fan is crucial for designing a high-efficiency fan. Dawes et al.[2] studied the complex three-dimensional unsteady flow caused by the interaction between the splitter centrifugal impeller and the vaned diffuser. DV Bhope and PM Padole[3] used experimental and finite element methods to study centrifugal ventilation distribution and impeller stress distribution. They found that the stress of the impeller can be effectively reduced by setting a reinforcing ring, and the fan's air volume and air pressure can be appropriately reduced. EE Morfiadakis et al.[4] used a meshless (Lagrangian) numerical method to analyze the internal flow field of a centrifugal fan. Sakaguchi et al.[5] significantly improved the flow characteristics of centrifugal fans at small flow rates by optimizing the blade structure of centrifugal fans. Kishokanna[6] simulated the centrifugal fan by three-dimensional separation eddy simulation to predict the noise level in the far field.

This paper will use the steady-state multiple reference frame (MRF) method and the transient sliding mesh method (SMM) to conduct a numerical investigation of the near-stall conditions of a specific model of a medium-pressure centrifugal fan. The study will compare the prediction accuracy of wind pressure characteristics and the simulation accuracy of flow details within the internal flow field using different simulation methods, as well as investigate phenomena such as flow separation occurring inside the centrifugal fan.

2. Computational models and research methods

2.1. The governing equations

Because the numerical simulation of the centrifugal fan in this paper does not involve energy loss, the control equations of flow are the continuity equation and momentum equation in the Cartesian system:

$$\frac{\partial u}{\partial x}+\frac{\partial v}{\partial y}+\frac{\partial w}{\partial z}=0 \tag{1}$$

$$\frac{\partial(\rho u_i)}{\partial t}+\frac{\partial(\rho u_i u_j)}{\partial x_j}=-\frac{\partial p}{\partial x_i}+\frac{\partial(\tau_{ij}+\tau_{ij}^T)}{\partial x_j} \tag{2}$$

Where ρ is density, u_i is velocity component, p is pressure, τ_{ij} is molecular viscous stress and τ_{ij}^{T} is turbulent viscous stress.

2.2. Turbulence model

The Realizable $k-\varepsilon$ model adopted in this paper uses the empirical coefficient and strain rate C_μ to constrain the normal stress, which can better deal with the simulation of rotating flow. The transport equation is:

$$\frac{\partial(\rho k)}{\partial t}+\frac{\partial\left(\rho k u_j\right)}{\partial x_j}=\frac{\partial}{\partial x_j}\left[\left(\mu+\frac{\mu_t}{\sigma_k}\right)\frac{\partial k}{\partial x_j}\right]+G_k-\rho\varepsilon \tag{3}$$

$$\frac{\partial(\rho\varepsilon)}{\partial t}+\frac{\partial\left(\rho\varepsilon u_j\right)}{\partial x_j}=\frac{\partial}{\partial x_j}\left[\left(\mu+\frac{\mu_t}{\sigma_\varepsilon}\right)\frac{\partial\varepsilon}{\partial x_j}\right]+\rho C_1 E\varepsilon-\rho C_2\frac{\varepsilon^2}{k+\sqrt{v\varepsilon}} \tag{4}$$

Where μ_t is turbulent viscosity coefficient, μ is molecular viscosity, k is turbulent energy generation rate, ε is turbulent energy dissipation rate, G_k is turbulent kinetic energy generation term caused by average velocity gradient, σ_k, σ_ε, C_1, C_2 and E are empirical coefficients.

2.3. Method of calculation

The MRF (Multiple Reference Frame) method is a steady-state approximation approach in which different rotational or translational speeds are assigned to a specific region within a system. The flow in this region is computed using the appropriate moving reference frame equations. A reference frame transformation is applied at the interfaces between different regions, enabling the flow variables from one region to be utilized in the calculations of neighboring regions. It is important to note that MRF does not take into account the relative motion of the moving region concerning the adjacent regions, so the grid remains unchanged during the computation. Therefore, the MRF method is also known as the "frozen rotor method."

The sliding mesh method is a specific instance of the dynamic grid method in which all boundaries and cells within the mesh region move together as a rigid body. In this scenario, the grid nodes move in space, but the cells defined by the nodes remain undeformed. Additionally, adjacent moving grid regions can be connected across one or more non-conformal interfaces. So long as the interfaces remain in contact, the non-conformal interfaces can dynamically update with the movement of the mesh, allowing fluid to flow from one region to another.

Both MRF and SMM can deal with the rotating flow problem effectively. MRF can simplify the setup of the problem and improve the efficiency of calculation; SMM is more suitable for complex motions with high accuracy requirements, can better capture the dynamic changes of the flow, and can effectively deal with the physical differences between regions. The schematic diagrams are presented in Figure 1.

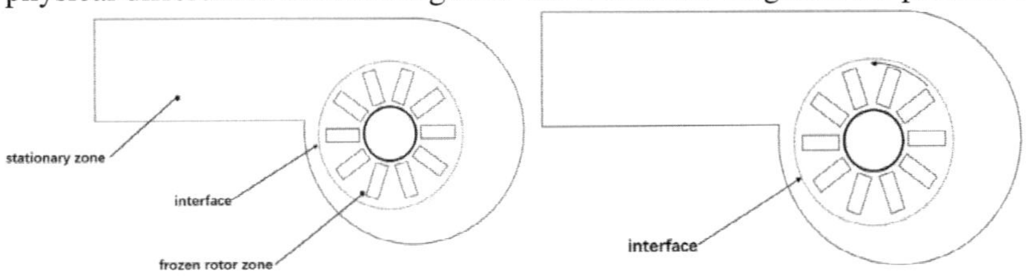

(a)The MRF schematics (b)The SMM schematics

Figure 1. Schematic diagram of calculation methods

2.4. Computational model and mesh-independence validation

The object of study in this paper is a certain type of medium-pressure centrifugal ventilator. The whole impeller consists of 12 blades. The diameter is 620mm, the inlet and outlet diameters are 1120mm, and the set rotation speed is 1255rpm. In the experiment, air is used as the fluid medium, with a density of 1.205kg/m^3 and a viscosity of 1.79×10^{-5}kg/(m*s). The atmospheric pressure is 101325Pa, and the temperature is 293.15K.

To facilitate the fluid entering the flow field better, we repair the geometry of the volute and turbine, respectively, and externally connect the inlet and outlet with 2m guide pipes. Because this calculation uses MRF and SMM, it is necessary to create a new geometric surface of the interface. The interface is a surface of revolution around the axis of rotation, which is located between the volute and the turbine and divides the internal area of the fan into two parts, namely, the volute area composed of the volute and the interface and the turbine area consisting of the turbine and the interface. The centrifuge provides two fluid domains: the wrap region of the volute and the interface, which is taken as the static domain, and the wrapping region of the interface, and the turbine generates the rotating domain.

To avoid wasting computational resources and improve accuracy, this paper divides the grid into four sets and performs a sensitivity test using the MRF method with an inlet flow rate of 1076.8 m³/min, comparing the total pressure rise with the experimental value of 2095.5 Pa to assess their closeness. The computational results are shown in Figure 2, from which we can see that the computed values converge more and more to the experimental values with the grid count increases. Still, when the grid count reaches 8 million, the computed values improve very little, and at this time, the relative error is 2.97%. The increase in the grid count can not significantly reduce the calculation error, and therefore, with a combination of computational requirements and simulation accuracy, we choose the 8 million grid as the calculation grid for this simulation. The height of the first layer grid at the blade is 0.8mm, the height of the first layer at the wall is 1mm, and the number of boundary layers is 15, with a growth rate of 1.1. The blade wall is encrypted, the final number of grids in the rotating computational domain is 3 million, and the number in the stationary computational domain is 5 million. The grid diagram is shown in Figure 3.

2.5. The boundary condition

This calculation is completed in Fluent, and the turbulence model selects the Realizable $k-\varepsilon$ model and standard wall function selection for near-wall conditions. The steady-state MRF and the transient SMM are selected for the calculations. Take the negative direction of the Z axis as the direction of the rotation axis, and set the rotation speed to 1255rpm. The inlet boundary conditions set the flow as independent variables, the outlet boundary is characterized as the static pressure outlet, the pressure was 0Pa, the impeller wall rotated with the rotation domain, the outer wall was set as the non-slip wall, and the interface was set for data transmission. Select coupled solver SIMPELEC, second-order upwind scheme. The calculation is considered complete when the residual converges to 10^{-5}.

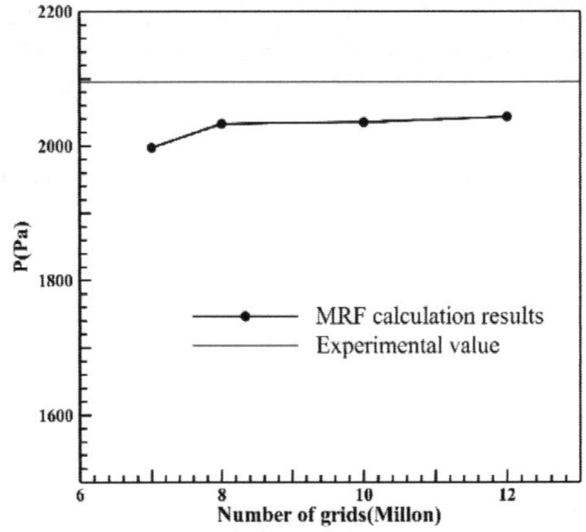

Figure 2. Pressure Difference vs. number of grids

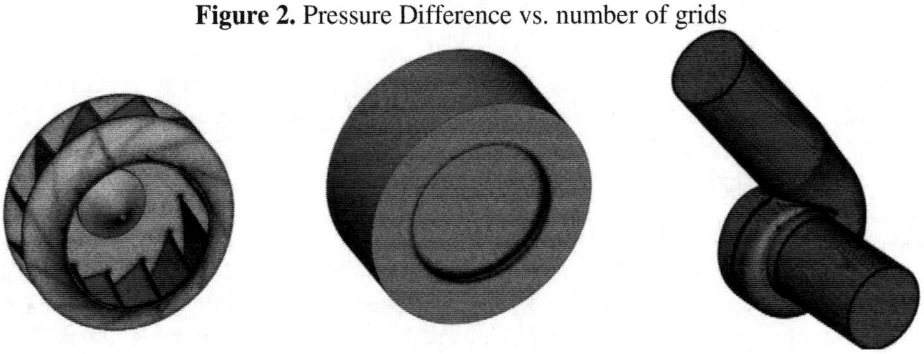

(a)Rotor (b) Rotating computational domain (c) Stationary computational domain

Figure 3. The grid diagram

3. Calculation results and analysis

3.1. Comparison of wind pressure characteristic values

We selected the standard working condition point (Q=943.7 m^3/min) and the near-stall point (Q=797.2 m^3/min) for simulation, respectively, and Table 1 exhibits the calculation results. The table shows that under the standard working condition flow rate, the results obtained from MRF and SMM are nearly identical to the experimental values. Still, at the point of near-stall, the computed results of MRF have a significant deviation of up to 9.5%, and the calculated results of SMM are obviously better than those of the steady-state, but they are still 3.9%. This is due to the fact that when at the near-stall point, the flow rate of the turbine is too low, which can lead to unstable airflow, coupled with the fact that due to the limitations of the turbulence modeling, high curvature may occur locally in such extreme cases, which can lead to extreme points of the strain rate, which affects the accuracy of the simulation.

Table 1. Numerical Table of Wind Pressure Characteristics.

Standard operating point	MRF	SMM	Experimental value
P (Pa)	2433.6	2453.4	2474.5
Relative error(%)	1.6	0.8	/
Near stall point	MRF	SMM	Experimental value
P (Pa)	2693.7	2860.6	2977.3
Relative error(%)	9.5	3.9	/

3.2. Flow field detail comparison

In order to better investigate the flow characteristics at the centrifuge stall point and to visually study the difference between the steady-state and transient calculation results, we performed a comparative analysis of the flow field details at the near-stall point, and the results are presented below.

Figure 4 shows the cloud diagram of static pressure in the meridional plane, from which it can be observed that the pressure reaches its minimum value at the inlet of the impeller channel and then gradually increases outward with the rotation of the impeller, and attains its highest value at the impeller outlet. The spread of pressure of the transient flow field has a more even distribution than that of the steady-state, and the negative pressure at the inlet is larger, and an apparent symmetrical distribution can be observed in the impeller channel, which is due to the fact that the transient calculation can describe the real dynamic behavior of the fluid system more comprehensively so that the results of the calculation approach the experimental data more closely. Under the effect of this high-pressure difference mentioned above, the gas will be pressed into the impeller channel again through the gap between the cover plate and the volute, accompanied by a large number of vortices appearing to produce internal backflow. This phenomenon causes leakage losses in the flow and reduces the static pressure conversion efficiency. This flow separation phenomenon can be observed in both flow fields, and the pressure gradient changes in the transient flow field are more complex, better capturing the unsteady flow conditions above the cover and producing more pronounced swirls.

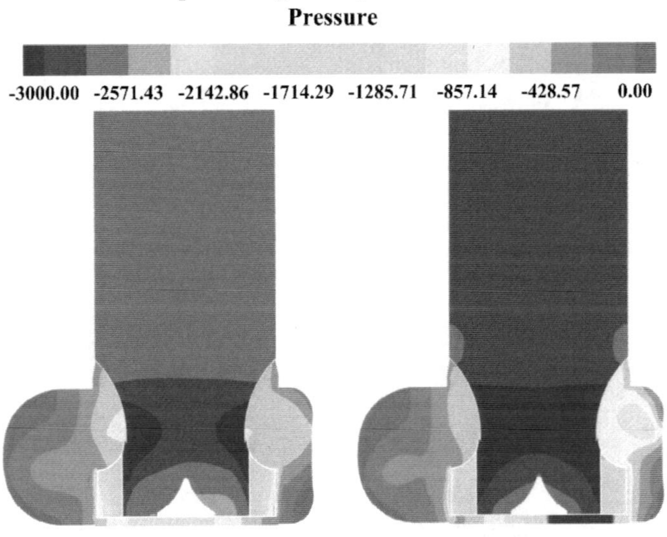

(a)The steady-state MRF (b) The transient SMM

Figure 4. Meridional Static Pressure Cloud Map.

Figure 5 shows the velocity cloud in the meridional plane, from which it can be found that the velocity change in the steady-state field is relatively smooth. Compared with the steady-state flow field, the transient flow field appears at the exit of the high-speed flow of the gas and produces a larger vortex scale, a larger gradient in the velocity change, and a more turbulent flow. In addition to this, the gas flow rate below the impeller channel produces a drastic change, which is closely related to the pressure change in Fig. 4. The scale of the vortex above the cover varies in different flow fields, and the velocity change of the vortex in the steady-state flow field is smoother, and the flow trend is more obvious; Transient flow field in the vortex flow velocity changes than the steady-state flow field is violent, and the flow is also more turbulent, due to the irregular changes in pressure, there are different pressure gradients in the same region, this phenomenon makes the gas is pushed back and forth, always in the flow channel back and forth circulation, in the large-scale vortex and the formation of a number of small-scale vortices.

Figure 6 presents the static pressure cloud map of the rotary surface, from which it can be observed that the lowest static pressure occurs at the center of the impeller, gradually increasing from the rotation center outward and reaching its highest value at the volute shell outlet, stabilizing at approximately 0Pa.

Journal of Physics: Conference Series **3041** (2025) 012027 doi:10.1088/1742-6596/3041/1/012027

In the impeller passage, as the gas rotates with the impeller, the pressure on the blade surface increases with the rotation, ultimately reaching its highest value at the impeller channel outlet. The static pressure on the pressure surface is consistently higher than on the aspiration surface. Compared with the steady flow field, the pressure transition of the transient flow field is smoother and shows noticeable periodic changes within the impeller channel, which is more in line with the actual flow. In addition, the transient flow field can be observed more clearly as a pressure rise of smaller magnitude at the outlet of the guide tube, which is caused by the small flow velocity near the wall and the internal friction between the different speeds of the airflow, which generates the pressure rise.

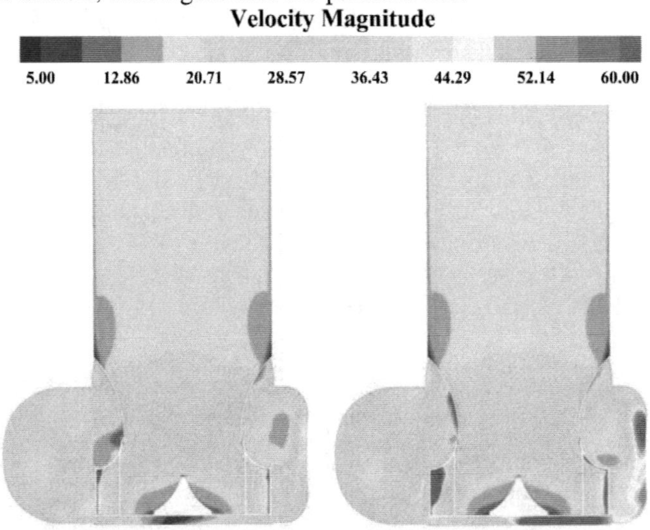

(a)The steady-state MRF (b) The transient SMM

Figure 5. Meridional Velocity Cloud Map.

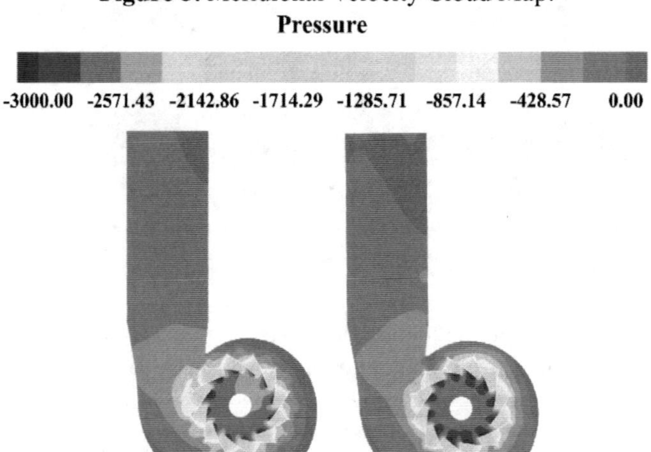

(a)The steady-state MRF (b) The transient SMM

Figure 6. Rotational Surface Static Pressure Cloud Map.

Figure 7 for the rotary surface velocity cloud diagram from the figure can be obtained in the entrance of the gas at the lowest speed, in the impeller rotating through the impeller channel by the impeller rotational work speed has been significantly increased after the flow through the centrifugal device at the exit of the guide tube, the fluid kinetic energy is again converted into pressure energy, the speed is gradually reduced. In agreement with Fig. 6, the velocity distribution of the transient flow field is also more regular and has a good transition. In addition to this, the near-wall effect of the above can be observed more clearly in this figure.

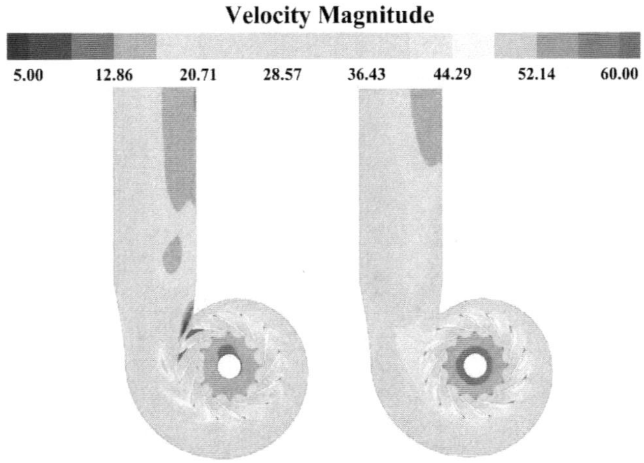

(a)The steady-state MRF (b) The transient SMM

Figure 7. Rotational Surface Velocity Cloud Map.

4. Conclusion

In this paper, the steady-state multiple reference coordinate system method, as well as the transient slip grid method, are used to numerically simulate a certain type of medium-pressure centrifugal ventilator and compared and analyzed with the experimental results, and the following conclusions are obtained:

- The wind pressures of centrifugal ventilators at the standard operating point and near-stall point were calculated, and it was found that at the standard operating point, the steady-state MRF and transient SMM have good accuracy, but at the near-stall point, the steady-state MRF has a large error.
- The flow characteristics near the point of the stall were analyzed, showing the distribution of static pressure and velocity inside the centrifuge and describing the process of fluid energy conversion. It was found that on both sides of the cover with high differential pressure, the fluid is highly susceptible to being re-pressed into the flow channel through the gap, and flow separation and leakage losses occur.
- By comparing the difference between the steady-state flow field and the transient flow field at the near-stall point, it is found that the transient flow field can better present the phenomenon of flow separation, can better capture the dynamic characteristics and local flow characteristics, and can more comprehensively describe the internal flow of centrifugal ventilators.

5. Reference

[1] Bin Y, Elhadi E E and Junlong X 2003 Three dimensional numerical analysis of multiblade fan. *Journal of Engineering Thermophysics*.(03):419-422.

[2] Dawes, W. N. 1995 A simulation of the unsteady interaction of a centrifugal impeller with its vaned diffuser: Flow analysis. *Journal of Turbomachinery* 117.

[3] Bhope, D. V. and Padole, P. M. 2004 Experimental and theoretical analysis of stresses, noise and flow in centrifugal fan impeller.*Mechanism and Machine Theory* 39, 1257–1271.

[4] Morfiadakis, E. E., Voutsinas, S. G. and Papantonis, D. E. 1991 Unsteady flow calculation in a radial flow centrifugal pump with spiral casing. *International Journal for Numerical Methods in Fluids* 12, 895–908.

[5] Sakaguchi, D., Sakue, D. and Tun, M. T. 2018 Global Search of a Three-dimensional Low Solidity Circular Cascade Diffuser for Centrifugal Blowers by Meta-model Assisted Optimization. *Journal of Thermal Science* 27, 111–116.

[6] Paramasivam, K., Rajoo, S., Romagnoli, A. and Yahya, W. J. 2017 Tonal noise prediction in a small high speed centrifugal fan and experimental validation. *Applied Acoustics* 125, 59–70

Journal of Physics: Conference Series **3041** (2025) 012028 doi:10.1088/1742-6596/3041/1/012028

Research on flow force of check valve for space station based on the theory of flow around motion

F F Wang, D Guo*, L Zhang, Z H Qiao, Z Wang and L W Fan

Key Laboratory of Space Utilization, Technology and Engineering Center for space Utilization, Chinese Academy of Sciences, No 9. Deng Zhuang South Rd, Hai Dian Dist, Beijing, China

* guodong@csu.ac.cn

Abstract. Check valve is an important component of the fluid drive unit used in the Chinese space station (CSS). At the minimum working flow rate, the valve core of the check valve overcomes the spring force to fully open and maintain a fully open state, ensuring high reliability during the operation of check valve. Through establishing force and parametric model on the valve core and spring of the check valve, deriving the mathematical model and carrying out the flow rate test. Test results match theoretical results quite well with low errors of 5.50%. The effectiveness of the theory has been confirmed, which can provide design references for valve cores and springs.

1. Introduction

With the successful launch of the Mengtian Lab Module in October 2022, China's manned spaceflight program completed the construction of the CSS. The designed operating life of the CSS is 15 years. A series of scientific experiments such as microgravity biology experiment and microgravity fluid physics have been carried out or will be carried out on the CSS [1~5]. A fluid drive unit is designed to drive the fluid circulation and provide thermal support and management for a series of scientific experiments [6].

Check valve is an important component of the fluid drive unit, which can prevent the reverse flow of liquid and ensure the safety of the fluid drive unit. The pump and check valve, as the core components of the fluid drive device, adopt a dual backup configuration, as shown in figure 1, to improve system reliability. At the minimum working flow rate, the valve core of the check valve overcomes the spring force to fully open and maintain a fully open state, which can reduce the reciprocating motion of the mating surface between the valve core and the outlet valve body and prevent wear of the mating surface during this process. It is a key design indicator to ensure high reliability during the operation of check valve. Traditional check valves pay less attention to this indicator.

Based on the check valve configuration shown in figure 2, this paper carried out parameterized design of the valve core and spring. By calculating the flow force around the valve core and matching the requirement for the valve core to fully open at the minimum working flow rate, high reliability was ensured.

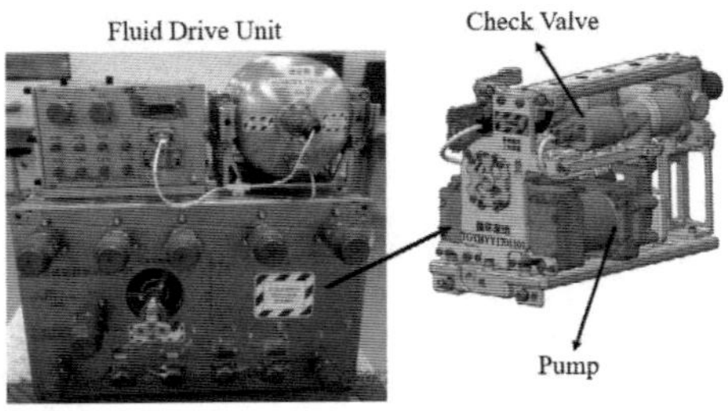

Figure 1. The fluid drive device and dual backup configuration

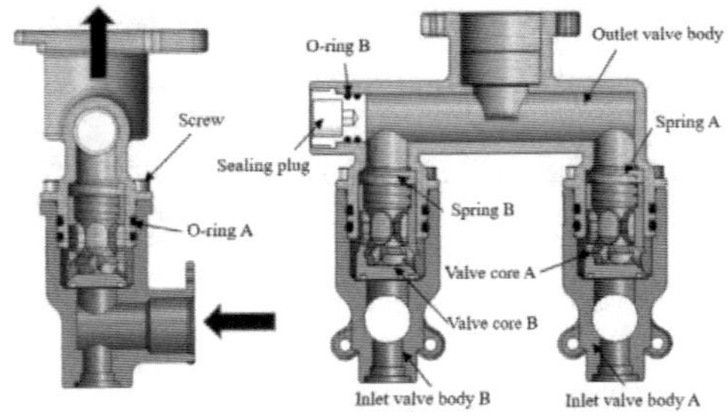

Figure 2. The check valve configuration

2. Model and Theoretical Analysis

2.1. Design Criterion

At the minimum working flow rate, the fluid driving force acting on the valve core matches the spring force. The specific criteria are as follows: one is that at the minimum working flow rate, the fluid driving force acting on the valve core in the working position cannot be lower than the spring force to ensure that the valve core can maintain a fully open state, the other is that at the minimum flow rate, the fluid driving force acting on the valve core in the working position should not be too large, as this can cause excessive flow resistance in the system.

2.2. Model

Based on the check valve configuration shown in figure 2, in the fully open position, the force on the valve core and spring is shown in figure 3, and the parameterized model of the check valve is shown in figure 4.

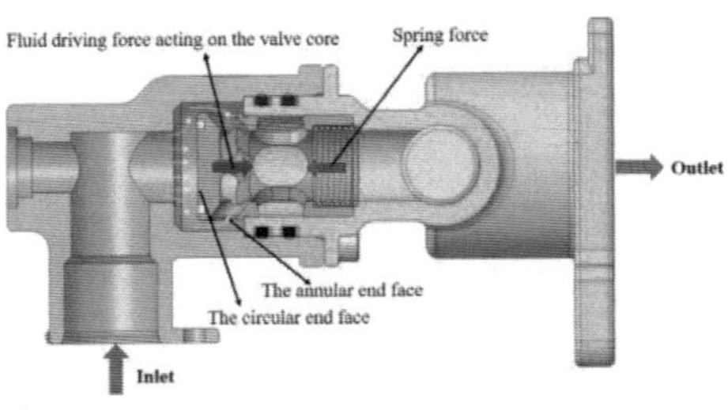

Figure 3. The force on the valve core and spring in the fully open position

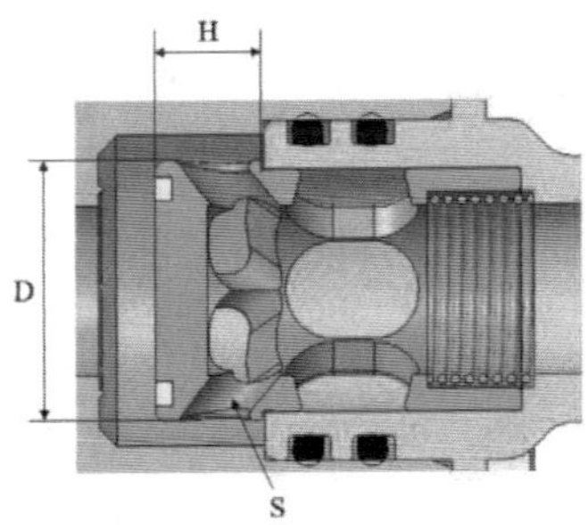

Figure 4. The parameterized model of the check valve

2.3. Force Analysis

According to the theory of flow around motion, the force of an object's flow can be obtained

$$F_D = C_D \rho \frac{V^2}{2} A \tag{1}$$

Where, C_D is the drag coefficient; ρ is the fluid density; V is the relative velocity; A is the projection area of incoming flow direction.

According to Figure 3, the fluid driving force of the valve core is decomposed into two parts: one is the fluid driving force F_1 acting on the circular end face, and the other is the fluid driving force F_2 acting on the annular end face.

Journal of Physics: Conference Series **3041** (2025) 012028 doi:10.1088/1742-6596/3041/1/012028

The fluid driving force F_1 and F_2 can be obtained.

$$\begin{cases} F_1 = C_1 \rho \dfrac{V^2}{2} A_1 \\ F_2 = C_2 \rho \dfrac{V^2}{2} A_2 \end{cases} \tag{2}$$

Where, C_1 is the drag coefficient on the circular end face; C_2 is the drag coefficient on the annular end face; A_1 is the projection area on the circular end face; A_2 is the projection area on the annular end face.

According to Figure 4, the following equations can be obtained

$$\begin{cases} A_1 = \pi \left(D/2 \right)^2 \\ A_2 = \pi DH - S \\ V = Q/S \end{cases} \tag{3}$$

Where, D is the diameter of valve core; H is the height of the valve core in contact with the fluid; S is the projection area of throttle hole in the flow direction; Q is the flow rate through valve.

So

$$\begin{cases} F_1 = C_1 \dfrac{\rho \pi D^2 Q^2}{8S^2} \\ F_2 = C_2 \dfrac{\rho Q^2}{2S^2} \left(\pi DH - S \right) \end{cases} \tag{4}$$

In the actual design process, when the valve core is in the fully open position, there will be an additional force F_0 to ensure that the valve core can be reliably maintained in the fully open position.

So

$$F_1 + F_2 - F_0 = F_{spring} \tag{5}$$

And the following equation can be obtained

$$C_1 \frac{\rho \pi D^2 Q^2}{8S^2} + C_2 \frac{\rho Q^2}{2S^2} \left(\pi DH - S \right) - F_0 = F_{spring} \tag{6}$$

According to the current design, the values of the above variables are shown in the table 1.

Table 1. The values of the different variables

Variables	D(m)	ρ (kg/L)	H (m)
Value	0.0214	1.11	0.0094

So

$$0.0002 \times \frac{C_1 Q^2}{S^2} + 0.55 \times \frac{C_2 Q^2}{S^2} \left(0.000366 - S \right) - F_0 = F_{spring} \tag{7}$$

By conducting processing tests to obtain different S, Q and F_{spring} values and confirming the values of C_1, C_2 and F_0, the fluid driving force of the check valve with the configuration shown in Figure 4 can be obtained.

3. Test

The test sample of check valve is manufactured and assembled as shown in figure 5. The inlet valve body is designed as a transparent component to facilitate observation of the position of the valve core. As shown in figure 6, valve cores with different parameters have been processed.

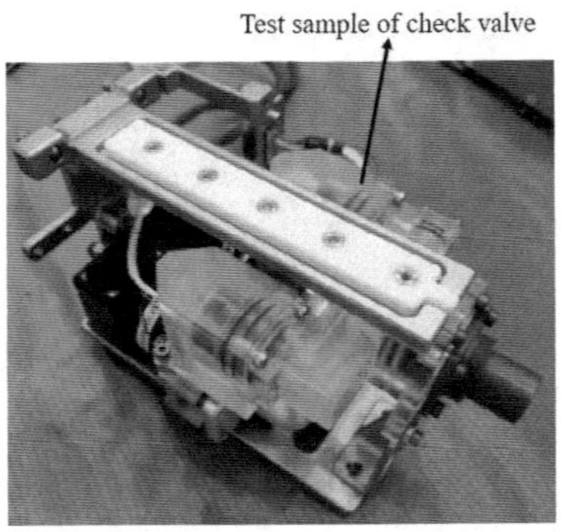

Figure 5. The test sample of check valve

Figure 6. Valve cores with different parameters

The ground testing system, as shown in figure 7, has been built. The flow rate values when the valve core is fully opened under different valve core and spring states are obtained, as shown in the table 2.

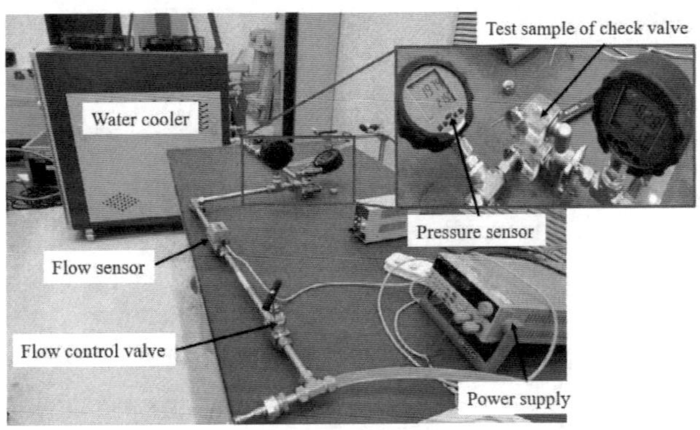

Figure 7. Ground testing system

Table 2. The ground test results

Number	Q(L/h)	S(mm^2)	F_{spring} (N)
1	430	41.6	1.5
2	800	77.12	1.5
3	390	41.6	1.2
4	430	46.9	1.2
5	420	36.6	1.8
6	530	46.9	1.8

By combining Equation (7) with the first three rows shown in Table 2, the following formula can be obtained

$$\begin{cases} 0.00165C_1 + 0.00147C_2 - F_0 = 1.5 \\ 0.00166C_1 + 0.00132C_2 - F_0 = 1.5 \\ 0.00136C_1 + 0.00121C_2 - F_0 = 1.2 \end{cases} \tag{8}$$

By solving a cubic linear equation, the values of C_1, C_2 and F_0 are calculated as follows

Journal of Physics: Conference Series **3041** (2025) 012028 doi:10.1088/1742-6596/3041/1/012028

$$\begin{cases} C_1 = 976.139 \\ C_2 = 65.076 \\ F_0 = 0.2063 \end{cases} \tag{9}$$

So, the following formula can be obtained

$$0.195 \times \frac{Q^2}{S^2} + 35.792 \times \frac{Q^2}{S^2}\left(0.000366 - S\right) - 0.2063 = F_{spring} \tag{10}$$

By combining Equation (10) and Table 2, the theoretical results and the error with the spring force can be obtained, as shown in Table 3.

Table 3. The theoretical results and the error with the spring force

Number	Q(L/h)	S(mm^2)	F_{spring} (N)	Theoretical results (N)	Error (%)
1	430	41.6	1.5	1.499	-0.07
2	800	77.12	1.5	1.501	0.07
3	390	41.6	1.2	1.196	-0.33
4	430	46.9	1.2	1.134	-5.50
5	420	36.6	1.8	1.897	5.39
6	530	46.9	1.8	1.845	2.50

From the analysis of the Table above, the theoretical results and testing results have good consistency and less error. The error mainly comes from F_0 in formula 5, and each test result of F_0 has subjective judgment bias of the testing personnel.

4. Conclusions

In this paper, the mathematical model of the flow force of the valve core was established and the flow rate and valve core position tests were carried out. Test results match theoretical results quite well with low errors of 5.50%. The effectiveness of the theory has been confirmed, which can provide design references for valve cores and springs.

5. References

[1] Cui JJ, Yi ZH, et al. Microgravity stress alters bacterial community assembly and co-occurrence networks during wheat seed germination. Science of The Total Environment 2023; 890 (9): 164147.

[2] Richard N, Paul L, et al. Materials research conducted aboard the International Space Station: Facilities overview, operational procedures, and experimental outcomes. Acta Astronautica 2008, 62, pp.491-498.

[3] Smirnova MN, Nikitin VF, Skryleva EI, Weisman YG. Capillary driven fluid flows in microgravity. Acta Astronautica 2023, 204, pp.892-899.

[4] Andreas G, Ulrich K, Andreas G, et al. A Fast and Self-Acting Release-Caging-Mechanism for Actively Driven Drop Tower Systems. Microgravity Sci. Technol. 2017, 29, pp.403-414.

[5] Wang F F, Zhang L, et al. Development of the on-orbit maintenance and manipulation workbench (MMW) for the Chinese space station. Acta Astronautica 214C (2024) pp. 366-379.

[6] Guo D, Xie M, et al. Design and On-orbit Performance of The Payload Rack Thermal Management System for China Space Station Experimental Lab Module. Applied Sciences 2024,14(7),2938

Acknowledgments

This work was supported by the national project (Y5140621PN).

Research on transient characteristics of the pintle injector

Hu Haifeng*, Zhu Wei, Zhang Xiao

National Key Laboratory of Aerospace Liquid Propulsion, Xi'an 710100, China

*Correspondence: huhaifeng_rocket@126.com

Abstract. In order to analyze the transient performance of start-up for the pintle injector, the equations were derived using mass and momentum conservation, ideal gas functions and many complex resistances. Based on the MFC framework, simulation software has been compiled, which can realize multi-parameter input and visualization output. To investigate the numerical method, both simulations and experimental tests were carried out on a specific pintle injector. The simulation results agree well with the test data, and the response time and test deviation of start-up is 3.535%, respectively, which confirms the validity of the model. The calculation examples show that the model is available and practical.

1. Introduction

The most difficult component in developing liquid rocket engines is the thrust chamber, as it is the part where the propellant is concentrated and mixed for combustion. The injector determines the maximum achievable combustion efficiency, the heat transfer rates to the combustion chamber walls, and whether or not high frequency combustion instabilities will occur. Since the development of the V-2 rocket, liquid rocket injection technology has so far developed, producing a variety of injection types, such as: the impinging stream injector, centrifugal swirl injector, and pintle injector. Compared with other types of injectors, the pintle injector has the following characteristics: deep throttling, variable thrust, fast response, small after-effect pulse, simple structure, simpleness but high reliability.etc. [1]. The main engine of the US SpaceX Dragon spacecraft [2] and the thrust chamber of the RL10 landing stage [3] all use this type of injector.

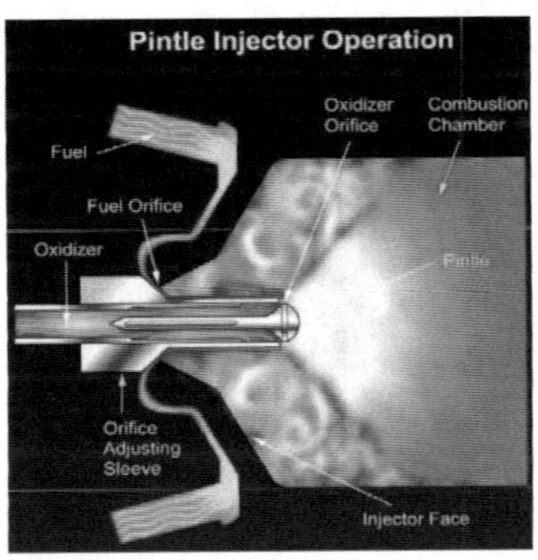

Figure 1. Schematic illustration of the typical pintle injector

In the typical bipropellant pintle injector is shown in Figure 1. One propellant, here represented as an oxidizer, enters the injector body through a separate, centrally located passageway and flows axially through the injector's central journal sleeve before being transformed into a uniform radial flow by the journal tip's contoured inner surface. The other propellant, which is fuel in this instance, is directed into a circumferential annular space that forms between the central injector element and the injector body's "snout" and meters the flow into the combustion chamber through the injector's external flow channels. This propellant exits the injector as an annular layer flowing axially, arriving at the point of impact with a consistent velocity profile circumferentially [4]. Both collide crosswise at a 90° angle, so that the propellants are atomized and mixed.

The concept of the pintle injector first appeared in the 1950s, a special structural testing device developed by the Jet Propulsion Laboratory in the United States[5]. After decades of research, current research on pintle injectors mainly focuses on their performance improvement, low cost and reusability, etc. When designing a pintle injector, the filling and discharging process of the pintle control chamber should be calculated to meet the requirements of a rapid technical index. With the development of simulation methods, it is possible to predict the response of the pintle using numerical simulation [6-12].

In the field of engineering application, it is necessary to quickly estimate the design scheme, obtain the transient performance parameters of the engine as a theoretical guide for optimal design, and then carry out numerical simulation to further verify the design scheme after optimizing the design parameters. For timeliness, it is necessary to apply different simulation models at different stages to improve simulation efficiency while ensuring simulation accuracy. Numerical simulation methods in this area are not time efficient. In this work, rapid prediction of the start-up response of a liquid rocket engine with a typical two-component pintle injector is considered as a research objective, and a physical and mathematical model based on the start-up process of a typical two-component pintle injector is created. Based on appropriate simplification, the equations of the zero-dimensional mathematical model are established by comprehensively applying engineering experience models, energy conservation, mass conservation, journal motion equation, ideal gas equation of state, and consideration of various complex resistances, friction forces and elastic forces. The system simulation software is written based on the classic MFC framework. By solving the above mathematical model, a system is created that can quickly predict the transient response of the pintle injector starting.

2. Analysis and model derivation of starting up process of pintle injector

Table 1. Nomenclature

k	specific heat ratio	V	volume
S_{nozzle}	minimum throttle area	V_0	initial volume
P_0	total local pressure	m	mass
R	spectral constant	m_0	initial mass
T_0	overall temperature	m_{kp}	mass change
Cd	correction factor	R	spectral constant
P	local static pressure	T	temperature
P_{high}	equivalent pressure sum of the external resistance	S	Surface of the pintle
Pc	thrust chamber pressure	L	pintle valve operating displacement
Pf	equivalent friction pressure	φ	quality of the virtual coefficient
P_u	equivalent seal pressure	g	acceleration due to gravity
P_{ihf}	pilot injection pressure	mass	mass of the pintle

2.1. Mathematical model of pintle injector starting up

Figure 2 shows a schematic diagram of a typical structure of a pintle injector. The pintle injection structure is very simple, and the heart is the needle valve with densely perforated surface. By designing the specification, angle, distribution and other characteristic parameters of the inner channel of the needle valve, the pressure flow adaptive zone is formed and the combustion instability is eliminated from the mechanism. Adjust the injection range to achieve a stable adjustment of the thrust. This article focused on the start-up process of the pintle injector corresponds to the model shown in Figure 3 by simplifying. In a simplified analysis, the thrust chamber is viewed as a downstream component of the injector. Fuel atomization occurs through a process where the switching motor

Journal of Physics: Conference Series **3041** (2025) 012029 doi:10.1088/1742-6596/3041/1/012029

activates or expels high-pressure gas via an externally connected pneumatic valve. When the engine is started, the thrust chamber propellants are filled in liquid state, and the two types of self-combustion fuels complete the atomization and combustion in the thrust chamber, meanwhile accelerate the exhaust gas through the nozzle to produce thrust. The entire start-up process mainly analyzes the internal movements and pneumatic parameter changes of the injector.

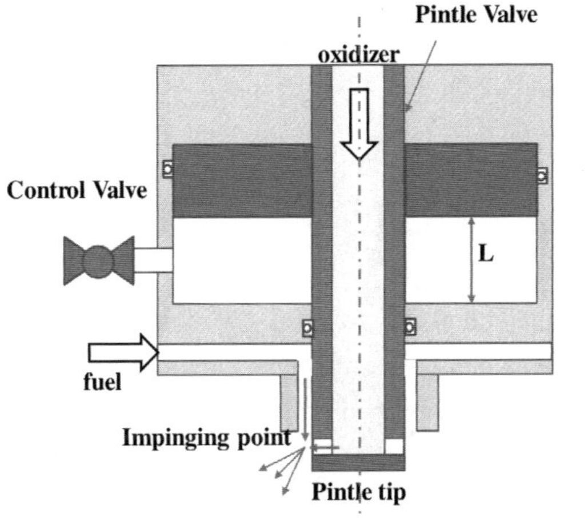

Figure 2. Schematic diagram of working principle of a typical bipropellant pintle injector

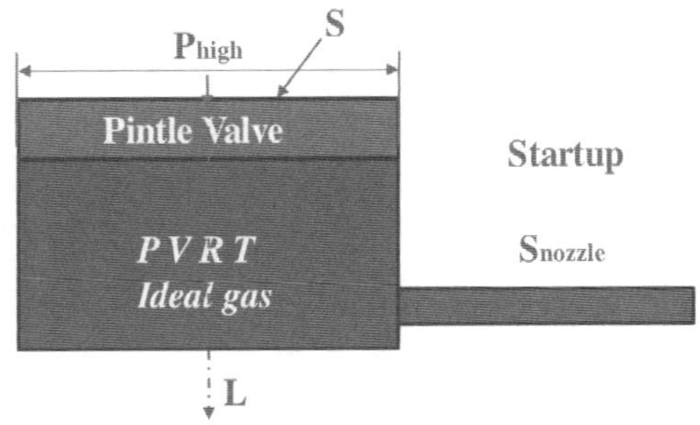

Figure 3. Diagram of the simple model for the start-up of the pintle injector

During the starting process, the high-pressure pneumatic valve sends high-pressure gas into the chamber and under the action of the high-pressure gas, the propellant moved through the needle valve enters the thrust chamber. After moving to the end position, the fuel no longer penetrates the thrust chamber and the thrust chamber switches off. The specific mathematical model is described as follows: The gas in the cavity is selected as the research object:

Mass conservation equation: During the entire start-up process, the gas in the needle valve cavity is analyzed. It is assumed that the minimum area in the flow process is the throttle point where the gas reaches the speed of sound, so the mass flow of the nozzle is:

$$\frac{dm_{kp}}{dt} = C_d \cdot K_0 \cdot S_{nozzle} \cdot P_0 / \sqrt{RT_0} \quad (1)$$

$$K_0 = (\frac{2}{k+1})^{\frac{k+1}{2(k-1)}} \cdot \sqrt{k} \quad (2)$$

Where k represents the specific heat ratio of the gas and *Snozzle* represents the minimum throttle area in the system (the coil diameter of the control valve in the analysis object). P_0 represents the total local pressure; R and T_0 represent the spectral constant and the overall temperature of the gas, respectively. *Cd* represents the correction factor and the second half of the gas cavity filling is in the subsonic state. Adopting the speed of sound in its entirety can result in inaccuracies; therefore, a correction factor is employed to mitigate these errors.

Equation of state for the gas in the cavity:

$$PV = mRT \quad (3)$$

P represents the local static pressure, V the volume of the gas, m the mass of the gas in the cavity and R and T the spectral constant or the temperature of the gas.

$$m = m_0 - m_{kp} \quad (4)$$

m_0 represents the initial mass of the gas in the cavity.

Cavity volume equation:

$$V = V_0 - SL \quad (5)$$

V_0 is the initial volume of the gas; *S* is the surface of the pen; *L* represents the trace of pen movement; if the pressure P in the chamber is less than the external pressure, P_{high} is the equivalent pressure sum of the external resistance (fluid pressure, sealing force, etc.).

$$P_{high}=P_C+P_f+P_u \quad (6)$$

Pc stands for the thrust chamber pressure, *Pf* for the equivalent friction pressure and P_u for the equivalent seal pressure. In the actual analysis process, the pilot injection pressure P_{ihf} is usually monitored, and the pintle injection pressure is equal to the pressure in the thrust chamber plus the pressure drop of the injector.

Select pintle valve as research objects:

Newton's equation of motion:

$$\varphi \cdot mass \cdot \frac{dv}{dt} = S \cdot P + \varphi \cdot mass \cdot g \quad (7)$$

While φ represents the quality of the virtual coefficient and the correction parameters, which is corrected by test results; Mass is the mass of the pintle and *g* is the acceleration due to gravity. The friction force and the elastic force during the pintle movement correspond to the pressure *P*.

Kinematic equation:

$$v = \frac{dL}{dt} \quad (8)$$

L is the pintle valve operating displacement. *V* represents velocity of motion.

2.2. The simulation system of start-up and shutdown process of pintle injector is realized

The simulation software for analyzing the start-up process of a pintle injector features a dialog-based interface for parameter input and output visualization through both charts and numerical displays. The specific software organization is shown in Figure4. Enter the related actuation cavity, cavity medium, calculation setting and other parameters in the input field to complete the initial state input. Fourth order Runge-Kutta algorithm is used to solve the above formula. For the equivalent *Pf* friction pressure, the existing empirical value is used to calculate the specified value. *Pu* represents the equivalent sealing pressure, which is calculated based on the design value. Click "Calculate" to complete the entire parameters of the pintle moving process of the pintle injector and output the pintle moving time, pintle moving speed and other curves and parameters.

```
[ def RK4(f,y0,t0,h,) // f mean the function; y0 initialization value;
  n=int(tf-t0/h)+1// the number of the iteration
  y=y[0]//
 for i in ragne(1,n): //start iteration
 k1=h*f(t0,y[i-1])
 k2=h*f(t0+h/2,y[i-1]+k1/2)
 k3-h*f(t0+h/2,y[i-1]+k2/2)
 k4=h*f(t0+h,y[i-1]+k3)
 y[i]=y[i-1]+(k1+2k2+2k3+k4)/6
 t0+=h//
 return y, t0
 ]
```

3. Calculation examples and analysis

Utilization of a system analysis program to assess pintle responsibility. The graphic user interface (GUI) was developed for the user controller to generate the input files and the operation of the program. The simulation calculation of a pintle engine is carried out and compared with the test data. The interface organization of the simulation system for starting up the pintle injector is shown in Figure 4. The specific calculation results and test data are compared and analyzed as follows:

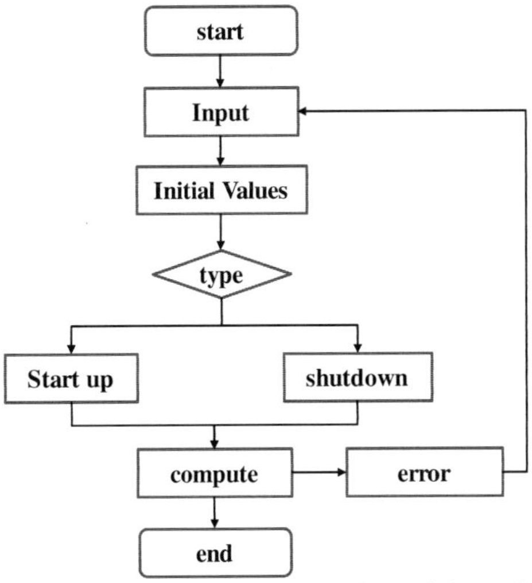

Figure 4. Schematic illustration of the software

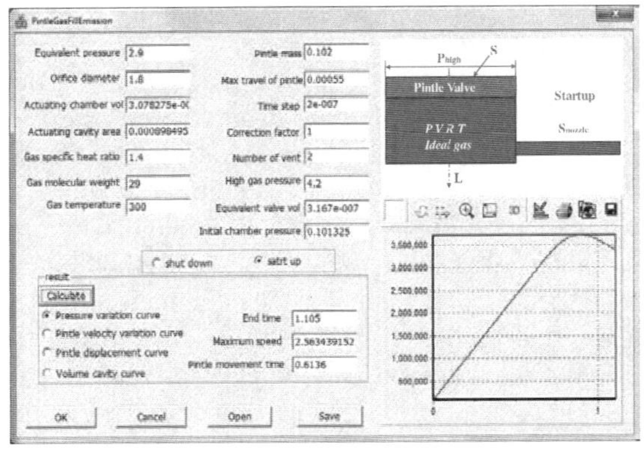

Figure 5. The simulation system of start up for the pintle injector

The response time of the P_{in} engine is shown in Figure 6. As you can see from the figure, the starting response time is about 0.021 s. The starting process of the engine can be understood as follows: at the moment "0", the valve begins to be energized, and when the valve opens at 0.014 s, the gas enters into the pintle actuation chamber. Almost simultaneously, the pressure P_{ihf} begins to change from the static state before fuel injection. At this point it is assumed that the pintle begins to act and the fuel is ejected. At 0.0185 s the chamber pressure is set and at t_{90} the chamber pressure is 0.021 s.

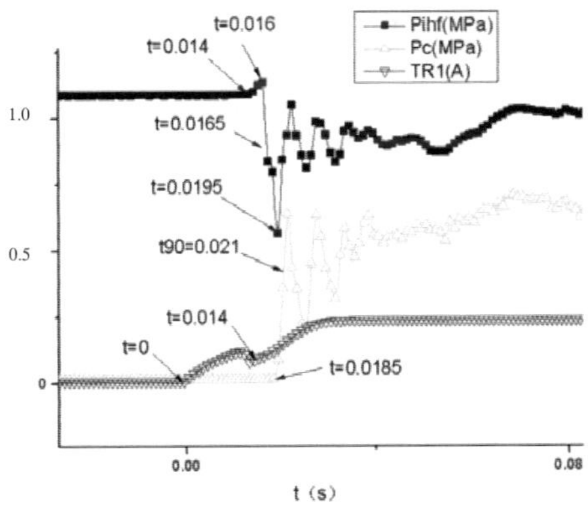

Figure 6. The response time of pintle injector during the start up phase

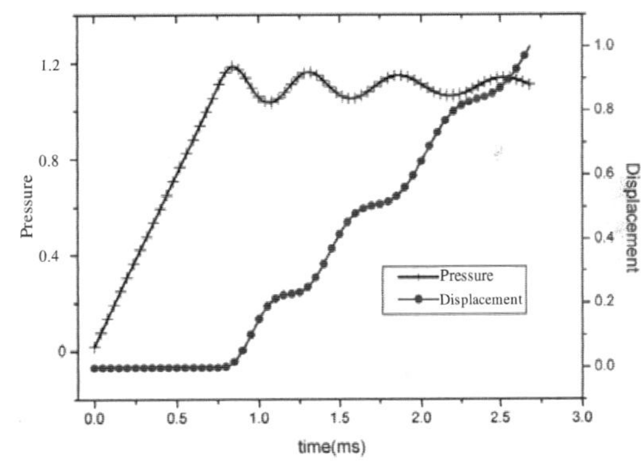

Figure 7. The simulation response time of pintle injector during the start up phase

According to the above analysis, the engine start response time t_{90} can be divided into the following parts: 0s to 0.014s is the opening time of the valve, 0.014s to 0.0185s is the time from the start of the pintle action to the start of the chamber pressure, and 0.0185-0.021 s is the time from the start of chamber pressure to reaching t_{90}. Figure7 shows a dimensionless schematic diagram of the simulated start-up response time of the pintle engine (where the pressure is dimensionless with respect to the chamber pressure and the needle valve lift is dimensionless with respect to the design value of the needle valve). The curves in figure 6 are normalized. The simulation results show that the pen starts moving at 0.7534 ms and the corresponding pressure reaches 88% of the inflation pressure, which is less than the inflation pressure but higher than the equivalent resistance load pressure. The maximum travel time is 2.6827 ms. 0s is defined as high pressure pneumatic inflation when the control valve is opened, the movement time of the pen is 1.9293 ms, and the deviation from the test is 3.535%.

4. Conclusion

In order to analyze the transient performance of start-up for the pintle injector, based on the analysis of the start-up process of a typical pintle injector, the equations were derived using mass and momentum conservation, ideal gas functions and many complex resistances. The mathematical model fully takes into account the typical voltage environment of the pintle injector. Based on the MFC framework, simulation software has been compiled, which can realize multi-parameter input and visualization output. To investigate the numerical method, both simulations and experimental tests were carried out on a specific pintle injector. The simulation results agree well with the test data, and the response time and test deviation of start-up is 3.535%, respectively, which confirms the validity of the model. The calculation software can quickly predict the start-up response of each pintle injector and the calculation result is right and the calculation speed is fast, which makes it easier to compare the advantages and disadvantages of different structural parameters in the demonstration phase. The pintle injector scheme enables the design process to be quickly iteration and has a strong technical application and popularization value.

5. References

[1] Matthew J. Casiano, James R. Hulka, Vigor Yang. Liquid-Propellant Rocket Engine Throttling: A Comprehensive Review[C]. 45th AIAA/ASME/SAE/ASEE Joint Propulsion Conference & Exhibit 2-5 August 2009.

[2] Max Vozoff and John Couluris, SpaceX Products – Advancing the Use of Space[C]. AIAA SPACE 2008 Conference & Exposition 9 - 11 September 2008, San Diego, California. AIAA 2008-7836.

[3] Tom Mueller, et al. TRW 40K1bf LOX /RP-1 low cost pintle engine test results. AIAA 2000-3863.

[4] Mishra A K, Kavipriya J. A Technical Review on effect of spray angles and characteristics for a pintle injector[J]. Journal of Advanced Engineering and Technology, 2021, 7: 30-36.

[5] Dressler G, Bauer J. TRW pintle engine heritage and performance characteristics[C]//36th AiAA/ASME/SAE/ASEE joint propulsion conference and exhibit. 2000: 3871.

[6] ZHANG Botao, LI Ping, WANG Kai, YANG Baoe, Review on Pintle Injector of Throttling Liquid Rocket Engine[J]. Journal of Astronautics 2020, 41(12):1481-1489. DOI: 10.3873/j.issn.1000-1328.2020.12.001. (in Chinese)

[7] Nardi R, Perez V, Pimenta A. Experiments with pintle injector design and development[C]//51st AIAA/SAE/ASEE joint propulsion conference. 2015: 3810.

[8] Rios A, Chowdhury M, Khan M M I, et al. Design and test of a pintle injector for oxy-coal combustion[C]//AIAA Propulsion and Energy 2019 Forum. 2019: 4325.

[9] Fang X, Shen C. Study on atomization and combustion characteristics of LOX/methane pintle injectors[J]. Acta Astronautica, 2017, 136: 369-379.

[10] Lee I, Shim D, Kim I, et al. Effect of the Flow Passage Area Ratio on a Double Slit-Type Pintle Injector[J]. International Journal of Aeronautical and Space Sciences, 2024, 25(2):540-554. DOI:10.1007/s42405-023-00683-4.

[11] Zhao F, Zhang H, Zhao L, et al. Effects of Liquid Film Instability on Spray Features of Liquid–Liquid Pintle Injectors[J]. AIAA Journal, 2024, 62(11):18. DOI:10.2514/1.J064017.

[12] Jiao B, Yu N, Zhou C, et al. Combustion flow field and thermal protection performance of a liquid oxygen/kerosene engine with pintle injectors[J]. Acta Astronautica, 2024, 225:928-941. DOI:10.1016/j.actaastro.2024.09.065.

An automatic inspection method for space launch sites

Jing Lan, Bingjie Wu, and Qiang Ma

Jiuquan Satellite Launch Center, Jiuquan, Gansu732750, China.

864475961@qq.com

Abstract. To promote the intelligent and unmanned construction of space launch sites and achieve automatic inspection of launch sites, a motion planning and control method for quadruped robots based on nonlinear model predictive control is proposed. The global body motion trajectory is quickly constructed in the launch site environment by motion planning method, and the motion control method combining Nonlinear Model Predictive Control (NMPC) and Whole-Body Control (WBC) tracks the motion trajectory, so that the quadruped robot can move autonomously in the launch site and realize the purpose of automatic inspection. Simulation validation using the quadruped robot Go1 in complex scenarios commonly found in space launch sites shows that the proposed method allows the robot to be autonomous, safe, and flexible for automated inspections.

1. Introduction

The space launch site system is a complex that provides security for the successful launch of carrier rockets and contains a series of ground facilities and equipment, which is characterized by a dangerous working environment, a large volume of facilities and equipment, and difficulties in inspection and maintenance. In recent years, the application of robots in space launch sites has become more and more widespread, and a variety of robots are used in launch sites for inspection, equipment maintenance and special operations[1].

The traditional inspection method is manual visual inspection, which is characterized by low detection accuracy, low efficiency and long time-consuming. In order to solve the shortcomings of manual inspection, robot intelligent inspection is being explored instead of manual inspection. The use of robots carrying various types of sensors to realize the site situation survey, temperature and humidity detection, smoke concentration alarm, leakage detection, fault location and disposal and other functions, to complete the inspection of special environments and special equipment[2-4]. Such as wall-climbing robots for large cranes inspection, hanging track inspection robots for underground comprehensive pipe corridors, pump rooms, substations inspection, wheeled tracked robots for hazardous environments, inspection of equipment. These robots have great motion defects when facing the complex rugged terrain and scattered facilities and equipment in the launch site, and are unable to complete all-round automatic inspection. Legged robots have higher motion mobility and flexibility[5-8], and can realize highly dynamic motion in complex and diverse environments, traversing terrains that cannot be traversed by wheeled and tracked robots, and have a broad application prospect in automatic inspection of space launch sites[9].

In order to realize the intelligent safety inspection of the space launch site, a quadruped robot space launch site automatic inspection method based on nonlinear model predictive control is proposed for the complex environment in the space launch site. Relying on the cameras and sensors carried by the quadruped robot itself, the automatic inspection of the space launch site is realized through the safe and autonomous movement of the quadruped robot in the launch site.

2. Overview of quadruped robot automated inspection system

The overall scheme of the quadruped robot for automatic inspection in a space launch site is shown in Figure 1, which mainly contains two parts: motion planning and motion control in complex environments. Motion planning is responsible for creating the body's global motion trajectory, and the global body motion planner (10Hz) plans the long-period collision-free motion trajectory of the quadruped robot in the launch site to guide the robot from the initial position to the target position, given the launch site terrain map and the target position. In motion control, a nonlinear model predictive control (50 Hz) is used to determine the quadruped robot foot-end contact forces and joint velocities to track the planned motion trajectory. Finally, the state estimation and whole body controller are executed at a rate of 400 Hz. A high frequency whole-body controller is used to convert the optimized NMPC trajectory into torque commands, and the state estimation module estimates the robot's position, velocity, direction and angular velocity in the world coordinate system.

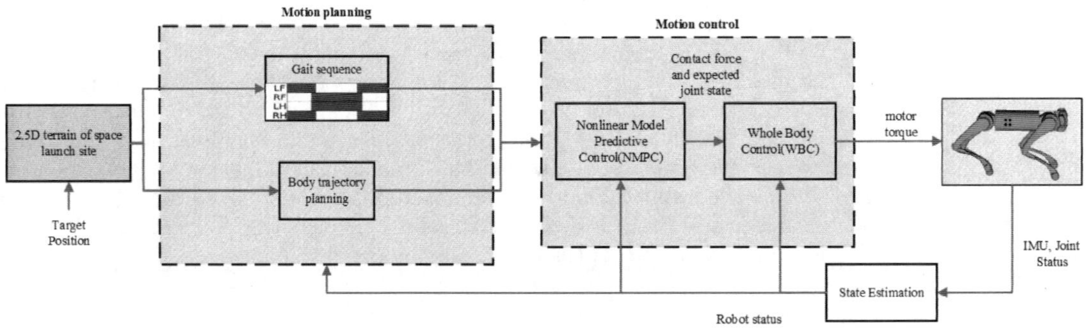

Figure 1. Block diagram of quadruped robot control system

3. Motion planning for automated inspection of quadruped robots

In wheeled and tracked robots, path planning algorithms are commonly used for motion planning, and these motions are usually restricted to 2D environments. In quadruped robot motion planning, if the same motion planning methods as wheeled robots are used, quadruped robots are unable to realize a series of motions, such as climbing and going up and down the stairs, which greatly limits the application of these motion planning algorithms to quadruped robots. Therefore, based on a 2.5D height map of the launch site terrain, a RRT*-Connect algorithm[10] is used to calculate the collision-free trajectory of the body of the quadruped robot by a low-order dynamics model at a frequency of 10 Hz under the condition of satisfying the dynamics and kinematics of the quadruped robot, and to rapidly plan the body's long-period motion trajectory during the automated inspection process of the quadruped robot[11], so that the quadruped robot to inspect the launch site quickly and safely by walking, climbing and jumping movements.

4. Motion control for automated inspection of quadruped robots

A hierarchical motion control framework combining a nonlinear model predictive controller and a whole-body controller is used to execute agile motions of a quadruped robot designed by a motion planner. The input to the motion control frame is the body motion trajectory and the output is the motor torque of the robot. The task of the state estimation module is to estimate the position, velocity, orientation and angular velocity of the robot body in the world coordinate system[12], using the state estimation method of Cheetah 3[13] to estimate the position and velocity of the quadruped robot.

A quadruped is a multi-jointed floating base system with all four legs attached to the base $\{B\}$ .Figure 2 represents the main reference system (inertial reference system $\{W\}$, body coordinate system $\{B\}$) used in deriving the equations of motion of our robot, as well as the contact forces acting on the legs of the quadruped robot.

Define the quadruped robot generalized coordinates and generalized velocity as:

$$\boldsymbol{q} = [\boldsymbol{r}_B, \boldsymbol{p}_B, \boldsymbol{q}_j],\ \ \dot{\boldsymbol{q}} = [\boldsymbol{\omega}_B, \boldsymbol{v}_B, \dot{\boldsymbol{q}}_j] \tag{1}$$

where $\boldsymbol{r}_B \in \mathbb{R}^3$ is the orientation of the quadruped robot body, and a ZYX-Euler Euler angle is assumed to represent the orientation of the robot body, $\boldsymbol{p}_B \in \mathbb{R}^3$ is the position of the body in the inertial reference system, $\boldsymbol{\omega}_B$ and $\boldsymbol{v}_B$ are the angular and linear velocities of the body in the body coordinate system, $\boldsymbol{q}_j$ and $\dot{\boldsymbol{q}}_j$ indicates the position and velocity of the joint.

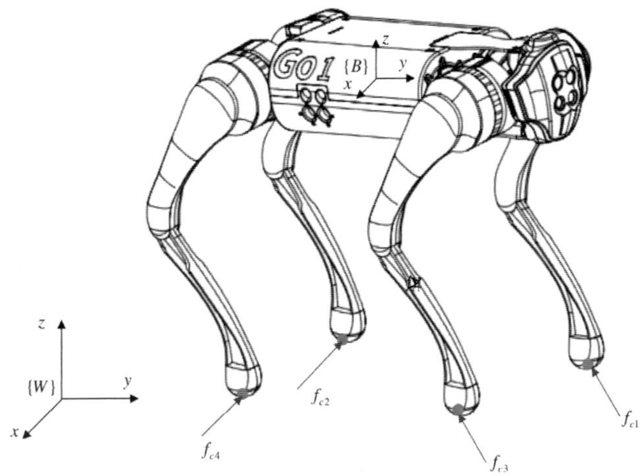

Figure 2. Floating Base System for Quadruped Robot

The rigid body dynamics of a quadruped robot, a typical floating base system, can be described as:

$$\boldsymbol{M}(\boldsymbol{q})\ddot{\boldsymbol{q}} + \boldsymbol{b}(\boldsymbol{q},\dot{\boldsymbol{q}}) = \boldsymbol{S}^T\boldsymbol{\tau} + \boldsymbol{J}_s^{\ T}\boldsymbol{F}_c \tag{2}$$

$\boldsymbol{M}$ is a generalized mass matrix, $\boldsymbol{b}$ is the vector containing all generalized forces (Coriolis force, centrifugal force, gravity), $\boldsymbol{\tau}$ is the joint moment vector, $\boldsymbol{S}$ is a selection matrix reflecting the underactuated characteristics of the robot, $\boldsymbol{J}_s = [\boldsymbol{J}_{C_i}^T \ \dots \ \boldsymbol{J}_{C_{n_c}}^T]^T$ is the foot-end contact Jacobi matrix, $\boldsymbol{F}_c$ represents the sum of the force and moment vectors applied to the robot foot end.

A quadruped robot can be appropriately modeled as an underactuated three-dimensional rigid body. Left-multiplication of equation (2) by $\boldsymbol{Q}^T$, The dynamics equations can be categorized into underactuated and actuated parts, thus the quadruped robot dynamics are represented by the following set of equations:

$$\boldsymbol{Q}_c^T[\boldsymbol{M}(\boldsymbol{q})\ddot{\boldsymbol{q}} + \boldsymbol{b}(\boldsymbol{q},\dot{\boldsymbol{q}})] = \boldsymbol{Q}_c^T\boldsymbol{S}^T\boldsymbol{\tau} \tag{3}$$

$$\boldsymbol{Q}_c^T[\boldsymbol{M}(\boldsymbol{q})\ddot{\boldsymbol{q}} + \boldsymbol{b}(\boldsymbol{q},\dot{\boldsymbol{q}})] = \boldsymbol{Q}_c^T\boldsymbol{S}^T\boldsymbol{\tau} + \boldsymbol{R}\boldsymbol{F}_c \tag{4}$$

From equation (3) and (4), the motion of the quadruped robot can be represented by two dynamics equations that include the foot-end contact force and do not include the foot-end contact force, which can be selected according to the actual needs when designing the robot motion control. (4) describes the dynamics of the entire robotic system as long as the motion of the quadruped robot is in the contact-constrained zero space. The computational expression for the foot-end contact force can be obtained by transforming (4):

$$\boldsymbol{F}_c = \boldsymbol{R}^{-1}\boldsymbol{Q}_c^T[\boldsymbol{M}(\boldsymbol{q})\ddot{\boldsymbol{q}} + \boldsymbol{b}(\boldsymbol{q},\dot{\boldsymbol{q}}) - \boldsymbol{S}^T\boldsymbol{\tau}] \tag{5}$$

Journal of Physics: Conference Series **3041** (2025) 012030 doi:10.1088/1742-6596/3041/1/012030

Equation (5) allows the foot-end contact force to be calculated with known quadruped robot joint torque $\boldsymbol{\tau}$ and robot dynamics parameters.
Equations (3) and (4) are simplified [14].

$$\boldsymbol{M}_u(\boldsymbol{q})\ddot{\boldsymbol{q}}+\boldsymbol{b}_u(\boldsymbol{q},\dot{\boldsymbol{q}})=\boldsymbol{J}_{su}{}^T(\boldsymbol{q})\boldsymbol{F}_c \tag{6}$$

$$\boldsymbol{M}_a(\boldsymbol{q})\ddot{\boldsymbol{q}}+\boldsymbol{b}_a(\boldsymbol{q},\dot{\boldsymbol{q}})=\boldsymbol{\tau}+\boldsymbol{J}_{sa}{}^T(\boldsymbol{q})\boldsymbol{F}_c \tag{7}$$

The equations represented by subscripts a,u correspond to driving and non-driving dynamics, respectively. The quadruped robot center-of-mass dynamics drive the kinetic (7) that control the range of motion that can be achieved. A transformation of (7) yields the Newton-Euler equations or center-of-mass dynamics applied to a robotic center of mass (CoM), i.e., (7) is equivalent to the Newton-Euler equations:

$$\begin{bmatrix}\dot{\boldsymbol{I}}_{com}\\ \dot{\boldsymbol{K}}_{com}\end{bmatrix}=\begin{bmatrix}\boldsymbol{F}_c+m\boldsymbol{g}\\ \sum_{i=1}^{n_c}(\boldsymbol{r}_{c_i}-\boldsymbol{c}_{com})\times\boldsymbol{f}_{c_i}+\boldsymbol{\tau}_{c_i}\end{bmatrix} \tag{8}$$

where $\boldsymbol{I}_{com}$ and $\boldsymbol{K}_{com}$ are the linear and angular momentum in the center-of-mass coordinate system, respectively. The center-of-mass coordinate system is a reference system attached to the robot's center of mass and aligned with the world coordinate system, which is consistent with the body coordinate system due to the uniform distribution of the robot's mass and shape. m is the total mass of the quadruped robot, $\boldsymbol{r}_{c_i}$ represents the position of the contact point c_i, $\boldsymbol{c}_{com}$ represents the position of the centre of mass, $\boldsymbol{f}_{c_i}$ and $\boldsymbol{\tau}_{c_i}$ are the contact forces and moments that the environment applies to the robot. The effect of the rate of change of the generalized coordinates on the center of mass momentum $\boldsymbol{A}_{com}$ is obtained through the mapping of the center of mass momentum matrix $\boldsymbol{N}(\boldsymbol{q})=\boldsymbol{N}_b(\boldsymbol{q})\,\boldsymbol{N}_j(\boldsymbol{q})$ [15]:

$$\boldsymbol{A}_{com}=\boldsymbol{N}(\boldsymbol{q})\begin{bmatrix}\dot{\boldsymbol{q}}_b\\ \dot{\boldsymbol{q}}_j\end{bmatrix} \tag{9}$$

Equation (9) can be re-expressed as:

$$\dot{\boldsymbol{q}}_b=\boldsymbol{N}_b^{-1}(\boldsymbol{A}_{com}-\boldsymbol{N}_j\dot{\boldsymbol{q}}_j) \tag{10}$$

where $\boldsymbol{q}_b$ is the body position relative to the inertial system. Having completed the definition of physically consistent dynamics, the state vector of the quadruped robot is now defined as $\boldsymbol{x}=(\boldsymbol{A}_{com},\boldsymbol{q}_b,\boldsymbol{q}_j)\in\mathbb{R}^{12+n_s}$, The input vector is $\boldsymbol{u}=(\boldsymbol{f}_{c_i},...,\boldsymbol{f}_{c_{n_c}},\boldsymbol{k}_j)\in\mathbb{R}^{3n_l+n_S}$.

$$\dot{\boldsymbol{q}}_j=\boldsymbol{k}_j \tag{11}$$

The contact moments are ignored in the control inputs, as shown in Figure 2 The quadruped robot system consists of 12 actuated joints (3 joints per leg) and 4 foot-end contact points, which requires a total of 24 state variables and 24 inputs. System status $\boldsymbol{x}\in\mathbb{R}^{24}$, input $\boldsymbol{u}\in\mathbb{R}^{24}$ 。

$$\boldsymbol{x}=[\boldsymbol{A}^T{}_{com},\boldsymbol{q}_b^T,\boldsymbol{q}_j^T]^T,\quad \boldsymbol{u}=[\mathbf{f}_c^T,\boldsymbol{k}_j^T]^T \tag{12}$$

$\mathbf{f}_c\in\mathbb{R}^{12}$ consists of contact forces at four points of contact, i.e., four ground reaction forces at the foot end.
Therefore, combining equation (8), (10), and (11) the continuous-time system of center-of-mass dynamics for the quadruped robot is:

$$\frac{d}{dt}\begin{bmatrix} \boldsymbol{I}_{com} \\ \boldsymbol{K}_{com} \\ \boldsymbol{q}_b \\ \boldsymbol{q}_j \end{bmatrix} = \begin{bmatrix} \boldsymbol{F}_c + m\boldsymbol{g} \\ \sum_i^4 (\boldsymbol{r}_{c_i} - \boldsymbol{c}_{com}) \times \boldsymbol{f}_{c_i} + \tau_{c_i} \\ \boldsymbol{N}_b^{-1}(\boldsymbol{A}_{com} - \boldsymbol{N}_j \dot{\boldsymbol{q}}_j) \\ \boldsymbol{k}_j \end{bmatrix} \tag{13}$$

Equation (13) can be discretized and the dynamics of the quadruped robot expressed in discrete time form.

Using center-of-mass dynamics it is possible to generate foot-end contact forces and joint velocities for a robot tracking a motion trajectory by solving the optimal control problem (OCP).

The real-time iterative multiple targeting method is utilized to solve the nonlinear model predictive control problem for quadruped robots[16].Then, the results of the model predictive control optimization are fed into the hierarchically optimized whole-body controller, which decomposes the control problem of a quadruped robot in a complex environment into multiple tasks based on the actual motion characteristics, expresses these tasks with linear equations or inequalities, and determines the priority of these tasks. The Whole Body Controller finds optimal joint torques and joint accelerations by solving a set of quadratic programming problems with prioritization. By defining multiple tasks (equal or unequal constraints), the WBC solves the quadratic programming problem in a linearly constrained zero space of high-priority tasks and attempts to minimize the slack variables of the unequal constraints. The method can fully account for nonlinear rigid body dynamics and ensure strict task prioritization. We use the Pinocchio rigid body library[17] to solve the robot dynamics, the necessary kinematic transformations and Jacobi matrices, and the qpOASES QP solver[18] to implement a whole-body controller for a quadruped robot.

5. System simulation verification

Motion planning and motion control algorithms are verified in the Gazebo simulation environment, and a quadruped robot Unitree Go1 from Yu Shu Technology is used as the simulation platform to verify the designed automatic inspection method of the launch site. The parameters used in the simulation are the real parameters of the quadruped robot Go1, and the proposed motion control and planning method can be directly loaded on the real robot after simulation verification.

5.1. Validation of global motion planning for automated inspection of quadruped robots

In the space launch site environment, there are various steps, slopes, gullies and various obstacles, and the quadruped robot needs to be fully adapted to these terrains in order to accomplish the inspection task. Various terrains that may exist in a launch site are created as shown in Figure 3, and the effectiveness of the proposed global motion planning method is verified through simulation. The red line segment in the figure indicates the trajectory of the body when the robot is walking, the purple trajectory indicates the trajectory of the robot in the jumping phase, and the green trajectory indicates the trajectory of the body before jumping and during the landing of the jumping movement. Simulation results show that under conditions of intermittent robot contact, underactuated, and restricted kinematics and dynamics, global motion planning can create long-period feasible body trajectories for robots to autonomously execute dynamic motions in these environments, and that the speedy and global nature of the motion planning solution enhances robot motion flexibility.

Journal of Physics: Conference Series **3041** (2025) 012030 doi:10.1088/1742-6596/3041/1/012030

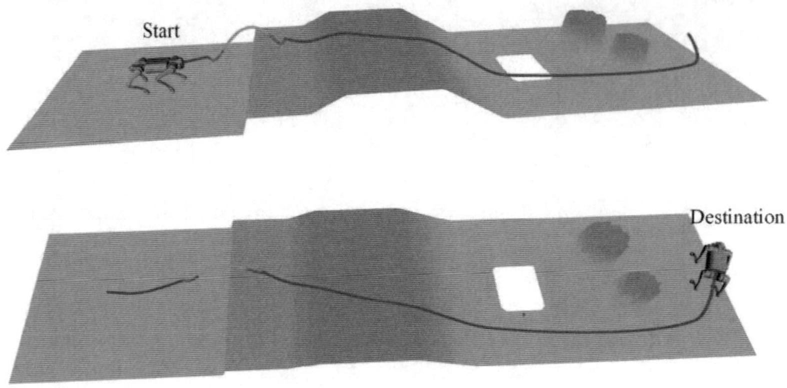

Figure 3. Robot global motion planning

5.2. Quadruped Robot Automated Inspection Flexibility Verification

For quadruped robots, locomotion in discrete terrains in a space launch site is a challenge because the robot's motion is highly dynamic. A quadruped robot was allowed to pass through discrete terrains with gap distances varying between 10cm and 15cm to validate the passability of the quadruped autorobot during inspections. Trot gait is used to allow the robot to pass through such terrain at 0.5m/s for simulation of non-regular terrain. Figure 4 shows the variation of robot body height, foot end and X-direction motion velocity when the quadruped robot passes through the discrete terrain with a speed of 0.5 m/s. In the simulation, the robot's foot landing points all fall on the safe motion plane, and there is no stepping out of the air.

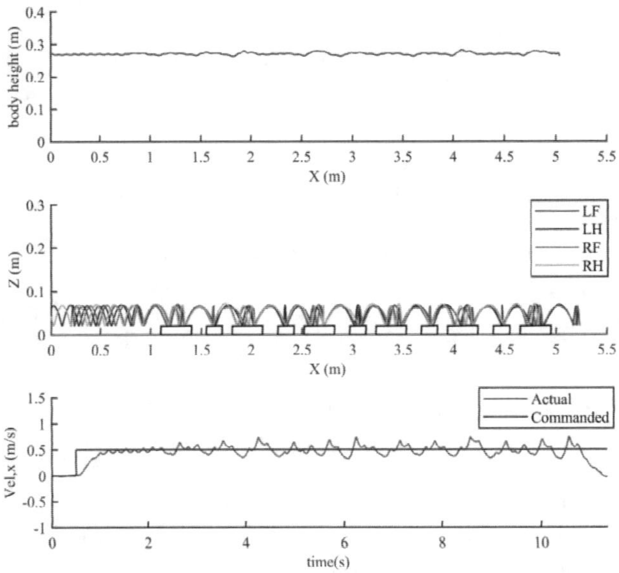

Figure 4. Robots passing through discrete terrain at 0.5m/s in simulation

In order to further validate the passage capability of the quadruped robot in a space launch site, stair climbing simulations were performed. The staircase was a standard staircase with dimensions of 280 mm wide and 150 mm high, and the staircase used for the experiment consisted of 5 steps. The robot moves at 0.5m/s using a Trot gait with a gait period of 0.5s. As can be seen in Figure 5, the robot's swinging legs did not collide with the stairs during the stair climbing process, and it was able to climb the stairs under the premise of ensuring stability and safety.

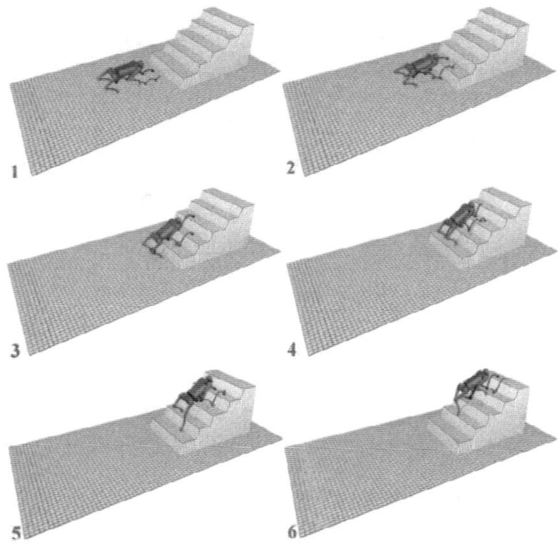

Figure 5. Quadruped robot simulation climbing stairs

Gullies are a common type of terrain for quadruped robots moving through a space launch site, and robots often pass through such terrain by straddling. The motion planning and motion control method proposed in this paper enables the quadruped robot Go1 to cross a 40cm gully that is 62% of the overall robot length (0.645m). As shown in Figure 6, the robot adopts Trot gait to pass through such a terrain at a speed of 0.5 m/s in the simulation, and in the process of striding across the gully, the swing leg distance of the quadruped robot becomes larger, and the fluctuation of the pitch angle, roll angle, and yaw angle of the quadruped robot in the process of the movement is small, so that it can complete the spanning action in a stable way.

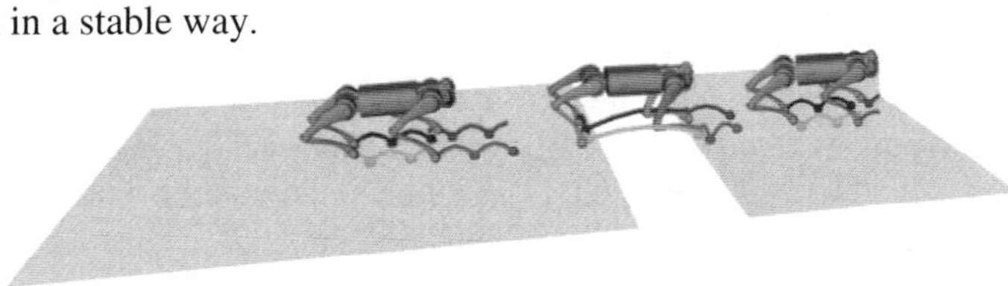

Figure 6. Quadruped robot crosses 40cm ravine

5.3. Robustness validation of a motion control framework for automated inspection of quadruped robots

Quadruped robots may collide with obstacles or encounter some sudden dynamic moving objects colliding with the robots while patrolling in a space launch site. Thus, We tested the robustness of the motion controllers for motion on flat terrain in simulation. When the robot patrols the launch site, we suddenly apply a large impact force in the Y direction of the robot body. As shown in Figure 7 for the motion of the robot when the quadruped robot is subjected to a lateral impact, the legs of the quadruped robot are able to will swing outward in the same direction, and the swinging frequency of the swinging legs is increased in order to resist the force of that impact. In the simulation, the robot deviates from the traveling direction, and both the center of mass trajectory and the swing leg trajectory move toward the method of the impact force, but after the impact force disappears, the robot returns to the original motion direction after about 1s. After the robot regained stability, the trajectory position of the center of mass was not shifted and the direction of robot motion was shifted by a small amount.

Journal of Physics: Conference Series **3041** (2025) 012030 doi:10.1088/1742-6596/3041/1/012030

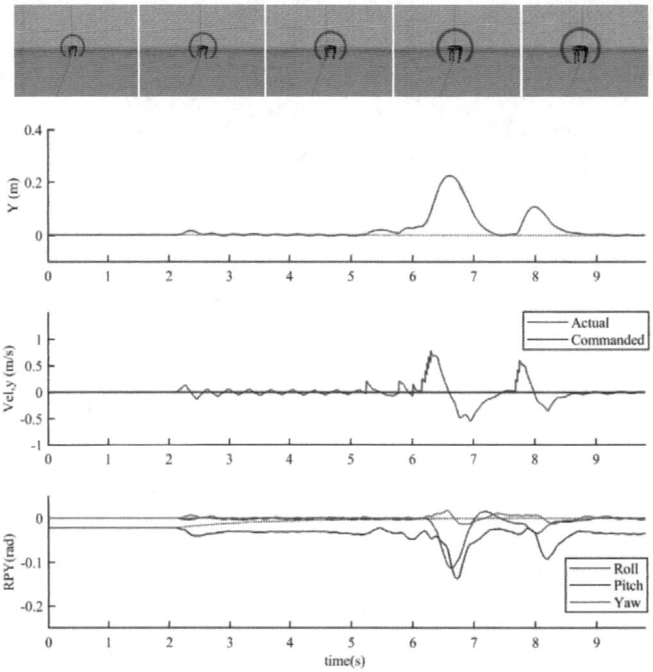

Figure 7. Quadruped robot is subjected to impact force in the y -direction

6. Summary

This paper proposes a quadruped robot-based automatic inspection method for space launch sites, integrating the space launch site terrain to design the motion planning and motion control method for the quadruped robot, which can enable the quadruped robot to move safely and autonomously in the space launch site. Simulation on the quadruped robot Go1, in which the quadruped robot is able to pass through discrete terrains with varying gap distances, 40cm gullies, multistage standard staircases, and resistance to lateral impacts, allows the robot to autonomously, flexibly, and safely realize the task of automated inspection of the launch site.

Future work is to equip robots with vision or laser sensors to accurately map the robot's environment instead of using known terrain information, improving the robot's adaptability to the environment and range of applications.

7. References

[1] Wang X M, Liu Z Q, Teng F, Hou K W 2021 *Manned Spaceflight* **27** 522-9.
[2] Sun Y, Li F, Hao M, Hong X W, Chen Y 2020 Manufacturing Automation **42** 69-72.
[3] Liu W, Wang Y B, Liu G Q, Liu F, Chen L J, Yu W J, Xu Ming 2019 *Water and Wastewater Engineering* **55** 122-8.
[4] Jiang X J, Sha W L, Rui J, Wang Y 2018 *Automation and Instrumentation* **09** 153-5.
[5] Hooks J, Ahn M S, Yu J, Zhang X G, Zhu T Y N, Chae H 2020 *RAL* **5** 5409-16.
[6] Hutter M, Gehring C, Jud D, Lauber A, Bellicoso C D, Tsounis V 2016 *IROS* 38-44.
[7] Bouman A, Ginting M F, Alatur N, Palieri M, Fan D D, Touma T, Pailevanian T, Kim S K, Otsu K, Burdick J 2018 *IROS* 2518-25.
[8] Raibert M, Blankespoor K, Nelson G, Playter R 2008 *IFAC* **41** 10822-5.
[9] Lee J, Hwangbo J, Wellhausen L, Koltun V, Hutter M 2020 *Science Robot* **5**.
[10] Norby J, Johnson A M 2020 *IROS* 3829-36.
[11] Norby J, Yang Y, Tajbakhsh A, Ren J M, Yim J K, Stutt A, Yu Q S, Flowers N, Johnson A M 2022 *In ICRA Workshop on Legged Robots*.
[12] Hosseini M, Rodriguez D, Behnke S 2022 *IRC* 103-10.
[13] Bledt G, Powell M J, Katz B, Carlo J D, Wensing P M, Kim S 2018 *IROS* 2245-52.

Journal of Physics: Conference Series **3041** (2025) 012030 doi:10.1088/1742-6596/3041/1/012030

[14] Sleiman J P, Farshidian F, Minniti M V, Hutter M 2021 *RAL* **6** 4688-95.
[15] Orin D E, Goswami A, Lee S-H 2013 *Autonomous Robots* **35** 161-76.
[16] Grandia R, Jenelten F, Yang S, Farshidian F, Hutter M, 2023 *TRO* 1-20.
[17] Carpentier J, Saurel G, Buondonno G, Mirabel J, Lamiraux F, Stasse O, Mansard N 2019 *SII* 614-9.
[18] Ferreau H J, Kirches C, Potschka A, Bock H G, Diehl M 2014 *Mathematical Programming Computation* **6** 327-63.

Journal of Physics: Conference Series **3041** (2025) 012031 doi:10.1088/1742-6596/3041/1/012031

Intelligent path programming approach for autonomous ecological vehicles based on fuzzy logic control

Biyogo Nchama Vicente Angel Obama[1,5], Hossen MD Akram[2,5], Ye Yukun[3], Han Haoyu[4], Zhao Shuang[5*], Shi Peng[1*]

[*1] School of Astronautics, Beihang University, Beijing 102206, China
[2] School of Automation Science and Electrical Engineering, Beihang University, Beijing 100191, China
[3] Sino-French Engineer School, Beihang University, Beijing 100191, China
[4] School of Mechanical Engineering & Automation, Beihang University, Beijing 100191, China
[*5] International Relations Department, Beihang University, Beijing 100191, China

E-mail: shipeng@buaa.edu.cn, vaobsendi@buaa.edu.cn

Abstract. This paper proposes an intelligent path-planning approach for autonomous vehicle driving based on fuzzy control. Compared to the traditional control method, this approach is applied directly to the nonlinear system, additionally, this approach is simple to implement compared to the modern adaptive and predictive nonlinear controllers. This research finding aims to be compatible with the existing global satellite positioning systems and the emerging green ecological technologies, equipping the vehicle with reliable technology to track target trajectories autonomously and accomplish multiple tasks in autonomous driving and green agriculture. The paper focuses on ground vehicles' nonlinear dynamics control and analysis in different topological scenarios. The simulation results prove the reliability of using the fuzzy logic-based intelligent controller to train the system to track the desired trajectory and orientation autonomously, providing a great reference for designing an intelligent multi-use vehicle and a valuable reference for nonlinear system control.

1. Introduction

In parallel with the rapid development of modern science and technology, developing new technologies or analyzing the reliability of using other fields' technology to achieve a green ecological environment is an essential topic[1,2]. Thereby, many countries have set the realization of a green environment as one of the main goals in their development objective[1]. Furthermore, many researchers have delved into the design of intelligent[3] and ecological[4] vehicle technologies, such as automatic farming machines and so on[5-8]. The development of the next generation of autonomous vehicles needs to satisfy three basic properties[9], i.e., intelligent, green propellants[7], and advanced positioning. With the development of global satellite positioning systems, such as BeiDou (BDS) and GPS, and with the current advances in electric vehicle technology, the lack is the development of a technology-algorithm capable of combining existing technologies to accomplish autonomous driving on a small and industrial scale.

Motivated by the lack of joint technologies for intelligent autonomous driving, this research aims to design a fuzzy-logic-based intelligent control algorithm, which aims to be compatible with the existing global satellite positioning system and green propellant technologies to accomplish autonomous vehicle driving. This paper is structured as follows: Section 1 gives a general overview of the importance of the

Journal of Physics: Conference Series **3041** (2025) 012031 doi:10.1088/1742-6596/3041/1/012031

research topic. Section 2 delves into the mathematical formulation of the system motion, focusing on ground vehicles. Section 3 presents the mathematical implementation of the proposed controller. And finally, Section 4 conducts simulations and analyses of the system's performance. Figure 1 shows the potential application of solar-energy-sourced BDS-guided vehicles in farmer-agriculture labours.

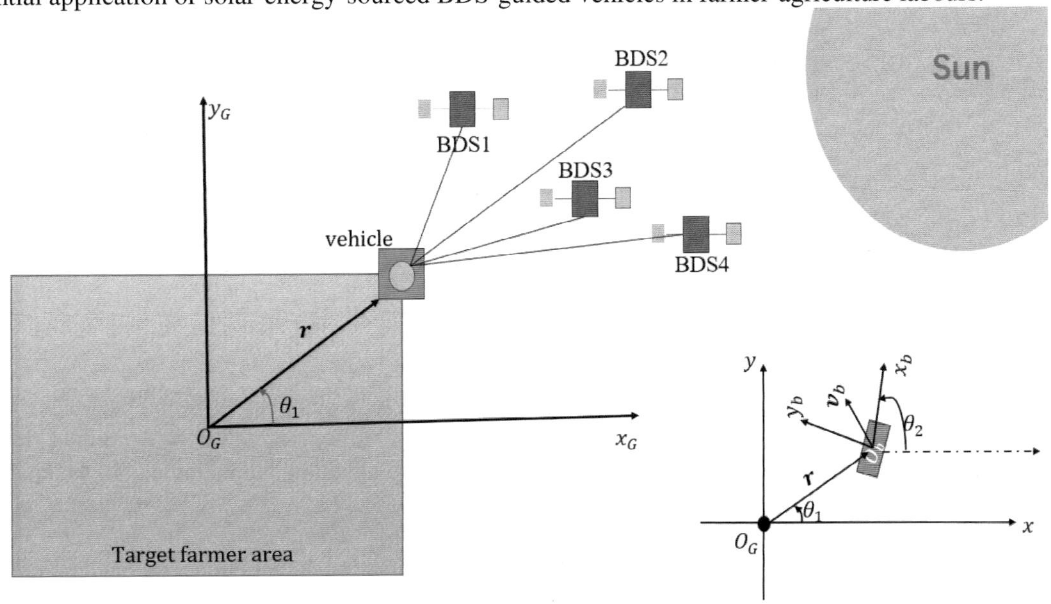

Figure 1. The conceptual idea of a solar-sourced electrical farm vehicle guided by BDS

2. Mathematical preliminaries and system analysis

From a ground vehicle activities perspective, the Local Ground-Inertial Frame $S_G(O_G xyz)$ is defined as shown in Figure 1, the vehicle is assumed to be a rigid body, therefore, we define the vehicle Body coordinates system $S_B(O_B x_B y_B z_B)$ to point to vehicle's inertial axes. The rotational angle $\theta = \theta_1 + \theta_2$ is the parameter that describes the orientation of the vehicle w.r.t the local ground frame. Since the sensors can measure the vehicle state in the body frame, the body frame-based dynamics equation is

$$\begin{cases} \dfrac{\mathrm{d}(\boldsymbol{r})_G}{\mathrm{d}t} = \boldsymbol{L}_{BG}^{-1}(\boldsymbol{v})_B \\ \dfrac{\mathrm{d}(\boldsymbol{v})_B}{\mathrm{d}t} + [\boldsymbol{\omega}]^{\times} \cdot (\boldsymbol{v})_B = \dfrac{1}{m}\boldsymbol{T} + \boldsymbol{g} + \dfrac{1}{m}\boldsymbol{D} + \dfrac{1}{m}\boldsymbol{F}_{\mathrm{C}} \\ \dfrac{\mathrm{d}\boldsymbol{h}}{\mathrm{d}t} + [\boldsymbol{\omega}]^{\times} \cdot (\boldsymbol{h})_B = \boldsymbol{\tau} + \boldsymbol{\tau}_C \\ \dfrac{\mathrm{d}m}{\mathrm{d}t} = f(t) \end{cases} \tag{1}$$

where the sub-indexes refer to the frame in which each term is expressed, some sub-indexes have been omitted considering the vectorial operations should be in the same frame. m is the vehicle's mass. $\boldsymbol{r}$ is its position vector concerning the ground point O_G, $\boldsymbol{v}$ is its velocity, $\boldsymbol{T}$ is the Thrust vector, $\boldsymbol{g}$ is the gravitational acceleration vector, $\boldsymbol{D}$ is the friction force, and $\boldsymbol{F}_C$ is the active control force. The angular velocity's cross matrix, the coordinates transformation matrix, and the momentum of inertia are

$$[\boldsymbol{\omega}]^{\times} = \begin{bmatrix} 0 & -\omega & 0 \\ \omega & 0 & 0 \\ 0 & 0 & 0 \end{bmatrix};\ \boldsymbol{L}_{BG}^{-1} = \begin{bmatrix} \cos\theta & -\sin\theta & 0 \\ \sin\theta & \cos\theta & 0 \\ 0 & 0 & 1 \end{bmatrix};\ \boldsymbol{h} = \boldsymbol{J}\cdot\boldsymbol{\omega} = \begin{bmatrix} J_x & 0 & 0 \\ 0 & J_y & 0 \\ 0 & 0 & J_z \end{bmatrix}\begin{bmatrix} 0 \\ 0 \\ \omega \end{bmatrix} \tag{2}$$

Another reason to describe the dynamic motion in the vehicle's body frame is because the angular momentum is constant on it. Thus, the transform matrix can be used to simplify the calculations.

Journal of Physics: Conference Series **3041** (2025) 012031 doi:10.1088/1742-6596/3041/1/012031

3. Fuzzy-Learning-Based Trajectory planning

Figure 2 shows the conceptual implementation of an autonomous vehicle based on fuzzy control. The fuzzy controller is implemented in three steps: signal fuzzification, i.e., converting the control's inputs into fuzzy logic signals; control law computation; and signal defuzzification, i.e., converting the fuzzy control law into a deterministic signal to accomplish the control objective. In this research, we use the Mamdani-based[10,11] interference, or "*if-then*" combination, i.e., then $\boldsymbol{e} \wedge \boldsymbol{ec} \rightarrow \boldsymbol{u}$, to compute the control law. The median[11] approach is used for the defuzzification of the control law.

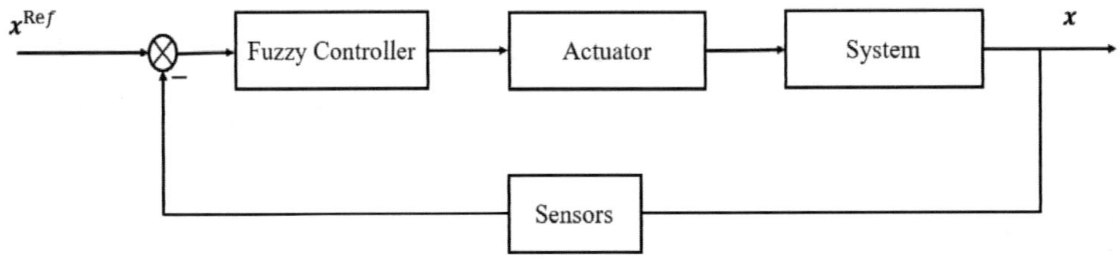

Figure 2. Conceptual implementation of an intelligent farming machine

Without loss of generality, focusing on planar motion, we simplify the nonlinear equation(1) into the nonlinear equation(3), where the first 4 equations are in the body frame, and the last two represent the vehicle position expressed in the ground frame. The sub-index B means Body-frame parameter. Figure 3 shows how this idea works from an axis perspective.

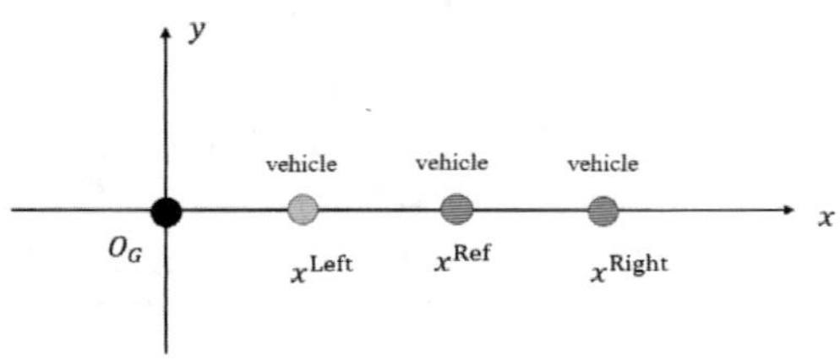

Figure 3. One-dimensional approach

$$\begin{cases} \dot{v}_{xB} = -cv_{xB} + \omega v_{yB} + u_{xB} \\ \dot{v}_{yB} = -cv_{yB} - \omega v_{xB} + u_{yB} \\ \dot{\omega} = \dfrac{1}{J_z} u_\tau \\ \dot{\theta} = \omega \\ x_G = x_B \cos(\theta) - y_B \sin(\theta) \\ y_G = x_B \sin(\theta) + y_B \cos(\theta) \end{cases} \tag{3}$$

where c is friction's constant, $u = \begin{bmatrix} u_x & u_y & u_\tau \end{bmatrix}^{\mathrm{T}}$ is the required intelligent control to guide the vehicle to the tracking position and orientation. Defining the vehicle's reference as $\boldsymbol{x}^{\mathrm{Ref}}$, and the sensor's measured as $\boldsymbol{x}$, i.e., $\boldsymbol{x}^{\mathrm{Ref}} = \begin{bmatrix} x^{\mathrm{Ref}} & y^{\mathrm{Ref}} & \theta^{\mathrm{Ref}} & v_x^{\mathrm{Ref}} & v_y^{\mathrm{Ref}} & \omega^{\mathrm{Ref}} \end{bmatrix}^{\mathrm{T}}; \boldsymbol{x} = \begin{bmatrix} x & y & \theta & v_x & v_x & \omega \end{bmatrix}^{\mathrm{T}}$, therefore, the position and velocity errors can be defined as:

$$\boldsymbol{e} = \begin{bmatrix} \boldsymbol{r}^{\mathrm{Ref}} - \boldsymbol{r} \\ \theta^{\mathrm{Ref}} - \theta \end{bmatrix} \Rightarrow \boldsymbol{ec} = \dot{\boldsymbol{e}} = \begin{bmatrix} \dot{\boldsymbol{r}}^{\mathrm{Ref}} - \dot{\boldsymbol{r}} \\ \dot{\theta}^{\mathrm{Ref}} - \dot{\theta} \end{bmatrix} \tag{4}$$

thereby, considering a one-dimensional motion along $O_G x$ direction, the vehicle can be in three possible positions: at the desired position x^{Ref} , at the left of the desired position x^{Left} , and at the right side of the desired position x^{Right} , hence, for $\boldsymbol{e} \in E$, $\boldsymbol{ec} \in EC$, and $\boldsymbol{u} \in U$, we can define two possible signs for the control action: the positive control $u_x > 0$ to drive the vehicle from left to the target position, and the negative control action $u_x < 0$ to push the vehicle from right to desired. Consequently, this analysis is feasible and general for any considered position vector $\boldsymbol{r}$, since we can always compute its projection along each axis and then proceed with the same analysis.

However, the above law is not feasible for numerical computation, thus we use fuzzy mathematics to quantize the above errors: for the state error e and the state error rate ec along each axis, we can define three-level linguistic values, i.e., N (Negative), Z (Zero), and P (Positive); Furthermore, for each axis control action u, we define five-level linguistic values, NB (Negative Big), NS (Negative Small), Z (Zero), PS (Positive Small), and PB (Positive Big); Figure 4 shows related memberships.

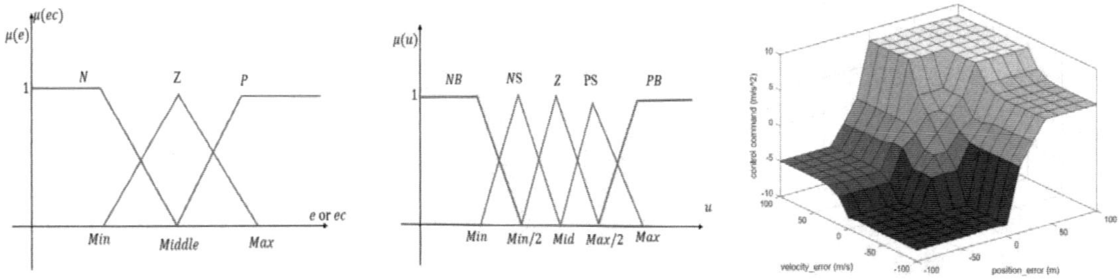

Figure 4. Membership functions of ***e*** and ***ec*** (left) and ***u*** (middle) and normalized control surface(right)

from the membership functions in Figure 4, we define a more precise control law in Table 1. Thus, the accuracy and robustness of the system depend on each defined trial ***e, ec, u***, Furthermore, the control system control surface also agrees with the basic engineering control definition.

Table 1. Fuzzy Control Rules

ec \ e	N	Z	P
N	NB	NB	PS
Z	NB	Z	PB
P	NS	PB	PB

4. Simulations

Consider a vehicle with mass $m = 100\text{kg}$, $J = 50\text{kgm}^2$; the boundaries of $\boldsymbol{e}$ and $\boldsymbol{ec}$ are ± 100, considering an aggressive control $\boldsymbol{u}$ between $\pm 10\text{m/s}^2$, and the rotation θ between $\pm 7\text{rad}$. The reference is

$$\boldsymbol{x}^{\text{Ref}} = \left[\boldsymbol{r}^{\text{Ref}} \quad \theta^{\text{Ref}} \quad \boldsymbol{v}^{\text{Ref}} \quad \dot{\theta}^{\text{Ref}}\right]^T = \left[\boldsymbol{r}^{\text{Ref}} \quad \theta^{\text{Ref}} \quad \boldsymbol{v}^{\text{Ref}} \quad \dot{\theta}^{\text{Ref}}\right] = \left[100 \quad 100 \quad \pi/2 \quad 0 \quad 0 \quad 0\right]^T$$

this means tracking position (-100,100) in the ground frame and the rotating 90deg. The performance of the vehicle along each axis is expressed in the body frame are shown in Figure 5 and Figure 6

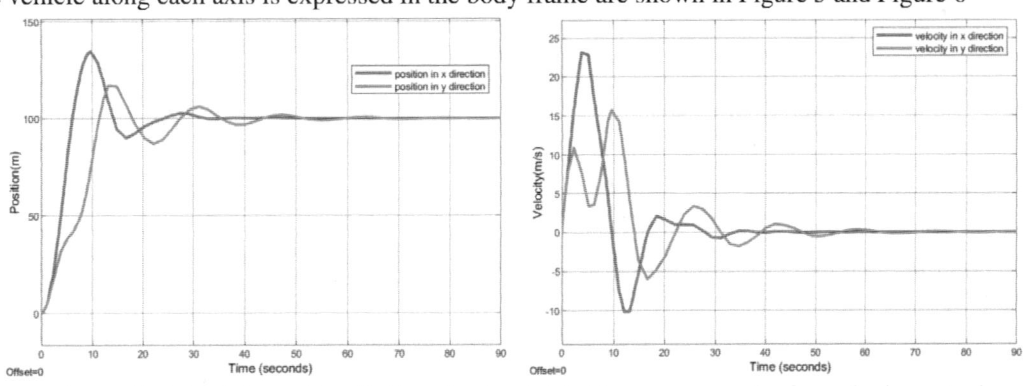

Figure 5. Vehicle's position(left) and velocity(right) response under fuzzy logic control

Journal of Physics: Conference Series **3041** (2025) 012031 doi:10.1088/1742-6596/3041/1/012031

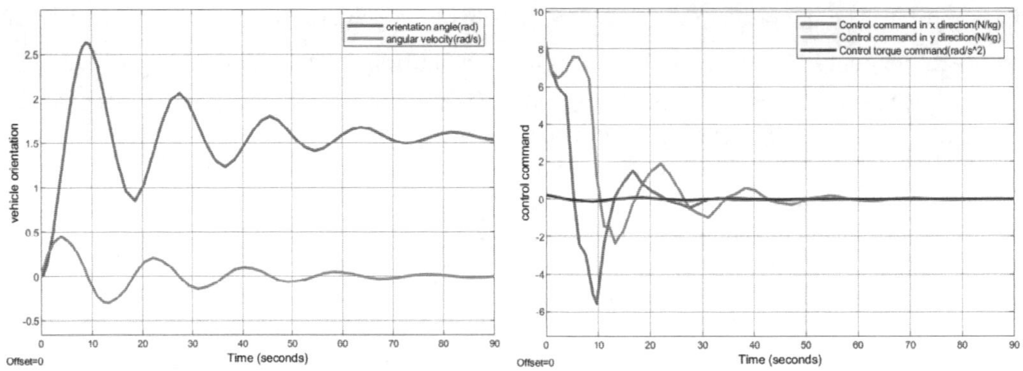

Figure 6. Vehicle's orientation response(left) and required control action (right) based on fuzzy control

The simulations reveal that the fuzzy controller drives the vehicle to the desired position and makes it rotate 90 degrees as desired. The time response on each axis also shows the great advantage of the proposed research idea for engineering implementation. Thereby, we can assert that the proposed research is in agreement with the engineering needs for autonomous vehicle driving. Furthermore, the robustness of the controller depends on the trials ***e, ec, u*** defined in Table 1, which can be adjusted according to the requirements.

5. Conclusions

This paper presented a feasible approach for implementing intelligent-autonomous vehicle driving. The research focused on ground vehicles, a detailed analysis of the system's dynamics was carried out, and the basic idea behind fuzzy control was described and implemented to design an intelligent controller to train the ground vehicle to track the desired trajectory and orientation autonomously, The simulation results revealed the robustness and intelligence performance of the system under any given tracking state, proving the feasibility of the research idea for autonomous driving and its potential compatibility with the emerging green and global positioning technologies to achieve large scale autonomous driving and its potential in nonlinear systems control.

Future works: The simplicity of the presented approach in achieving autonomous driving inspired us to go for the implementation process. In the next, we aim to apply electronics knowledge for hardware design and combine this controller with positioning and electrical technology for engineering uses.

References

[1] Y. Lin, F. Qian, J. Wu, et al, "Looking for the Green Future of Sustainable City: A Broad Perspective on Ecology Urban Design," *2010 International Conference on Management and Service Science*, Wuhan, China, 2010, pp. 1-4

[2] C. L. Fang, "The urbanization and urban development in China after the reform and opening- up, " Economic Geography, vol. 29(1), 2006, pp. 19-25.

[3] Y. Liu, Z. Xiaofe, W. Huang, D. He and X. Mi, "A Framework for Rare and Endangered Plants Protection Based on Beidou and Digital Twins Technology," *2021 International Conference on Electronic Information Technology and Smart Agriculture (ICEITSA)*, Huaihua, China, 2021, pp. 130-134

[4] Rashmi Priya and Dharavath Ramesh, "ML based sustainable precision agriculture: A future generation perspective", Sustainable Computing: Informatics and Systems, vol. 28, pp. 100439, 2020.

[5] Nadir Ali Bhangar and Abul Kashem Shahriyar, "IoT and AI for Next-Generation Farming: Opportunities Challenges and Outlook", International Journal of Sustainable Infrastructure for Cities and Societies 8, no. 2, pp. 14-26, 2023.

[6] A. Gahlot and M. Agarwal, "Towards Sustainable Agriculture: A Comprehensive Survey on Next Generation Smart Farming," 2024 Second International Conference on Intelligent Cyber Physical Systems and Internet of Things (ICoICI), Coimbatore, India, 2024, pp. 1148-1154.

[7] H. H. Vogt, D. Albiero, B. Schmuelling, "Electric tractor propelled by renewable energy for small-scale family farming," Thirteenth International Conference on Ecological Vehicles and Renewable Energies (EVER), Monte Carlo, Monaco, 2018, pp. 1-4

[8] E Mehrdad et al., Modern Electrical Hybrid Electric and Fuel Cell Vehicles, Boca Raton FL:CRC Press, 2010.

[9] X. Wang and X. He, "Evaluation of Multisignal and Multiorbit Multipath Reflectometry of BeiDou Navigation Satellite System," in IEEE Geoscience and Remote Sensing Letters, vol. 20, pp. 1-5, 2023.

[10] J. G. Juang, R. W. Lin, W. K. Liu Comparison of classical control and intelligent control for a MIMO system[J]. Applied mathematics and computation,2008,205(2)770-791.

[11] L. J. Kun, Intelligent Control Theoretical Basis, Algorithmic Design and Application[M]. Beijing: TSINGHUA University Press, 2022 (in Chinese).

Acknowledgments

The authors thank the Chinese Government Scholarship (CSC) “high-level education program” for supporting this research paper.

Journal of Physics: Conference Series **3041** (2025) 012032 doi:10.1088/1742-6596/3041/1/012032

Corrosion inhibition in electrochemical grinding of honeycomb seal using sodium lignosulfonate

J H Wang[1], L Wang[1], H S Li[1] and N S Qu[1,2]

[1] College of Mechanical and Electrical Engineering, Nanjing University of Aeronautics and Astronautics, Nanjing, 210016, China

E-mail: nsqu@nuaa.edu.cn

Abstract. Honeycomb seal is an important sealing component in aviation engines, which ensures the gas tightness between the rotor and stator, and it has been widely put into effect for the output efficiency of aero engines. Electrochemical grinding (ECG) is one of its efficient manufacturing methods. However, due to the small wall thickness of its structure, it is susceptible to corrosion and dissolution by electrolysis during the processing. Therefore, sodium lignosulfonate (SL), a green corrosion inhibitor, is added to the electrolyte of ECG to inhibit the corrosion of honeycomb seal and reduce environmental pollution during the machining process. Electrochemical tests and experimental machining with different concentrations of SL are conducted, and surface composition analysis is performed. The ECG experiments show that the efficiency of corrosion inhibition can reach 84.67% and the excessive corrosion can be reduced by 44.8% by adding 0.06 wt.% SL in 2 wt.% NaCl solution. The study provides an efficient and eco-friendly means for restraining the corrosion of honeycomb seal during ECG.

1. Introduction

Honeycomb seals have been implemented as critical sealing components in the turbine engines, exhibiting excellent performance in reducing leakage between different levels of turbine blades and improving engine efficiency [1].

Due to the thin-walled honeycomb structures, significant deformation occurs on the surface under mechanical stress during processing, and burrs are easily generated, resulting in lower processing quality. Qiu et al. [2] designed a low-temperature ice-solidified fixture to solve the deformation problem of aluminum honeycomb structural components during milling, achieving effective clamping of honeycomb cores. Sun et al. [3] designed a disk tool to cut aluminum honeycombs and used ultrasonic processing to reduce the machining deformation of it, which can keep the hexagonal honeycomb shape intact and improve manufacturing quality. The above studies effectively suppress the deformation of honeycomb core materials during processing by improving mechanical processing techniques. However, according to the research of Tian et al. [4], the honeycomb seal's surface, characterized by its small and closely packed apertures, is susceptible to significant deformation when subjected to mechanical stress during the manufacturing process. Furthermore, the burrs and chips produced in this process are inclined to accumulate within the honeycomb seal's perforations, posing potential challenges in the manufacturing and application of it. Therefore, a machining method with low mechanical stress is needed.

Gao et al. [5] conducted research on electrical discharge grinding (EDG) technology, using kerosene as the working fluid and the fan-shaped graphite electrode. The obtained workpiece has higher radial dimensional accuracy. However, the surface of the electrical discharge grinding has a thermal affected layer with a certain recast structure.

Electrochemical Grinding (ECG) is a technology that integrates electrolysis with mechanical grinding. Since electrolysis also acts a role in material removal during the machining process, it can minimize the plastic deformation of the seal surface caused by machining. At the same time, ECG doesn't have a discharge process, so it will not produce a remelted layer on the honeycomb surface. However, due to the corrosive influence of the electrolytic action on the wall surface of honeycomb seal, further research is needed to explore methods to suppress this process.

Sodium lignosulfonate (SL) is an anionic surfactant that is green and pollution-free and can be used as a corrosion inhibitor. It can be extracted from a natural lignocellulose waste [6]. Calcium lignosulfonate (CLS) was studied by Lin et al. [7] on the associated corrosion suppression effectiveness. They also found, in the alkaline solution, some different types of sodium salts (such as Na_2MoO_4, Na_2SnO_3, and $NaWO_4$) along with CLS molecules can make an adsorption onto the material surfaces and generate a metal oxides film, protecting the steel from dissolution from NaCl. Hajjari et al. [8] investigated a new green corrosion inhibitor compromising SL and zinc acetate, improving the chemical impedance of the steel in NaCl solution, and pointed out that SL can adhere on the metal-face and make up a protection film, forming chelate compounds. Ahangar et al. [9] also pointed out that the sodium lignosulfonate can cooperate with zinc-acetate and form a deposition film to constitute zinc ion-containing metal surfaces to block the erosion of the solution. During ECG, the surface of the honeycomb is positively charged due to its connection to the positive electrode of the power source, while SL dissociates into negatively charged sulfonate ions in the solution, which can combine with the positively charged metal-face. Meanwhile, during the ECG process, the surface of the honeycomb will dissolve material and release metal ions, and SL can chelate with them to form complexes. Therefore, SL can be used as a corrosion inhibitor for honeycomb seal structures in ECG processes.

This article introduces a method to reduce the corrosion of honeycomb seal structures during ECG using a green corrosion inhibitor, sodium lignosulfonate. Electrochemical impedance spectroscopy (EIS) tests, as well as Tafel, were studied for ascertaining the efficiency of corrosion inhibition. The experiments found SL can reduce depression of the honeycomb after ECG on the single-foiled segment. This article provides a green and pollution-free machining method for inhibiting corrosion of honeycomb structures during ECG process.

2. Materials and methods

2.1. Honeycomb seal and grinding wheel

To conduct machining experiments on test pieces, the honeycomb seal ring was cut into 30° fan-shaped segments, with a honeycomb layer width of 16 mm. The material of the honeycomb structure (Andertechs, China) is a nickel-based superalloys, GH3536. The NaCl and SL (Analytical pure, 99.8% and 99%) used in the electrochemical tests are all of analytical purity. As shown in figure 1, it shows the theoretical progress of the ECG process and the structure size of honeycomb core. The honeycomb layer is composed of a hexagonal hole structure, and the thickness of each segment is 50 μm. During ECG, the positive pole of the electrical supply is jointed to the seal, and the negative pole of it is linked to the conductive grinding wheel. In the processing area, the electrolyte is supplied from the nozzle. Due to the thin thickness of the single-foiled segment, it will be largely thinned caused by electrolysis in ECG manufacturing, resulting in the excessive corrosion as shown in figure 1.

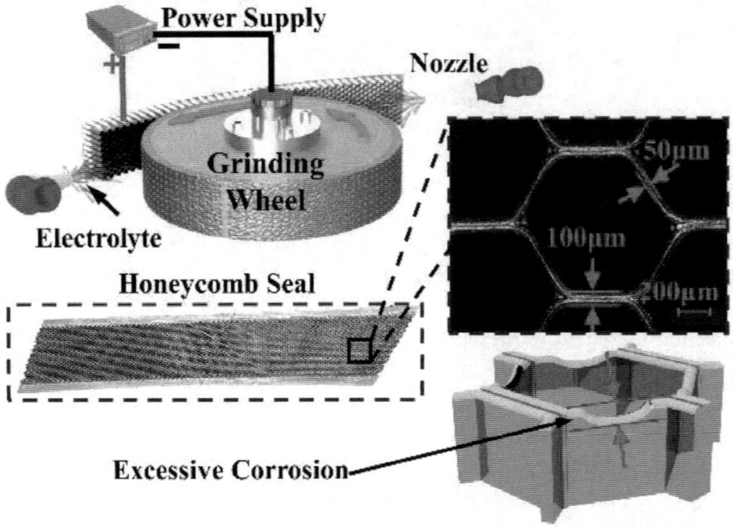

Figure 1. Electrolytic Machining of Honeycomb Seals.

2.2. Electrochemical test and morphological observation

At the electrochemical workstation (ZENNIUM pro, Zahner, Germany), electrochemical testing was conducted to study the corrosion reduction of SL in the 2 wt.% NaCl solution. Solution temperature is controlled using a water bath providing an intact environment of 30 ± 1 °C. A saturated calomel electrode worked as the reference pole in the solution, and the counter one was served by a platinum plate. Tafel curves are processed with a scanning speed of 0.5 mV/s. And the voltages were set at ± 200 mV around the open circuit potential. EIS progressed at frequency ranging from 100 kHz to 10 mHz, as well as an AC destabilization of 10 mV. The processed honeycomb seals were sent to scanning electron microscope (SEM), and surface scanning tests were proceeded by energy dispersive spectroscopy (EDS) at the same time.

2.3. Experimental setup

The CNC machine tool used in the experiment has an accuracy of 0.001 mm. The machine tool possesses four degrees of freedom, enabling linear motions of X, Y, Z, and rotational motion of spindle. In figure 2, the apparatuses include a data collector, a power supply (IT6500, ITCH), an electrolyte pool, a pressure filter, and a honeycomb clamping platform, as shown in figure 2. The electrolyte collected during the processing is in the electrolyte tanks, each equipped with a filter to continuously purify the electrolyte.

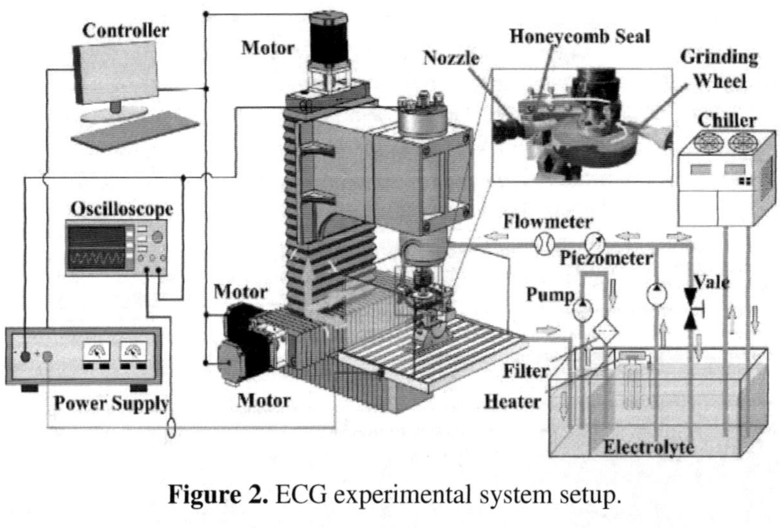

Figure 2. ECG experimental system setup.

Journal of Physics: Conference Series **3041** (2025) 012032 doi:10.1088/1742-6596/3041/1/012032

3. Results and discussion

3.1. Tafel and EIS

The experiment tested in the solution of 2 wt.% NaCl with SL for Tafel polarization and EIS. The results of the measurements are depicted in figure 3(a). By using the expanding methodology of the anodic and cathodic polarization curves, the corrosion potential (E_{corr}) as well as corrosion current density (I_{corr}) are obtained [10], shown in table 1. Equation (1) gives the calculation of inhibition efficiency:

$$\eta=\frac{I_{corr}-I_{SL}}{I_{corr}}\times 100\% \tag{1}$$

where I_{SL} represents the corrosion current density after adding SL, while I_{corr} represents the one without the participation of SL. From table 1, after the addition of SL, E_{corr} and I_{corr} have been obviously decreased. The inhibition efficiency η is 81.28% with a SL concentration of 0.06 wt.%. Therefore, SL can notably inhibit the anodic dissolution of GH3536 in the 2 wt.% NaCl solution [11].

In figure 3(b) and (c), the results of EIS are demonstrated, where all curves exhibit a convex arc in the Nyquist plot, which has relevance with the charge transfer resistance (R_{ct}), as well as double-layer capacitance (C_{dl}). This construction can be explained by the domination of charge transfer process in the dissolution [12]. In the 2 wt.% NaCl solution with SL, the Nyquist curve indicates a larger arc radius, indicating that the corrosion inhibition ability can be achieved with SL [13]. In the phase plots, electrolysis of GH3536 in 2 wt.% NaCl solution is controlled by only one charge transfer, as it has only one phase peak [14]. After adding the SL, the phase angle is enlarged within 1 Hz to 10 mHz, indicating that a protective film will be generated by SL on the metal surface [15].

The equivalent circuit fitting of EIS can be fitted using the structure shown in figure 3(b) [16]. The fitting results calculated from the matching operations are shown in table 2. The circuit is set that R_s means the solution resistance, R_f serves as the surface layer resistance, R_{ct} denotes the charge transfer resistance, CPE_f is the surface layer capacitance and CPE_{dl} is the double-layer capacitance [17]. The polarization resistance (R_p) has the function of evaluating the corrosion impedance of the metal, as $R_p = R_f + R_{ct}$ [18]. A larger polarization resistance (R_p) accompanies with a higher dissolution resistance. In equation (2), the impedance of CPE can be calculated by:

$$Z_{CPE}=\frac{1}{Y_0\left(j\omega\right)^n} \tag{2}$$

where Y_0 is the constant, $j^2 = -1$, n is the exponent of the CPE, ω is the frequency. If $n = 1$, Z_{CPE} characterizes an ideal capacitor, and it corresponds to a pure resistor when $n = 0$ [19].

In equation (3), the inhibition efficiency η (%) is illustrated as:

$$\eta=\frac{R_{ctSL}-R_{ct}}{R_{ctSL}}\times 100\% \tag{3}$$

where R_{ctSL} represents the charge transfer resistance of the solution added SL, and R_{ct} represents the charge transfer resistance of the one without SL. With the adsorption of SL adsorption onto the metal face, Y_{0dl} decreases consistently [20]. The inhibition efficiency η is magnified with SL mass, reaching 84.67% at 0.06 wt.% SL.

Based on Tafel and EIS results, the corrosion inhibition rate of SL can exceed 80%. According to research by Thabet et al. [21], SL exhibits a corrosion inhibition rate exceeding 50% in NaCl solution. This demonstrates that it possesses a high corrosion inhibition capability.

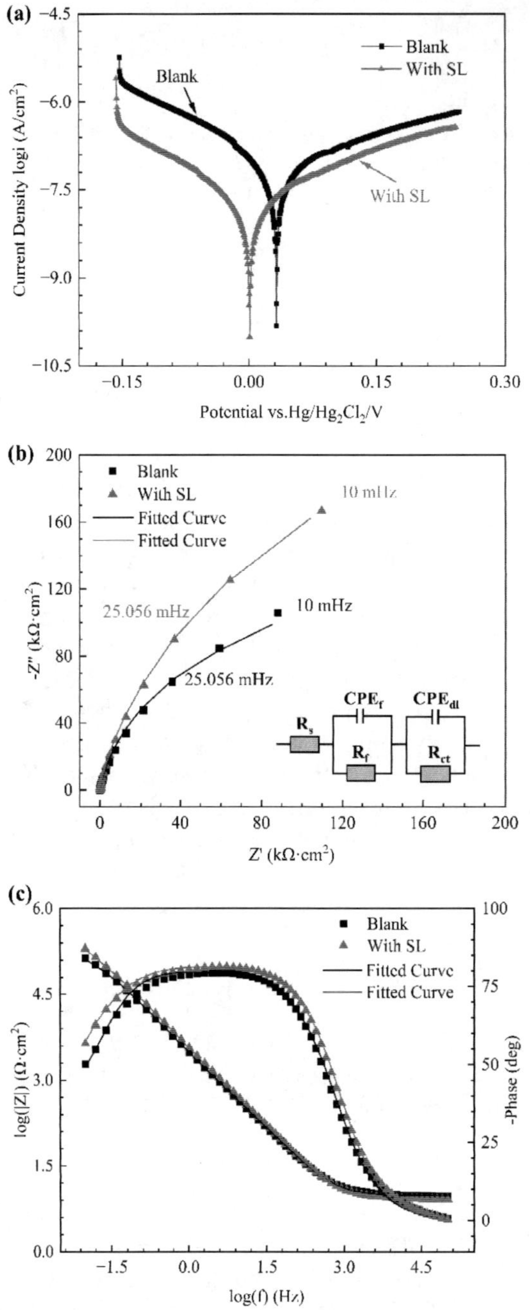

Figure 3. Results of electrochemical test: (a)Tafel plots, (b) Nyquist plots, (c) Bode plots.

Table 1. Tafel parameters with and without SL.

M(wt.%)	E_{corr}(mV/SCE)	I_{corr}(μA/cm^2)	η(%)
Blank	37.7840	0.0924	--
0.06	3.6130	0.0173	81.28

Table 2. The impedance parameters with and without SL.

M(wt.%)	Y_{0f} (10^{-4} Ω^{-1} cm^{-2} s^n)	n_f	Y_{0dl} (10^{-5} Ω^{-1} cm^{-2} s^n)	n_{dl}	Rs (Ω cm^2)	R_f (10^4 Ω cm^2)	R_{ct} (10^5 Ω cm^2)	R_p (10^5 Ω cm^2)	η(%)
Blank	1.1803	0.8417	12.6720	0.9665	9.6650	35.4170	6.6233	42.0403	--
0.06	1.0936	0.8669	8.9998	0.9549	8.4770	7.7458	43.2180	50.9638	84.67

3.2. The impact of SL on honeycomb seal in ECG

To validate the inhibitory effect of SL during the ECG process, machining experiments with different concentrations of SL were conducted, and the optical images of the machined surfaces are illustrated in figure 4.

As depicted in figure 4(a)-(c), during electrochemical grinding in the NaCl solution without SL, excessive corrosion came out on the machined seal due to the effects of electrolysis. This is because the thin-wall characteristic of single-foiled segment, making it susceptible to thinning under electrolysis, resulting in the formation of high current density areas and ultimately corrosion, as identified in figure 4(c) with depression. It is seen in figure 4(d)-(f), when 0.06 wt.% SL is involved in the solution, the average value of excessive corrosion is reduced from 58 μm to 32 μm, and an improvement in the machined surface of the honeycomb can be seen in figure 4(d) and (e), indicating the inhibition of electrolysis during the machining process. The corrosion depressions on the honeycomb surface are effectively reduced comparing figure 4(a) and (b) to the machining results with SL illustrated in figure 4(d) and (e).

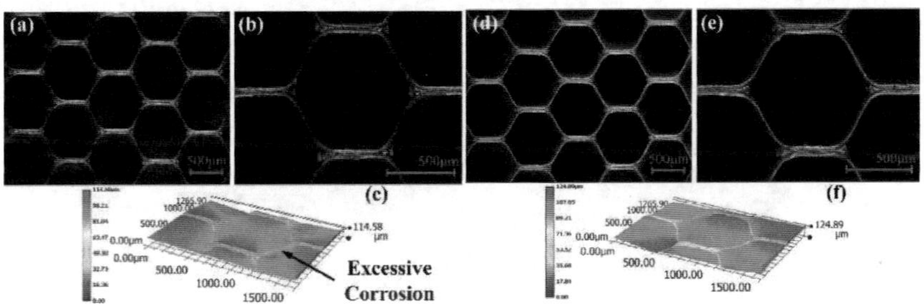

Figure 4. ECG results of honeycomb seals: (a), (b), (c) without SL; (d), (e), (f) with SL.

3.3. SEM results

The SEM results without SL are shown in figure 5, while those with 0.06 wt.% SL are depicted in figure 6. As shown in figure 5(a), severe corrosion is observed in the double-foiled segment of the processed surface, with the metal surface corroded to expose the intermediate brazing material. Such defects can impair the usability of the honeycomb structure. Additionally, figure 5(b) shows a sharp-edged structure on the honeycomb wall surface, which can lead to excessive corrosion under the electric field during ECG. However, no exposed brazing material is observed in the double-foiled region in figure 6(a). The surface of the double-foiled segment remains level and intact after processing. Moreover, in figure 6(b), no sharp structure is observed on the wall surface of the single-foiled segment. Elemental composition analysis using EDS in figure 5(a) and (b) reveals the presence of Si elements on the surfaces of both double-foiled and single-foiled segments after ECG processing without SL, suggesting the dissolution of honeycomb walls due to electrolysis, exposing the brazing material. Conversely, after adding SL, the Si content on the surface of the ECG processed wall decreases, indicating a reduction in electrolytic corrosion on the honeycomb wall surface and prevention of brazing material exposure. Moreover, the sulfur (S) content increases in figure 6(a) and (b) after adding SL, as SL contains sulfonic acid groups, indicating SL's ability to adsorb on the honeycomb surface during the ECG process, thereby reducing corrosion. Comparing the oxygen (O) content in figure 5 and figure 6, a decrease in O on the honeycomb surface after adding SL indicates a reduction in metal oxide formed in ECG.

Journal of Physics: Conference Series **3041** (2025) 012032 doi:10.1088/1742-6596/3041/1/012032

According to the study by Yin et al. on Hastelloy X's corrosion characteristics in $NaNO_3$ solution, it is known that the main components of this material's oxidation products are NiO, Cr_2O_3, and MoO_3, and the elements of Ni, Fe, and Cr decrease sharply and disappear, indicating their conversion into metal ions and hydroxides [22]. Therefore, during the processing of Hastelloy X honeycomb metal, when oxidation products are generated due to electrolytic action, the oxygen content on its surface increases, and the elemental content of Ni, Fe, and Cr decreases on the surface. From the SEM results in figure 5 and figure 6, it is evident that the presence of SL demonstrating that SL can inhibit the corrosion process of the honeycomb metal.

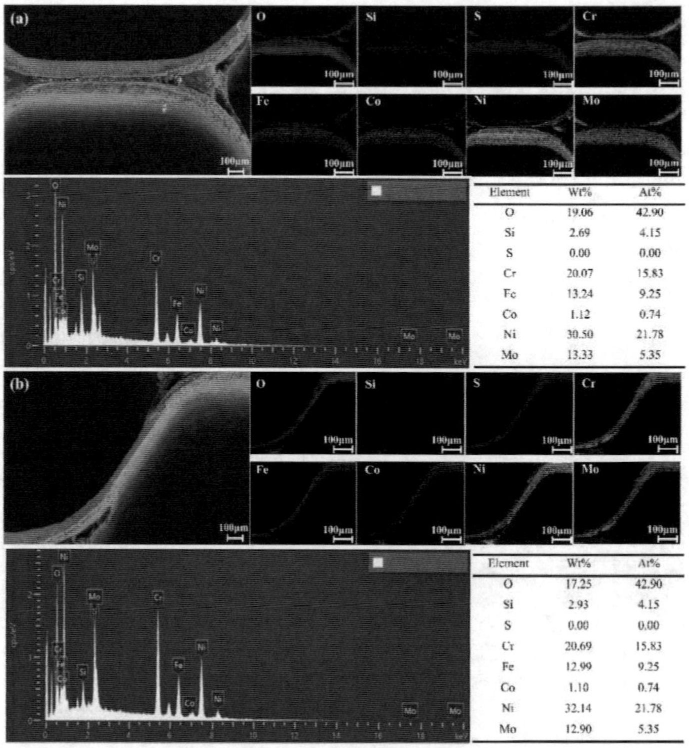

Figure 5. SEM and EDS images of honeycomb without SL.

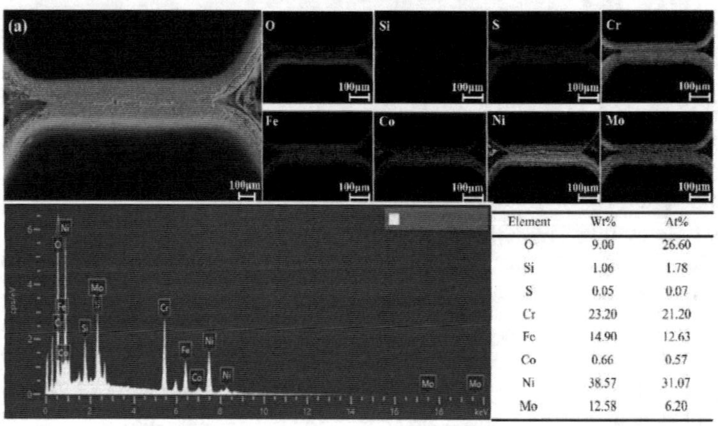

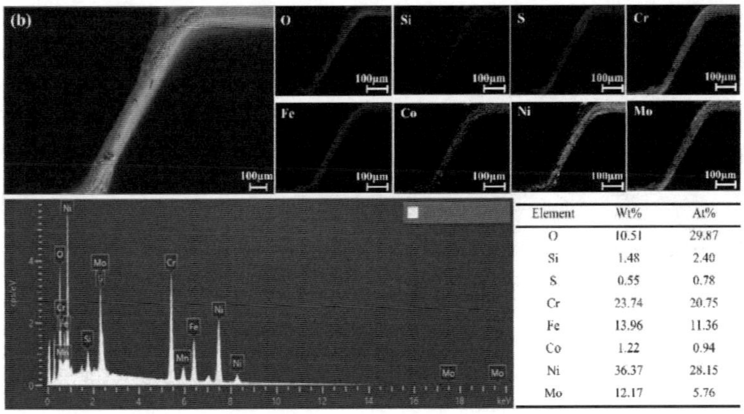

Element	Wt%	At%
O	10.51	29.87
Si	1.48	2.40
S	0.55	0.78
Cr	23.74	20.75
Fe	13.96	11.36
Co	1.22	0.94
Ni	36.37	28.15
Mo	12.17	5.76

Figure 6. SEM and EDS images of honeycomb without SL.

3.4. Limitation and future challenge
Ni^{2+} ions are generated during the dissolution process of honeycomb metal, and SL can complex with nickel ions [23]. Consequently, SL is consumed in the course of use. Due to SL consumption, additional SL needs to be introduced into the electrolyte after a period of use. Therefore, SL consumption should be taken into account in the ECG processing of honeycomb seals. As SL is an anionic surfactant, varying concentrations of SL in the electrolyte affect its surface tension [24]. Thus, the SL concentration can be monitored by measuring the surface tension of the electrolyte [25], allowing for timely replenishment of SL.

This paper focuses on only one corrosion inhibitor, and other green corrosion inhibitors, like sodium dodecylbenzene sulfonate [26] and sodium gluconate [27], have been proven to also inhibit metal corrosion. Therefore, subsequent research can focus on corrosion inhibitors with enhanced corrosion inhibition abilities, to mitigate corrosion of honeycomb seals during the ECG processing. As SL exhibits corrosion inhibition on Q235 steel [28], stainless steel [29], and zinc [30], it can be applied in ECG processing of these materials to control electrolytic corrosion. Therefore, the corrosion inhibition effect of SL in the electrochemical grinding process can be extended to the processing of more metallic materials in the future.

4. Conclusion
This article describes a means for inhibiting corrosion during ECG of honeycomb seal by using SL. The corrosion inhibition efficiency of it was obtained through electrochemical tests, and its corrosion inhibition effect on honeycomb structures was verified by machining experiments. The principal findings are summarized as follows:

(1) The data of Tafel curves and EIS tests indicate that SL can inhibit the corrosion of GH3536 in 2 wt.% NaCl solution. With 0.06 wt.% mass fraction of SL, the corrosion inhibition rates of Tafel and EIS reach 81.28% and 84.67%, respectively.

(2) In the solution with SL, the excessive corrosion of the seal is largely restrained. In solutions without SL, the honeycomb's single-foiled area is corroded by the electrolysis, forming an average excessive-corrosion depth of 58 μm. The excessive corrosion, after the addition of 0.06 wt.% SL, on the seal on the single-foiled segment, is inhibited, and the average depth is reduced to 32 μm.

(3) On the machined seal surface, SL is capable of diminishing the corrosion and lowering exposure of brazing material, evidenced by the EDS analysis which noted a decrease in the levels of O and Si. Additionally, the EDS data show that SL can adhere to the honeycomb during ECG machining, with the detection of S on the surface.

This article provides an environmentally friendly corrosion inhibitor for ECG processing of honeycomb structures to suppress corrosion during processing. In future work, further research could be done to explore more green dissolution inhibitors for ECG processing of honeycomb seal.

5. References

[1] Zaniewski D, Klimaszewski P, Klonowicz P, Lampart P, Witanowski Ł, Jędrzejewski Ł, Suchocki T and Antczak Ł 2021 Performance of the honeycomb type sealings in organic vapour microturbines *Energy* **226** 120242

[2] Qiu K X, Ming W W, Shen L F, An Q L and Chen M 2017 Study on the cutting force in machining of aluminum honeycomb core material *Compos Struct* **164** 58–67

[3] Sun J S, Dong Z G, Wang X P,Wang Y D, Qin Y and Kang R K 2020 Simulation and experimental study of ultrasonic cutting for aluminum honeycomb by disc cutter *Ultrasonics* **103** 106102

[4] Tian J C, Kang R K, Dong Z G, Liu Z Q, Guo D M and Bao Y 2023 Multi-scale machining damages of CFRP circular cell honeycomb during end face machining *J Manuf Process* **86** 282–293

[5] Gao J, Ma X F and Liu Q Y 2012 Experimental study of electrical discharge grinding honeycomb tori *Advanced Materials Research* **468** 924–7

[6] Abdelrahman N S, Galiwango E, Al-Marzouqi A H and Mahmound E 2022 Sodium lignosulfonate: a renewable corrosion inhibitor extracted from lignocellulosic waste. *Biomass Convers Bior* **14** 1–11

[7] Lin B, Tang J L, Wang Y Y, Wang H and Zuo U 2020 Study on synergistic corrosion inhibition effect between calcium lignosulfonate (CLS) and inorganic inhibitors on Q235 carbon steel in alkaline environment with Cl^- *Molecules* **25(18)** 4200

[8] Hajjari A, Shahrabi T and Mohammadi I 2023 Synthesis of a novel environmentally friendly hybrid pigment for effective corrosion control of mild steel *J Environ Chem Eng* **11(2)** 109383

[9] Ahangar M, Izadi M, Shahrabi T and Mohammadi I 2020 The synergistic effect of zinc acetate on the protective behavior of sodium lignosulfonate for corrosion prevention of mild steel in 3.5 wt% NaCl electrolyte: Surface and electrochemical studies *J Mol Liq* **314** 113617

[10] Qiang, Y J, Zhang, S T, Xu S Y and Yin L L 2015 The effect of 5-nitroindazole as an inhibitor for the corrosion of copper in a 3.0% NaCl solution *RSC Adv* **5(78)** 63866–73

[11] Zhao C Z, Yu Z and Heng J B 2020 Eco-friendly anticorrosion for carbon steel in artificial seawater: Effect of the ethylene oxide number of polyoxyethylene lauryl ether on its inhibition behaviour combining experimental and theoretical perspectives *Constr Build Mater* **260** 119950

[12] Qiang Y J, Zhang S T, Xu S Y and Li W P 2016 Experimental and theoretical studies on the corrosion inhibition of copper by two indazole derivatives in 3.0% NaCl solution *J Colloid Interf Sci* **472** 52–9.

[13] Gao C P, Qu N S, Ding B and Shen Y Z 2019 An insight into cathodic reactions during wire electrochemical micromachining in aqueous hydrochloric acid solution *Electrochim Acta* **295** 67–74

[14] Zhao R X, Xu W D, Yu Q and Niu L 2020 Synergistic effect of SAMs of S-containing amino acids and surfactant on corrosion inhibition of 316L stainless steel in 0.5 M NaCl solution *J Mol Liq* **318** 114322

[15] Fernandes C M, Alvarez L X, dos Santos N E, Maldonado Barrios A C and Ponio E A 2019 Green synthesis of 1-benzyl-4-phenyl-1H-1, 2, 3-triazole, its application as corrosion inhibitor for mild steel in acidic medium and new approach of classical electrochemical analyses *Corros Sci* **149** 185–194

[16] Hosseini S A, Shahrabi T and Ramezanzadeh B 2020 Synergistic effect of Black cumin extract and zinc cations on the mild steel corrosion resistance improvement in NaCl solution; Surface and electrochemical explorations *Colloid Surface* A **654** 130153

[17] Kong D C, Ni X Q, Dong C F, Zhang L, Yao J Z, Man C, Wang L, Xiao K and Li X G 2019 Anisotropic response in mechanical and corrosion properties of hastelloy X fabricated by selective laser melting *Constr Build Mater* **221** 720–9

[18] He X, Wang L, Kong D C, Li R X, Zhang W, Dai K J, Ni X Q, He K T and Dong C F 2023 Recrystallization effect on surface passivation of Hastelloy X alloy fabricated by laser powder bed fusion *J Mater Sci Technol* **163** 245–58

[19] Growcock F B and Jasinski R J 1989 Time-resolved impedance spectroscopy of mild steel in concentrated hydrochloric acid *J Electrochem Soc* **136(8)** 2310
[20] Qiang Y J, Zhang S T, Guo L, Xu S Y, Feng L, Obot B I and Chen S J 2017 Sodium dodecyl benzene sulfonate as a sustainable inhibitor for zinc corrosion in 26% NH4Cl solution *J Clean Prod* **152** 17–25
[21] Thabet M, El-Moselhy M M, Azooz R E and El-Zomrawy A A 2024 Rule of Lignosulfonate as a corrosion inhibitor for steel in neutral media, computational and reaction pathway *Results in Surfaces and Interfaces* **16** 100266
[22] Yin Y, Zhang J, Ma Y, Huo J X, Zhao K and Meng X Y 2020 Electrochemical dissolution behavior of nickel-based Hastelloy X superalloy at low current densities *IEEE Access* **8** 62714-62724
[23] Gao W, Wang L and Liu Y 2023 Assembly of sodium lignosulfonate-metal hybrids as highly efficient adsorbents for rapid removal of dye molecules *Desalin Water Treat* **294** 259-271
[24] Rana D, Neale G and Hornof V 2002 Surface tension of mixed surfactant systems: lignosulfonate and sodium dodecyl sulfate *Colloid Polym Sci* **280** 775-778
[25] Li R, Aghamiri S F, Yang D, Chen P and Qiu X Q 2013 Dynamic surface tension and adsorption kinetics of sodium lignosulfonate aqueous solutions *J Disper Sci Technol* **34(5)** 709-715
[26] Li Y, Lu X P, Wu K X, Yang L, Zhang T and Wang F H 2020 Exploration the inhibition mechanism of sodium dodecyl sulfate on Mg alloy *Corros Sci* **168** 108559.
[27] Ivušić F, Lahodny-Šarc O, Ćurković H O and Alar V 2015 Synergistic inhibition of carbon steel corrosion in seawater by cerium chloride and sodium gluconate *Corros Sci* **98** 88-97
[28] Li J, Liu W and Xie W 2020 Influence of sodium lignosulfonate on the corrosion-inhibition behavior of Q235 steel in simulated concrete pore solutions *Int J Electrochem Sc* **15(8)** 7136-7151
[29] Wu J, Dai N, Sun Y T, Tang Y W Jiang Y M and Li Jin 2019 Pitting inhibition of newly developed lean duplex stainless steel 2002 in NaCl solution by a green inhibitor *Mater Res Express* **6(7)** 076569
[30] Kolmachikhina E B, Sviridov A V and Naumov K D 2020 Investigation into the influence of sodium lignosulfonate, anionic surfactants, and their mixtures on the copper cementation rate by zinc *Russ J Non-Ferr Met+* **61** 488-493

Acknowledgments
This work was supported by The National Natural Science Foundation of China (No. 52275432), Defense Industrial Technology Development Program (No. JCKY2023205B001).

Numerical investigation on cooling characteristics of an additively manufacturable composite cooling structure with swirling outflow

Yutao Zhuang[1], Honglin Li[1], Timothy H Lee[1], Bangwei Liu[1], Jiahuan Cui[1,2,*]

[1] ZJU-UIUC Institute, Zhejiang University, Haining, 314400, China
[2] School of Aeronautics and Astronautics, Zhejiang University, Hangzhou, 310007, China

*Email: jiahuancui@intl.zju.edu.cn

Abstract. Innovative structural designs for cooling technologies play a crucial role in the advancement of next-generation aero-engines. This study proposes an additively manufacturable composite cooling structure featuring swirling coolant outflow, and its cooling characteristics are investigated using numerical simulation. The operating conditions considered include blowing ratios of 0.5 and 2.0, analyzed under conjugate heat transfer and adiabatic contexts. The findings indicate that the swirl composite cooling structure exhibits significantly higher cooling effectiveness in comparison of the traditional film cooling structure, with this superiority becoming more pronounced under high blowing ratio condition. When the blowing ratio is 2.0, the improvements of area-averaged cooling effectiveness reach 55.3% in conjugate heat transfer and 202.9% in adiabatic contexts. On one hand, the swirl composite cooling structure achieves excellent internal cooling due to coolant impingement and an enlarged heat transfer surface area. On the other hand, the swirling coolant outflow creates a unique velocity distribution within the film hole, leading to a top vortex that compresses the bottom vortex upon merging with the mainstream flow. As a result, the lift-off effect of coolant outflow is restricted and the cooling air coverage is significantly enhanced, causing preeminent cooling performance of the swirl composite cooling configuration.

1. Introduction

In response to modern energy deficits, the aviation sector is compelled to innovate aero-engines with higher efficiency. The aero-engine operates based on the conventional Brayton cycle, wherein elevating the turbine inlet temperature substantially enhances thermal efficiency[1]. As increasing gas temperatures in advanced aero-engines now exceed the limits of metallic materials, there is an urgent need for effective cooling technologies to safeguard the operational safety and longevity of high-temperature turbine components. Laminated cooling is widely recognized as a key technology for efficiently handling extreme heat. As one of the composite cooling technologies, it integrates film cooling externally and impingement cooling internally[2]. Film cooling reduces the temperature of adjacent airflow by forming a protective air film over external surfaces, while impingement cooling directs high-velocity jets at internal surfaces to create thin boundary layers that enable exceptional heat transfer—surpassing the effectiveness of other internal cooling methods [3].

Significant advancements in the cooling performance of laminated structures have focused predominantly on structural refinements. Extensive efforts have been directed toward augmenting the convective heat transfer coefficient of laminated cooling by optimizing rib turbulators, increasing the number of laminate layers, and designing uniquely shaped cooling chambers. Relevant findings are comprehensively documented in the literature[4-8]. Moreover, adjusting the layout and geometry of film holes can also markedly improve the film cooling performance of laminated cooling[9-12]. Research done Gu et al.[10], Rao et al.[12], and Liu et al.[11] have studied laminated structures with slotted, shallowly dimpled, and hexagonally arranged film holes, respectively. The aforementioned structural modifications not only boost traditional laminated cooling performance, but also offer important guidance for the design of advanced cooling strategies in next-generation aero-engines. However, achieving even higher cooling efficiencies requires further in-depth exploration of structural modifications of laminated cooling.

Swirling flow technologies, which thin the thermal boundary layer and amplify heat transfer through elevated radial pressure gradients, represent a promising cooling methodology[13, 14]. However, research on laminated structures incorporating impingement-induced swirling film cooling remains sparse. Han et al.[15] have explored a laminated design that combines swirl and film cooling, focusing more on augmenting internal convective cooling effects. Zhang et al.[16] have developed a laminated structure incorporating impingement-induced swirling film cooling, but its structural integrity is not sufficient.

In this paper, through further structural innovation in laminated cooling design, an additively manufacturable composite cooling structure with hemispherical cavity is proposed, which can generate swirling coolant outflow. CFD numerical simulation is conducted to investigate the cooling performance and the cooling mechanism of the swirl composite cooling structure in comparison to the traditional film cooling structure.

2. Numerical model

Computational Fluid Dynamics (CFD) simulation is a widely used method for investigating fluid behavior and heat transfer issues, as it offers detailed insights and high-precision results with high efficiency and low cost. This section details the geometric configuration of the swirl composite cooling structure and the numerical model for CFD simulation.

2.1. Geometric configuration

The geometric configuration of the additively manufacturable swirl composite cooling structure is illustrated in figure 1. This structure is composed of three primary components: a hemispherical cavity, a cylindrical film hole, and a cylindrical impingement hole. The radius of the hemispherical cavity is denoted as R, and the distance from its upper vertex to the upper surface of the cooling structure is defined as the top layer height h_f. The film hole, with a diameter of d_f, has an outer sidewall tangent to the hemispherical cavity, and the angle between the film hole's axis and the horizontal plane is defined as the film hole inclination angle α_f. The impingement hole, with a diameter of d_i, also has an outer sidewall tangent to the hemispherical cavity. In the projection of the hemispherical cavity's bottom surface, the line connecting the cavity center to the impingement hole center forms an angle with the axis of film hole (aligned with the direction of high-temperature mainstream), and this angle is defined as the impingement hole position angle θ_i. The total height of the cooling structure is denoted as H. The hemispherical cavity design not only serves as structural support to meet the requirements of additive manufacturing, but also enlarges the heat transfer area inside the cooling structure and induces an impingement cooling effect. And the primary purpose of staggered arrangement of the impingement hole and film hole is to generate a swirling coolant outflow.

In this study, the cooling characteristics of the swirl composite cooling structure are investigated compared to the traditional film cooling structure. The detailed parameter set of the two cooling structures is listed in table 1.

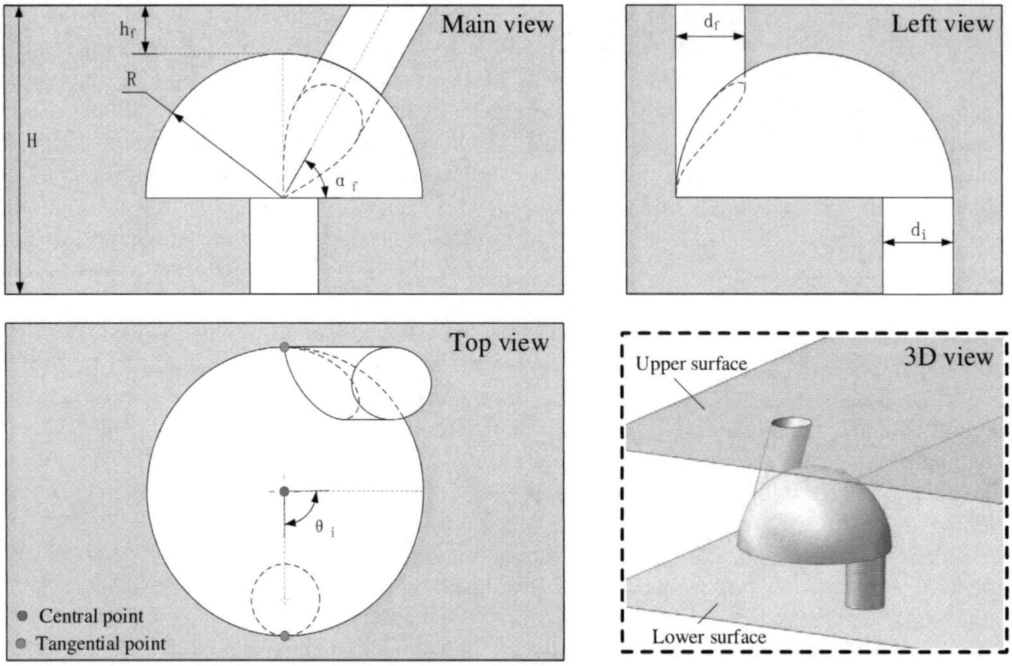

Figure 1. Geometric configuration of swirl composite cooling structure

Table 1. Parameter set of cooling structures

	d_f (mm)	H (mm)	R (mm)	α_f (°)	h_f (mm)	θ_i (°)	d_i (mm)
Swirl composite cooling	1.5	6	3	60	1	90	1.5
Traditional film cooling	1.5	6	-	60	-	-	-

2.2. Computational domain and boundary conditions

The computational domain employed for conjugate heat transfer simulation of the cooling structure includes both fluid and solid regions, as illustrated in figure 2. The mainstream flow proceeds along the positive X-axis, the lateral direction corresponds to the Y-axis, and the film hole exit is centered at the coordinate system's origin. Both the mainstream inlet and the coolant inlet are set as velocity inlet boundaries, while a pressure outlet boundary is applied to the mainstream outlet maintaining a constant pressure of 1 atm. Periodic boundaries are imposed on the sidewalls of both fluid and solid regions, and their interfaces are set as thermally coupled surfaces to ensure heat transfer while preventing mass flow. Air is defined as an incompressible ideal gas, with its dynamic viscosity calculated based on Sutherland's Law to accurately describe its properties. The material properties of the solid domain are set according to 316L stainless steel, with a thermal conductivity of 14.7 W/(m·K). The mainstream temperature is fixed at 373 K, the mainstream velocity at 20 m/s, and the coolant temperature at 303 K. The coolant inlet velocity is determined according to selected blowing ratios, which are 0.5 and 2.0 in this study.

Adiabatic cooling investigation is also conducted in this study, where the solid domain is removed and the corresponding interfaces are set as non-slip adiabatic walls.

2.3. Mesh generation and turbulence model

The poly-hexcore mesh for the computational domain generated using ANSYS Fluent Meshing is utilized for CFD simulation, as exhibited in figure 3. In the near-wall region of the fluid domain, 12 prism layers with an initial height of 0.03 mm are added to maintain the dimensionless y+ of the first mesh layer under 1, allowing for accurate capture of fluid flow characteristics near the wall.

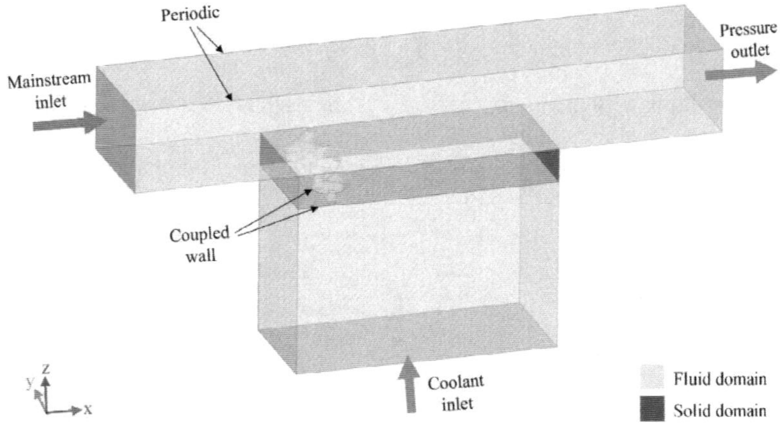

Figure 2. Computational domain

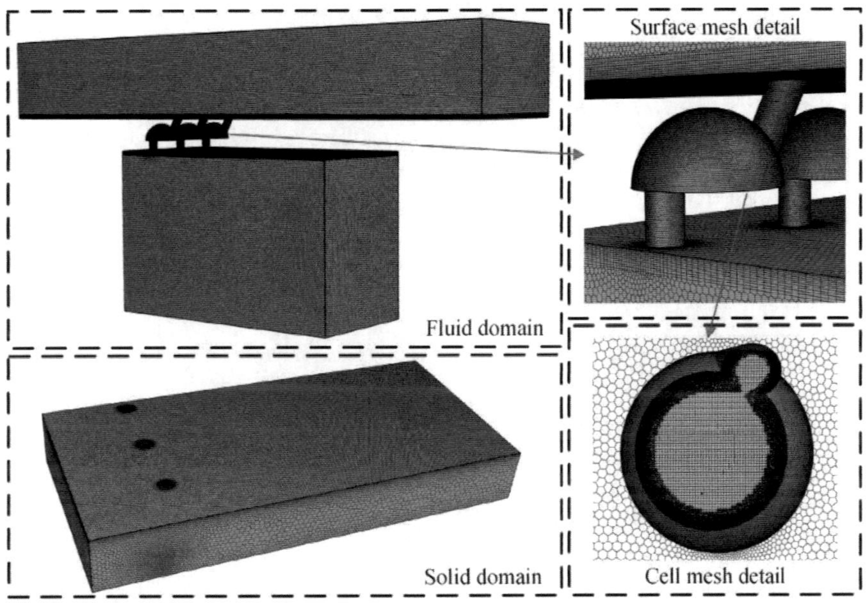

Figure 3. Mesh for the computational domain

In this study, numerical calculation is carried out using ANSYS Fluent. The Reynolds-Averaged Navier-Stokes (RANS) approach is applied to solve the governing equations, employing the realizable k-ε turbulence model. Coupled pressure-velocity coupling algorithm and Second Order Upwind discretization scheme are adopted. The convergence criterion for steady computation is that the residuals of energy, continuity, velocity, and turbulence variables be less than an order of 1e-6. The numerical model has been validated in author's previous work[17].

2.4. Data reduction

The blowing ratio (BR) is defined by the following equation:

$$BR = \frac{\rho_c V_c}{\rho_m V_m} \quad (1)$$

where ρ_c is the density of coolant, V_c is the average velocity of coolant film outflow, ρ_m is the density of mainstream air, V_m is the average velocity of mainstream flow.

Cooling effectiveness (η) is used to evaluate cooling performance in both conjugate heat transfer and adiabatic cooling contexts. The definition is given below:

$$\eta = \frac{T_m - T_w}{T_m - T_c} \tag{2}$$

where T_m is the temperature of mainstream flow, T_w is the surface temperature of the cooling wall, T_c is the temperature of coolant flow.

Lateral-averaged cooling effectiveness ($\bar{\eta}$) and area-averaged cooling effectiveness ($\bar{\bar{\eta}}$) are calculated by the following equations:

$$\bar{\eta} = \frac{\int \eta(y)dy}{L_y} \tag{3}$$

$$\bar{\bar{\eta}} = \frac{\int\int \eta(x,y)dxdy}{L_x L_y} \tag{4}$$

where L_x is the streamwise length of the cooling surface, L_y is the lateral length of the cooling surface.

2.5. Mesh independence check

In this study, mesh independence is checked for the swirl composite cooling structure, under $BR = 2.0$ operating condition, conjugate heat transfer. Through changing the setting of cell size, four kinds of mesh with varying cell counts are generated, including 3.7 million (Least), 4.36 million (Less), 6.43 million (Medium), and 14.87 million (Most). The simulation results of lateral-averaged cooling effectiveness using different meshes are shown in figure 4. The result indicates that the maximum error between cases with Medium mesh and with Most mesh is only 0.16%. Therefore, the Medium mesh setting is selected for further investigations to balance simulation accuracy and computation cost.

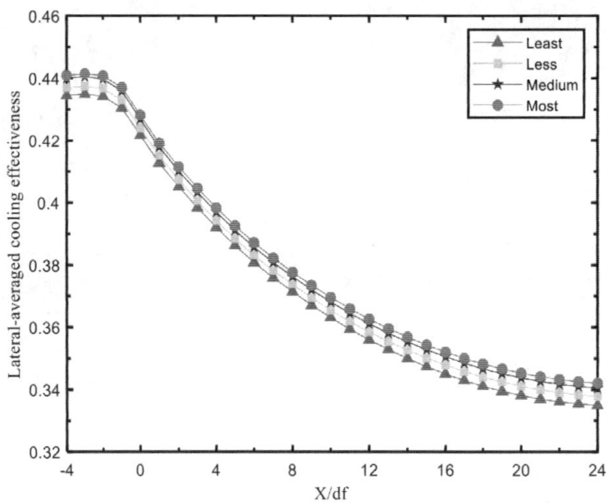

Figure 4. Mesh independence check

3. Numerical results and discussion

In this section, the cooling performance of the swirl composite cooling structure is evaluated in comparison to the traditional film cooling structure. Subsequently, flow behavior is investigated to reveal the underlying cooling mechanism.

3.1. Analysis of cooling effectiveness

The cooling effectiveness distribution for traditional film cooling and swirl composite cooling under the operating conditions of $BR = 0.5$ and $BR = 2.0$, conjugate heat transfer, are depicted in figure 5 and figure 6, respectively. The results indicate that, under both operating conditions, the cooling effectiveness of all cooling structures is highest in the upstream region and gradually decreases in the downstream direction, with a relatively uniform distribution along the lateral direction. This pattern is mainly attributed to the dominant role of internal heat exchange within cooling structures. As the swirl composite cooling induces impingement cooling and possesses a larger internal heat transfer area, its cooling effectiveness is significantly stronger than that of traditional film cooling. Additionally, for all cooling structures, increasements in cooling effectiveness are observed with a rise in the blowing ratio, primarily because of the enhanced internal heat exchange.

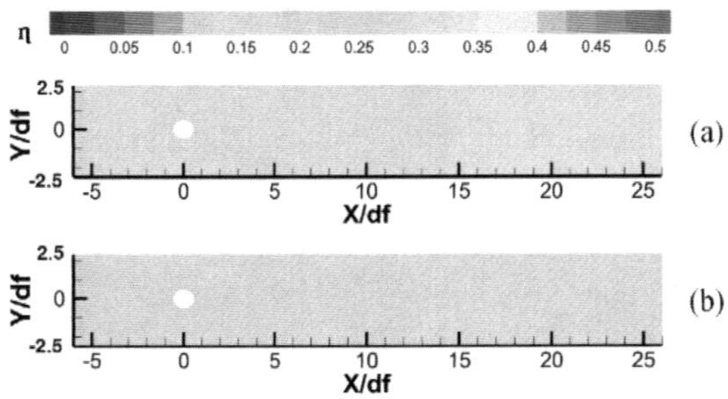

Figure 5.$BR=0.5$, conjugate heat transfer, cooling effectiveness distribution for (a) traditional film cooling and (b) swirl composite cooling

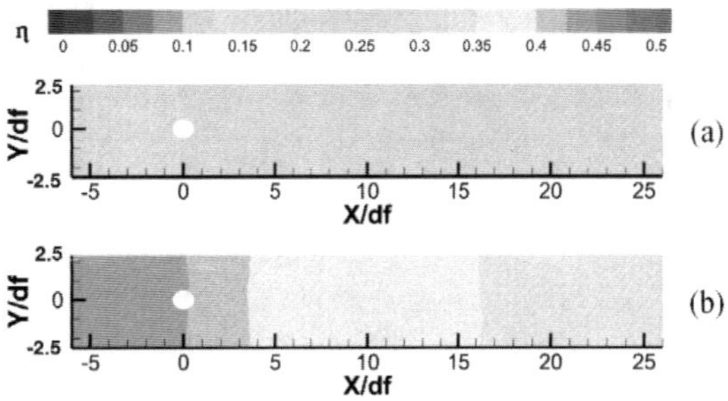

Figure 6.$BR=2.0$, conjugate heat transfer, cooling effectiveness distribution for (a) traditional film cooling and (b) swirl composite cooling

The cooling effectiveness distribution for traditional film cooling and swirl composite cooling under the operating conditions of $BR = 0.5$ and $BR = 2.0$, adiabatic, are depicted in figure 7 and figure 8, respectively. The results indicate that, under the operating condition of $BR = 0.5$, the cooling effectiveness distributions for the two cooling structures are nearly identical. Near the film hole exit, the adiabatic cooling effectiveness reaches its peak and progressively declines along the downstream path. On account of the relatively low momentum of the coolant outflow, the cooling air can adhere effectively to the hot-side surface after merging with the mainstream, thereby forming a protective film. However, under the operating condition of $BR = 2.0$, the cooling effectiveness distributions for the two cooling structures are obviously different. The cooling effectiveness of the traditional film cooling sharply decreases, as the increased outflow momentum causes a lift-off effect on cooling air, preventing the formation of an effective film coverage. In contrast, the coolant coverage of the swirl composite cooling is narrow near the film hole exit, with an approximate $0.5d_f$ deflection in the lateral direction. As it develops along the downstream direction, the cooling air noticeably reattaches to the surface around $X/d_f = 5$, accompanied by a substantial expansion in its lateral coverage. Consequently, it achieves a much higher film cooling effectiveness than traditional film cooling.

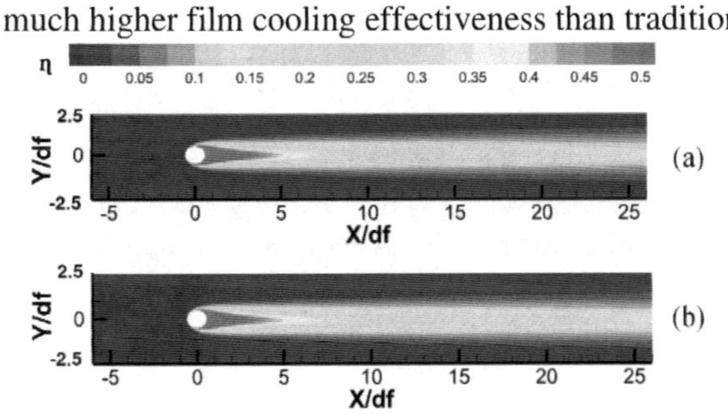

Figure 7.$BR=0.5$, adiabatic, cooling effectiveness distribution for (a) traditional film cooling and (b) swirl composite cooling

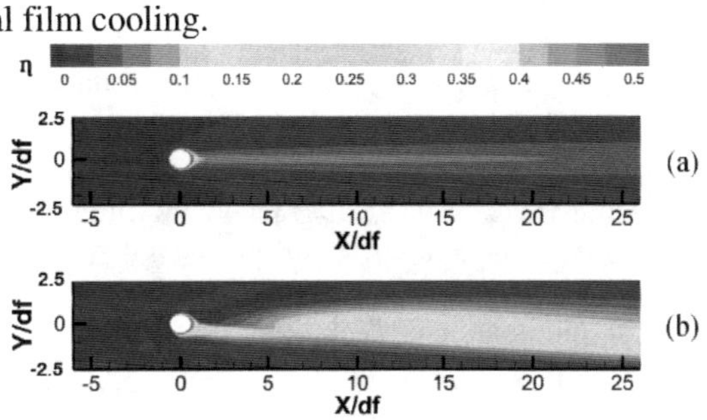

Figure 8.$BR=2.0$, adiabatic, cooling effectiveness distribution for (a) traditional film cooling and (b) swirl composite cooling

The conjugate heat transfer area-averaged cooling effectiveness for two cooling structures under different operating conditions, and the improvement percentage compared to traditional film cooling, are depicted in figure 9. The results demonstrate that under all operating conditions, swirl composite cooling significantly outperforms traditional film cooling in cooling performance. Moreover, cooling effectiveness increases with a higher blowing ratio, and swirl composite cooling shows a greater increment magnitude, leading to a more pronounced advantage under high blowing ratio conditions.

Journal of Physics: Conference Series **3041** (2025) 012033 doi:10.1088/1742-6596/3041/1/012033

The adiabatic area-averaged cooling effectiveness for two cooling structures under different operating conditions, and the improvement percentage compared to traditional film cooling, are depicted in figure 10. The results demonstrate that under $BR = 0.5$ operating condition, the two cooling structures exhibit parallel cooling performance. However, as the blowing ratio grows to 2.0, the cooling effectiveness of traditional film cooling dramatically decreases, while the swirl composite cooling structure evidently restrains the drastic deterioration of cooling performance, showing a 202.9% improvement over traditional film cooling.

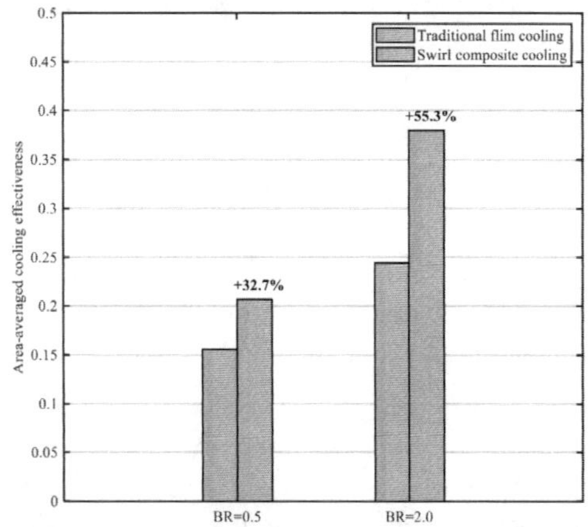

Figure 9. Conjugate heat transfer area-averaged cooling effectiveness for two cooling structures under different operating conditions

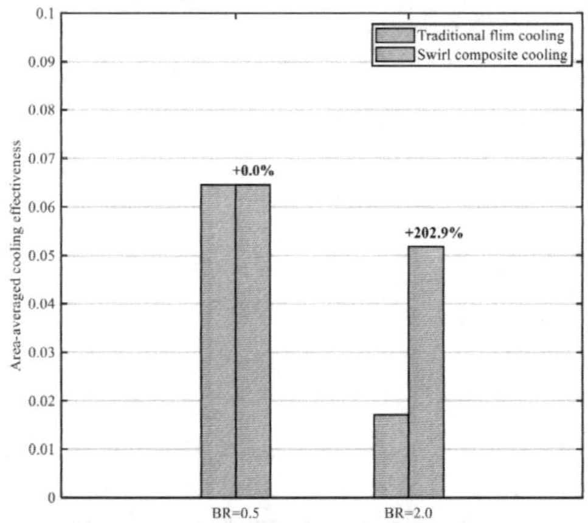

Figure 10. Adiabatic area-averaged cooling effectiveness for two cooling structures under different operating conditions

In summary, the swirl composite cooling structure outperforms the traditional film cooling structure in both conjugate heat transfer and adiabatic context. And this superiority grows larger under high blowing ratio operating condition.

3.2. Analysis of flow field

The flow behavior of the two cooling structures under the operating condition of $BR = 2.0$, adiabatic, is further investigated. Figure 11 presents the velocity distribution and streamline plots on axial cut plane of the film hole for the two cooling structures. The flow structure of traditional film cooling is depicted in figure 11 (a), which has a pattern similar to that observed in the research by Jiang et al.[18]. When the coolant enters the film hole from the cooling chamber, it accelerates and adheres more closely to the upstream side of the mainstream if it forms an acute angle with the film hole axis. Conversely, if the coolant forms an obtuse angle with the film hole axis, it decelerates and gradually compresses the airflow of high-speed region. Therefore, at the film hole exit, the downstream side becomes a low-velocity outflow region, while the upstream side turns into a high-velocity outflow region. This results in a noticeable lifting effect on the cooling air, preventing it from making an effective film coverage to the surface. The flow structure of swirl composite cooling is depicted in figure 11 (b). Due to the influence of the hemispherical cavity, the coolant generates a vortex before entering the film hole. This leads to a high-velocity outflow region on the downstream side and a low-velocity outflow region on the upstream side at the film hole exit. The high-speed airflow is compressed by the low-speed airflow, thereby reducing the lifting effect.

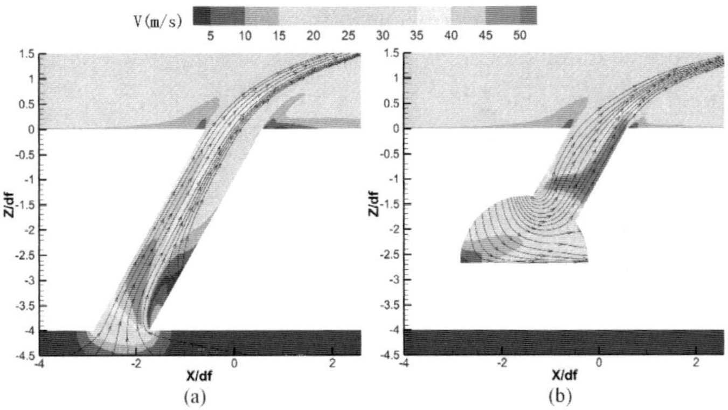

Figure 11. Velocity distribution and streamline plots on axial cut plane of the film hole for (a) traditional film cooling and (b) swirl composite cooling

Figure 12 presents the temperature distribution and streamline plots at various downstream sections of the film hole for two cooling structures. As shown in figure 12 (a), after merging with the mainstream, the cooling air forms a counter rotating vortex pair, which not only causes the cooling air to lift off from the hot-side surface but also draws high-temperature air from both sides into the lower region of the cooling airflow. By comparing results across different regions, it can be noticed that, the cooling air gradually separates from the wall along the mainstream direction, with its coverage area progressively narrowing, leading to a reduction in cooling effectiveness. As shown in figure 12 (b), the coolant outflow from the swirl composite cooling structure is deflected in the lateral direction. As it develops along the mainstream, the interaction of the high-velocity outflow and the mainstream forms a large vortex (bottom right in the figure), while the low-velocity outflow interacts with the mainstream generating a smaller vortex (top left in the figure). The smaller vortex at the top resists weakening by the mainstream and continues to squeeze the larger vortex below, causing the cooling air to stick closely to the hot-side surface and undergo significant lateral spread. This magnifies the coverage area of cooling air, thereby effectively enhancing the cooling performance.

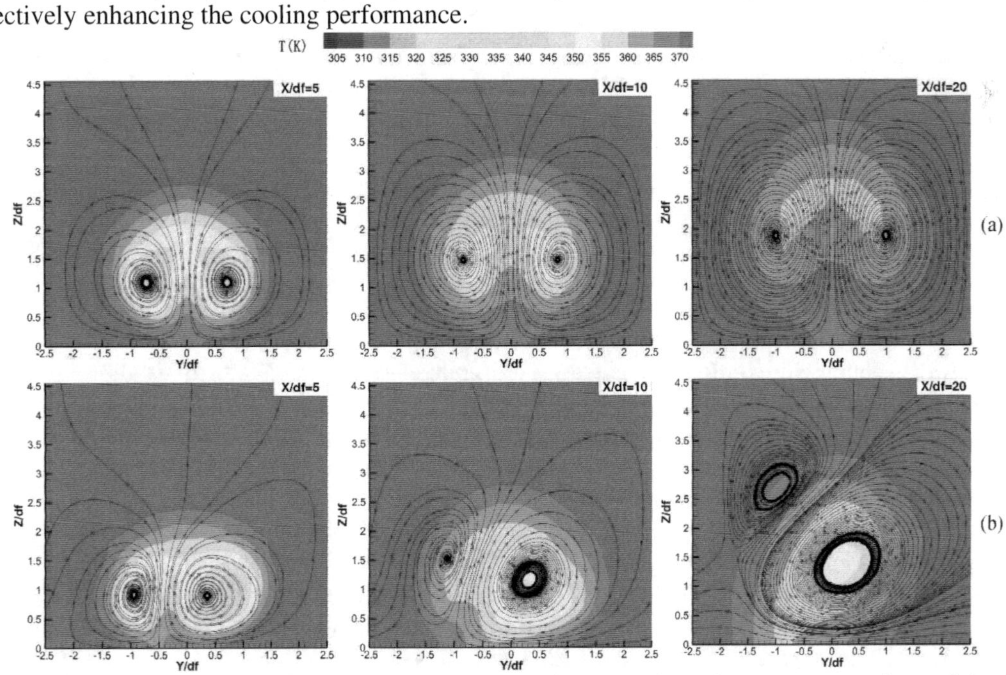

Figure 12. Temperature distribution and streamline plots at various downstream sections of the film hole for (a) traditional film cooling and (b) swirl composite cooling

In summary, on account of the swirling coolant outflow and the compressive effect of low-velocity airflow, the cooling air film coverage of the swirl composite cooling structure gets firm attachment and broad lateral spread, resulting in exceptional cooling performance.

4. Conclusion

In this paper, an additively manufacturable composite cooling structure with swirling outflow is proposed. Reliable numerical simulation is conducted to explore the cooling performance of the swirl composite cooling structure in comparison to the traditional film cooling structure. Subsequently, the flow behaviors of the two cooling structures are analyzed to uncover the cooling mechanism. The main conclusions are drawn as follows:

(1) The swirl composite cooling structure exhibits higher cooling effectiveness than the traditional film cooling structure in both conjugate heat transfer and adiabatic contexts, because of its stronger heat exchange inside the cooling structure and broader coolant coverage outside the cooling structure. Its superiority becomes more significant under high blowing ratio operating condition, with improvements of 55.3% and 202.9% in conjugate heat transfer and adiabatic context at a blowing ratio of 2.0.

(2) The swirling coolant outflow creates a distinct velocity distribution at the film hole exit compared to traditional film cooling. As a result, low-velocity cooling air interacts with mainstream flow and forms a vortex that continuously compresses another portion of cooling air. Accordingly, the lift-off effect is restrained and the lateral spread is enhanced for cooling air, causing a remarkable cooling performance improvement of the swirl composite cooling structure.

5. References

[1] Zamiri A, Barigozzi G and Chung JT Large eddy simulation of film cooling flow from shaped holes with different geometrical parameters *International Journal of Heat and Mass Transfer* 2022 **196** 123261

[2] Zhou W, Deng Q, He W, He J and Feng Z Conjugate heat transfer analysis for composite cooling structure using a decoupled method *International Journal of Heat and Mass Transfer* 2020 **149** 119200

[3] Wang K, Li H and Zhu J Experimental study of heat transfer characteristic on jet impingement cooling with film extraction flow *Applied thermal engineering* 2014 **70(1)** 620-9

[4] Zhang J, Han HZ, Li ZR and Zhong HG Effect of pin-fin forms on flow and cooling characteristics of three-layer porous laminate *Applied Thermal Engineering* 2021 **194** 117084

[5] Wenjing GA, Honglin LI, Lei LI, Zhe'nan ZH and Zhufeng YU Numerical simulation of broken pin effects on the flow field and cooling performance of a double-wall cooling configuration *Chinese Journal of Aeronautics* 2021 **34(2)** 358-75.

[6] Jaiswal AK, Mahapatra PS and Prasad BV Effect of microchannel on combined impingement and film cooling of a concave surface *International Communications in Heat and Mass Transfer* 2021 **126** 105441.

[7] Kong D, Guo T, Ma Z, Liu C and Isaev S Investigation of impingement heat transfer in double-wall cooling structures with corrugated impingement plate at small Reynolds numbers *Applied Thermal Engineering* 2023 **225** 120204.

[8] Bang M, Kim S, Park HS, Kim T, Rhee DH and Cho HH Impingement/effusion cooling with a hollow cylinder structure for additive manufacturing: Effect of channel gap height *International Journal of Heat and Mass Transfer* 2021 **175** 121420.

[9] Zhang XD, Liu JJ and An BT The investigations of slot film outflow used on the laminated cooling configuration *International Journal of Heat and Mass Transfer* 2019 **141** 1078-86.

[10] Gu H, Liang D, Duan P, Zhou D and Li W Aerothermal characteristics of thin double-wall effusion cooling systems with novel slot holes and cellular architectures for gas turbines *Aerospace Science and Technology* 2023 **140** 108441.

[11] Liu Y, Rao Y and Yang L Numerical simulations of a double-wall cooling with internal jet impingement and external hexagonal arrangement of film cooling holes *International Journal of Thermal Sciences* 2020 **153** 106337.

[12] Rao Y, Li Y, Liu Y and Yang L Experimental and numerical studies on enhanced effusion cooling with shallowly dimpled film holes on double-wall structure surface *International Journal of Heat and Fluid Flow* 2023 **101** 109135.

[13] Zhang R, Zhou L, Xing J, Luo C and Du X Numerical evaluation of film cooling performance of transverse trenched holes with shaped lips *International Communications in Heat and Mass Transfer* 2021 **125** 105326.

[14] Jia Y, Liu Y, Meng Z, Yin W and Hua W Numerical study on film cooling effectiveness from spiral-channel hole *International Communications in Heat and Mass Transfer* 2023 **143** 106716.

[15] Han S *et al.* Numerical simulation of composite swirl/film double-wall cooling structures and chamber designs for enhanced overall cooling effectiveness *International Journal of Heat and Mass Transfer* 2024 **228** 125664.

[16] Zhang J, Zheng Q, Xu J, Yue G and Jiang Y Conjugate heat transfer and flow analysis on double-wall cooling with impingement induced swirling and film cooling *Applied Thermal Engineering* 2023 **223** 120014.

[17] Li H, Xie F, Zhuang Y, Wang Y, Yan Y, Xu G and Cui J Numerical investigation of cooling performance of 3D-printable cooling structure for aero-engine combustor *Applied Thermal Engineering* 2024 **238** 122196.

[18] Jiang Y, Yue G, Dong P, Gao J and Zheng Q Investigation on film cooling with swirling coolant flow by optimizing the inflow chamber *International Communications in Heat and Mass Transfer* 2017 **88** 99-107.

AUTHOR INDEX